Molecular Cluster Magnets

World Scientific Series in Nanoscience and Nanotechnology

Series Editor: Mark Reed (*Yale University*)

Vol. 1 Molecular Electronics: An Introduciton to Theory and Experiment
Juan Carlos Cuevas (*Universidad Autónoma de Madrid, Spain*) *and*
Elke Scheer (*Universität Konstanz, Germany*)

Vol. 2 Nanostructures and Nanomaterials: Synthesis, Properties, and Applications, 2nd Edition
Guozhong Cao (*University of Washington, USA*) *and*
Ying Wang (*Louisiana State University, USA*)

Vol. 3 Molecular Cluster Magnets
edited by Richard Winpenny (*The University of Manchester, UK*)

Volume **3**

World Scientific Series in **Nanoscience and Nanotechnology**

Molecular Cluster Magnets

Editor

Richard Winpenny

The University of Manchester, UK

NEW JERSEY • LONDON • SINGAPORE • BEIJING • SHANGHAI • HONG KONG • TAIPEI • CHENNAI

Published by

World Scientific Publishing Co. Pte. Ltd.

5 Toh Tuck Link, Singapore 596224

USA office: 27 Warren Street, Suite 401-402, Hackensack, NJ 07601

UK office: 57 Shelton Street, Covent Garden, London WC2H 9HE

British Library Cataloguing-in-Publication Data
A catalogue record for this book is available from the British Library.

World Scientific Series in Nanoscience and Nanotechnology — Vol. 3
MOLECULAR CLUSTER MAGNETS

ISBN-13 978-981-4322-94-2
ISBN-10 981-4322-94-6

Typeset by Stallion Press
Email: enquiries@stallionpress.com

Printed in Singapore.

PREFACE

This book is intended as in introduction to certain aspects of molecular cluster magnets. There is a previous excellent book on this subject *Molecular Nanomagnets* by Dante Gatteschi, Roberta Sessoli and Jacques Villain. For this book, authors were chosen to cover areas not discussed in depth by this previous book — we intend this to be a complementary volume. For this reason we discuss very recent developments in single molecule magnets, rather than repeating a long discourse on theoretical aspects of these fascinating molecules. We also include a full review by Prof Laurie Thompson on the remarkable grid molecules that his group, and a few others, have been reporting. There is also a detailed review of the magnetism of the unique molecule $\{V_{15}\}$ — this was included as it is an excellent exemplar of what can be achieved by intelligent and extensive studies of one molecule; if you choose the correct molecule, a remarkable amount of physics can be extracted.

Important techniques for molecular nanomagnets include EPR spectroscopy and inelastic neutron scattering and we've included critical reviews of both these topics, written by two leading young scientists who utilise these techniques. Molecular magnetism also involves development of theoretical models, and the contribution from Schröder and Engelhardt was chosen because they use an approach slightly different to more conventional matrix-diagonalisation techniques. The approaches they have developed could be more applicable to large problems.

Richard Winpenny
Manchester, January 2011

CONTENTS

Chapter 1

SUPRAMOLECULAR POLYMETALLIC 2D [N × N] TRANSITION METAL GRIDS — APPROACHES TO ORDERED MOLECULAR ASSEMBLIES AND FUNCTIONAL MOLECULAR DEVICES

LAURENCE K THOMPSON, LOUISE N DAWE
and KONSTANTIN V SHUVAEV

Department of Chemistry, Memorial University, St. John's, Newfoundland, A1B 3X7, Canada

1. Convergent Self-assembly

1.1. *Introduction and overview*

The current 'top down' approaches to smaller functional subunits in the micro-electronics industry are based on making existing fundamental components smaller. Both CHIP circuitry and magnetic domain substructure are reaching critical limits because of charge leakage across CHIP circuit boundaries and thermal erasure in magnetic recording media. Typical limitations in size for thermal erasure in magnetic oxide domains are around 100 nm. An alternative way to approach the problem of miniaturization would be to use a convergent assembly or 'bottom up' strategy, where functional molecular based subunits can be constructed in a controlled way by the assembly of specific components. In addition, by tailoring the properties of the individual components, specific properties may be incorporated into the fundamental subunits. These could include magnetic, electronic, photophysical and other possible physical responses.

'LEGO' approaches to specific molecular construction require that elements of predictability be built into the fundamental components, which in the case of polymetallic assemblies would include the ligands and the appropriate metal ions. From the property perspective, transition metal ions offer the broadest range of potentially useful physical properties (redox, magnetic, electronic and photophysical) based in large measure on their complement of unpaired electrons in incomplete 'd' orbital subsets. The ligand is key in terms of its ability to organize metallic subunits into predictable assemblies, but has to be designed such that its coordination pockets create the 'right' environment to accommodate the metal ions, but also to have the correct geometrical disposition of the coordination pockets. In this context the preferred coordination 'algorithm' of the metal ion is also very important, in order to provide the right metal-ligand combination. This would include pocket size, donor identity (crystal field tuning), and donor disposition. The synthesis should be convergent in its approach such that the yield of the desired product is high, and that a minimum of other oligomeric components is produced.

From a practical perspective a collection of 'functional device' subunits must be organized in a regular fashion, and be accessible for stimulation and detection, based on a particular 'bistable' property. The 'bistable' nature of the subunit would allow for information storage at the subunit level, and rely critically on some fundamental property associated with the metal ions. The ligand may be benign in this context, and its fundamental role would simply be as a construction scaffold component, but of course it would influence the physical properties of the resulting complex to some extent. Since it would not be realistic to consider using an isolated single molecule or even a crystalline sample for information storage *per se*, by placing a uniform, regularly spaced arrangement of molecules, for example, on a flat surface, with some mechanism for molecular access and probing at the molecular level, a convenient medium for high density data storage could be created.

This chapter will focus on the design and synthesis of organic ligands encoded with the specific coordination information necessary to attract a group of metal ions, and also to create the conditions necessary for the formation of predetermined geometric 2D molecular arrays by spontaneous self-assembly. The ligand classes will include those which have small bridging subunits capable of bringing the metals into close proximity, such that intramolecular spin communication can be effected. This feature is considered to be fundamentally important in terms of the total pool of electronic charge associated with the collective group of paramagnetic metal ions. The metal ions will be considered on the basis of their coordination number and coordination geometry preferences, and also their general physical properties (e.g. electronic based properties). In addition, thermodynamic effects will be considered, particularly entropic effects, and the CFSE (crystal field stabilization energy) effects associated with the metal ions in their reacting form, in terms of their solvent medium coordination environments, and also in the self-assembled product. From an applications perspective studies of surface assemblies of polymetallic grids will be described, in addition to the consideration of possibly useful physical approaches for the encoding of information in individual surface bound molecules.

1.2. *Polytopic ligands for* $[n \times n]$ *square grids–design and self-assembly*

Metal-ligand interactions are driven largely by the formation of compatible ligand donor atom coordinate bonds with a metal ion, the formation of stable chelate rings, and the overall stability which these contribute to the complex which is formed. For simple mononuclear complexes the most effective metal-ligand interactions would be enhanced by the optimization of the chelate ring size, the choice of donors, and also the overall ligand denticity. The formation of polynuclear coordination complexes requires that some focal polynucleating (i.e. bridging) subunit be built into the ligand, which would provide the driving force for the attraction of more than one metal to a particular bridging site. This feature would then have to be incorporated into the more complex overall structure of the n-topic ligand (n represents the number of coordination pockets), such that the directed self-assembly

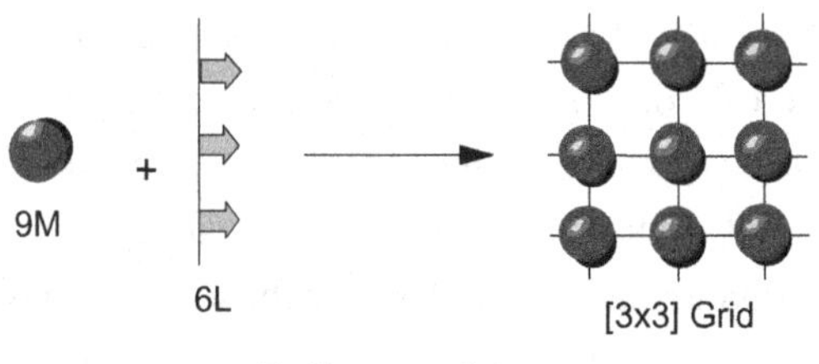

Scheme 1.

can proceed appropriately toward the desired product. This is illustrated in Scheme 1 for a hypothetical tritopic ligand, which would organize nine metal ions into a [3 × 3] square array as the optimum self-assembled product based on the match of donor (ligand) and acceptor (metal) properties. The denticity at each ligand coordination pocket must match that appropriate to the metal ion concerned, which will inevitably depend on the preferred metal coordination number and geometry. For typically tetrahedral metal ions, e.g. Cu(I), Ag(I), each pocket would of necessity be bidentate, while for typically octahedral metal ions, e.g. Mn(II), Ni(II), Co(II), it would have to be *mer*-tridentate. The size of the coordination sphere would of necessity be dependent on the nature of the metal ion, its oxidation state, and its preferred ionic radius. The disposition of the donor atoms in a particular ligand pocket would have to reflect this situation in an appropriate geometric way. The contiguous, linearly and evenly spaced arrangement of the pockets would then allow the necessary 90° twists to occur at each metal site based on the complementary 90° twists of the bidentate and *mer*-tridente ligand pockets (Scheme 2). Ideally this would then lead to the formation of a grid

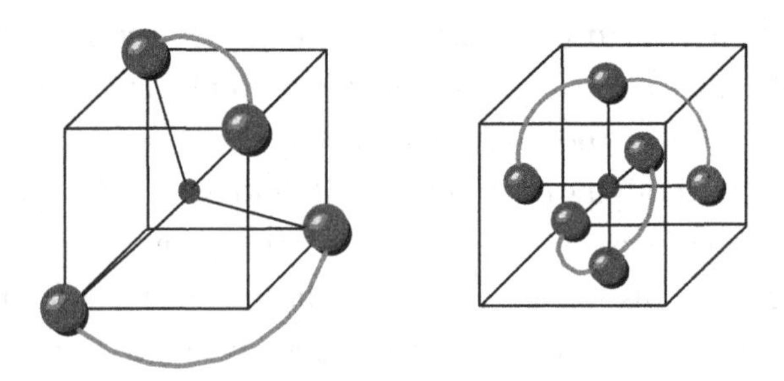

Scheme 2.

on the basis of the complementary arrangement of the n^2 metals and $2n$ ligands, and the 90° twists at all the ligand pockets. This is a complex set of encoded coordination instructions and requires that the ligand design is such that the yields of alternative non-grid arrangements are minimized. In reality one should expect a number of different equilibrium processes to occur in solution as the solvated metal ions and ligand react, and so other oligomers may also be anticipated (*vide infra*).

A comprehensive review of grid-type architectures published by Ruben (Ruben, 2004) examines square based $[n \times n]$ systems with nuclearities in the range 4 to 16. Some of the examples cited therein will be discussed in the present chapter, but in addition new ones will be discussed, along with other oligomeric non-grid assemblies, which occur as a result of the conformational flexibility of some ligands, and the coordination preferences of the metal ions, and other non-grid based equilibria.

1.3. *Thermodynamic aspects of the formation of convergent self-assembled grid architectures*

The synthesis of large, polymetallic architectures using a 'covergent' predictable approach presents the chemist with synthetic challenges in terms of producing suitable polytopic ligands. The extension of simple dinucleating fragments, e.g. secondary amino alcohols and thiols, phenols, and heterocyclic diazines (e.g. pyridazine, pyrimidine) has been a major focus, with considerable success (*vide infra*). The synthesis of self-assembled $[n \times n]$ (e.g. $n = 2-5$) grids is constrained by the normal ligand substitution issues (*vide supra*), as the ligand donor atoms compete for coordination sites on solvated metal cation precursors, but also by the organizational problems of aligning the requisite numbers of ligands (4, 8 and 10 respectively) and metal ions (4, 16 and 25 respectively) in an appropriate grid arrangement.

$$\begin{aligned} &4LH_2 + 4[M(II)(L')_6]^{2+} \\ &\quad \rightarrow [M(II)_4(L^{2-})_4] + 8H^+ + 24L' \quad (\Delta n = 25) \end{aligned} \tag{1}$$

$$\begin{aligned} &6LH_2 + 9[M(II)(L')_6]^{2+} \\ &\quad \rightarrow [M(II)_9(L^{2-})_6]^{6+} + 12H^+ + 54L' \quad (\Delta n = 52) \end{aligned} \tag{2}$$

$$8LH_2 + 16[M(II)(L')_6]^{2+} \rightarrow [M(II)_{16}(L^{2-})_8]^{16+} + 16H^+ + 96L' \quad (\Delta n = 89) \qquad (3)$$

$$10LH_2 + 25[M(II)(L')_6]^{2+} \rightarrow [M(II)_{25}(L^{2-})_{10}]^{30+} + 20H^+ + 150L' \quad (\Delta n = 136) \qquad (4)$$

Entropic considerations are clearly important, as part of the thermodynamic balance associated with the formation of intermediate species and fully assembled grids. Equations 1–4 highlight a set of idealized assembly reactions illustrating the entropy effects associated with the displacement of very large numbers of solvent molecules (ligands) from the metal ion coordination spheres as ligand substitution takes place in solution for hypothetical [2 × 2], [3 × 3], [4 × 4] and [5 × 5] grids respectively. The ligands in question are assumed to have readily dissociable protons (numbers actually based on the hydrazone ligands discussed in detail later). However, similar arguments would apply in the case of neutral ligands. The number of (idealized) independent particles released as a function of grid size goes up dramatically as the grid size itself increases, highlighting the importance of this contribution to the overall thermodynamic balance.

2. Ligands and Complexes

2.1. *Ditopic ligands and their complexes*

Typical ditopic ligands include those based on dinucleating diazine fragments, e.g. pyridazine, tetrazine, pyrazole, pyrazine, pyrimidine, open chain diazine etc. (e.g. L1–L10; Chart 1), in which the metals would ideally be bridged by the diazine groups, and so the metal separations would be limited by the geometrical projections of the nitrogen donor atoms at the bridges themselves. This would lead to close proximity of the metal ions in the case of e.g. pyridazine and pyrazole, but more remote positions in the case of pyrimidine and pyrazine. Any electronic communications between the metal ions would thus be dependent on the particular diazine fragment, and the interaction pathways involved. Molecular electronic properties are likely to be more useful if there is some degree of cooperativity between the paramagnetic unpaired spin centres, which could lead to single molecule magnetic

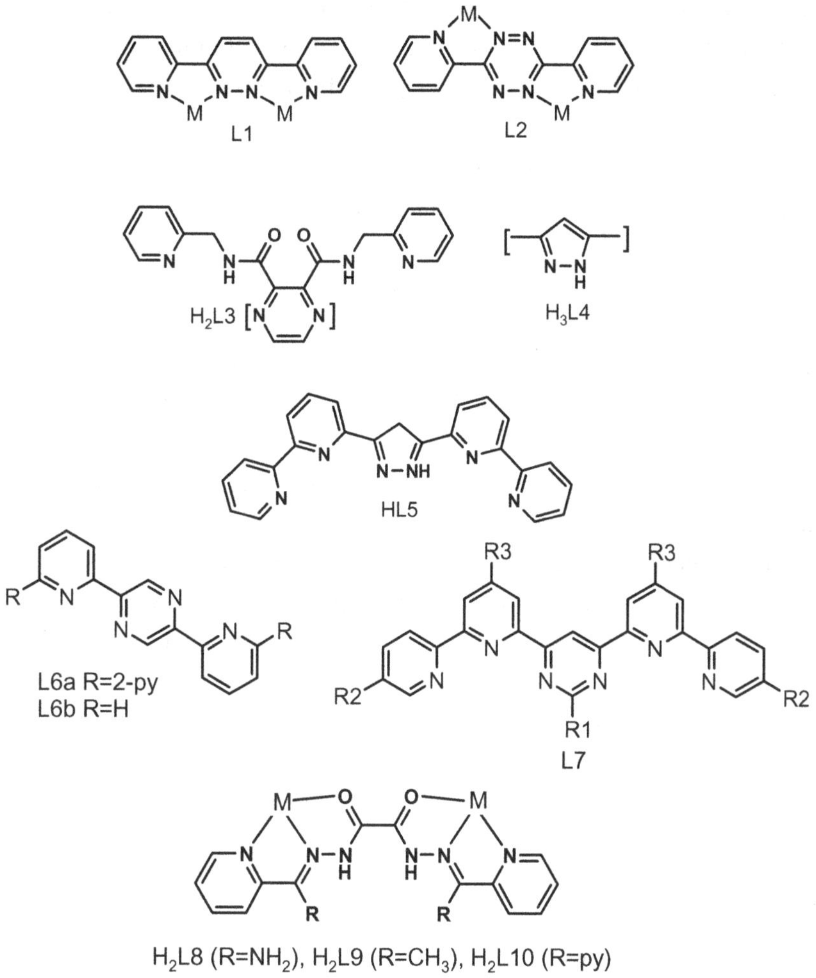

L1 L2

H_2L3 H_3L4

HL5

L6a R=2-py
L6b R=H

L7

H_2L8 (R=NH_2), H_2L9 (R=CH_3), H_2L10 (R=py)

Chart 1.

(SMM) behavior or functional and reversible metal based redox behavior, where charge redistribution/delocalization may lead to stability of intermediates, and overall bistability. This would be enhanced with smaller bridges between the metal ions. The classic magnetic iron oxides (e.g. Fe_2O_3) involve a network of single oxygen atom bridges between the iron centres. Spin exchange through such bridges in Mn and Fe polynuclear clusters often leads to SMM behavior. The now classic Mn_{12} carboxylate based cluster examples (Lis, 1980; Cornia *et al.*, 2002; Sessoli *et al.*, 1993a,b) display quantum tunneling of magnetization and SMM behaviour at low temperatures, as a result of the combination of mixed oxidation state metal sites (Mn(III) and Mn(IV)) and the specific orientation of the Mn(III) Jahn-Teller axes.

Single atom bridging subunits are limited to largely divalent group VI elements, typically oxygen, and less commonly sulphur. Halogens also act as bridging ligands in simple anionic form, but the chalcogenides can be integrated into a ligand framework and still function in a bridging bidentate capacity via lone or monovalent electron pairs. Substituted phenols or secondary alcohols, carbo- and thiocarbo-hydrazones fall into this class and provide a versatile group of organic ligands, which can be broadly functionalized to create grids as target molecules. Ligands H_nL11–15 (Chart 2) are typical examples, with potentially bridging single atoms e.g. O, S, which would of necessity produce short connections between metal ions, leading to relatively short metal-metal separations and more direct electronic interactions between the metals than would occur with diazine bridged systems (*vide supra*). Also, based on the proximity of the bidentate and tridentate coordination pockets, and the contiguous arrangement of five-membered chelate rings, these ligands are ideally constructed to lead to [2 × 2] grids via self-assembly (see Chart 1 for HL11 example showing the metal binding sites).

2.1.1. *Homometallic complexes*

2.1.1.1. [2 × 2] grids with heterocyclic diazine (N_2) bridging ligands

L1 is a bis-bidentate ditopic diazine ligand and was shown to be the first well documented example of a ligand which could direct the self-assembly of a [2 × 2] square grid on reaction with Cu(I). The [2 × 2] square, pyridazine bridged, homoleptic grid complex $[Cu(I)_4(L1)_4](CF_3SO_3)_4$ was produced in high yield, with four *syn*-ligands arranged in two parallel pairs above and below the plane of four tetrahedral Cu(I) ions (Youinou *et al.*, 1993). The four coordinate tetrahedral Cu(I) ions dispose the pairs of N_2 donor pockets in each intersecting ligand at 90°, which exactly matches the typical copper coordination sphere dimensions. The tetrazine ligand L2 has a similar topology, but forms the heteroleptic complex $[Zn(II)_4(L2)_4(H_2O)_4(CH_3CN)_4](ClO_4)_8$ with four Zn(II) ions held in a [2 × 2] square framework of four *anti*-ligands. In this and related cases the six-coordinate requirement of the Zn(II) ions dictates that eight co-ligands fill vacant sites (H_2O, CH_3CN), which result from the coordination mismatch between the ligand (bis-bidentate), and the six-coordinate Zn(II) ions (Bu *et al.*, 2000).

Self-assembly of L2 with Ni(II) salts appears to depend on the size of anion used with the [2 × 2] square grid $[Ni(II)_4(L2)_4(CH_3CN)_8](BF_4)_8$ and the pentagonal ring $[Ni(II)_5(L2)_5(CH_3CN)_{10}]$ $(SbF_6)_{10}$ forming in response to the size of the templating anion (SbF_6^- larger than BF_4^-) (Campos-Fernández *et al.*, 1999; Campos-Fernández *et al.*, 2005). Antiferromagnetic coupling between the Ni(II) centres through the tetrazine rings was observed in both cases. H_2L3 is based on a 1,4-pyrazine subunit, and provides the metal ion with a choice of N or O donors at the amide function. The complexes $[Cu(II)_4(HL3^-)_4](BF_4)_4$ (**1**) (Hausmann *et al.*, 2003) and $[Cu(II)_4(HL3^-)_4](ClO_4)_4$ (**1a**) (Cati *et al.*, 2004) involve self-assembled square arrangements of four six-coordinate Cu(II) ions four ligands completing *mer*-N_3 donor groupings at each metal ion. The structure of the cation in **1** is shown in Fig. 1. A similar structure was observed for **1a**. Weak antiferromagnetic exchange coupling was observed for **1a** ($J = -5.9\,cm^{-1}$)

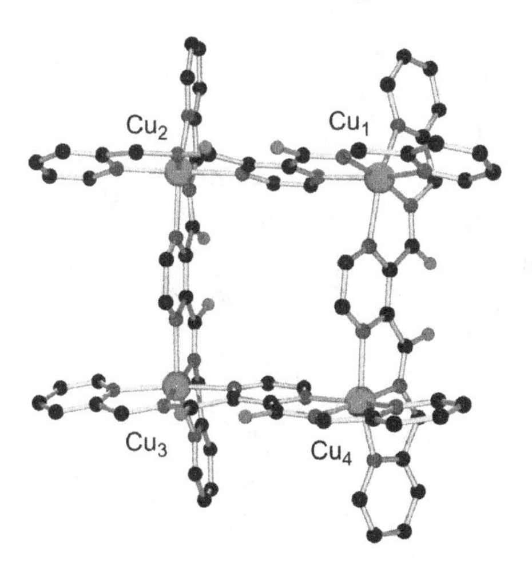

Fig. 1. Structure of Complex 1.

(Cati *et al.*, 2004), resulting from the remote disposition of the Cu(II) ions ($H_{ex} = -J\{S_1 \cdot S_2 + S_2 \cdot S_3 + S_3 \cdot S_4 + S_1 \cdot S_4\}$).

Substituting the central pyrazine with pyrazole (H_3L4) (Chart 1) creates a ligand with a bite comparable to H_2L3, except that the central pyrazole group constrains the ligand somewhat based on the angle projections of the pendant amide groups (vide infra). A [2 × 2] twisted square complex $[Cu(II)_4(HL4^{2-})_4]$ was obtained with just pyrazole bridges connecting the square-pyramidal Cu(II) ions, with terminal pyridine groups uncoordinated. Weak antiferromagnetic exchange was observed ($J = -8.2\,cm^{-1}$), consistent with the presence of non-orthogonal magnetic orbital connections (Klingele *et al.*, 2007). HL5 has a similar bite, but is a more rigid ligand. It forms [2 × 2] square based grids with Mn(II), Co(II) and Cu(II), with just pyrazolate bridges between metal ions (van der Vlugt, *et al.*, 2008). The structure of $[Cu(II)_4(L5)_4](ClO_4)_4$ (**2**) is shown in Fig. 2, and exhibits a pronounced tetrahedral distortion of the Cu_4 framework. All the copper ions are square pyramidal and linked by their basal planes via the pyrazole bridges. This leads to weak antiferromagnetic exchange between metal ions ($J = -18.7\,cm^{-1}$) (van der Vlugt, *et al.*, 2008).

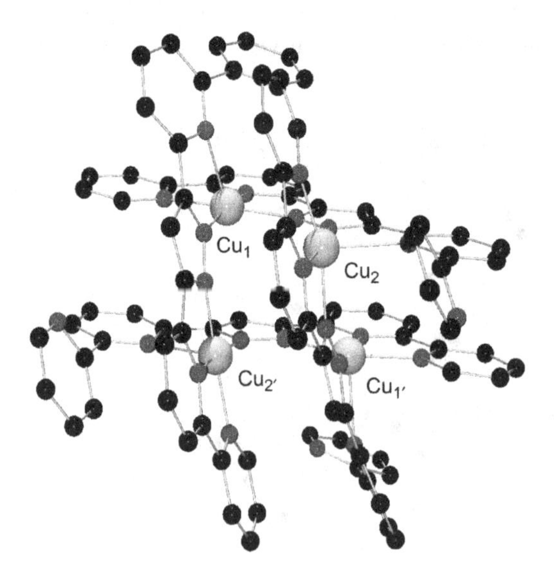

Fig. 2. Structure of Complex 2.

The ditopic bis-bidentate pyrazine ligand L6a is ideally suited to bind in a *mer* fashion to two six-coordinate metal ions, with an *anti*-arrangement of two N_3 donor pockets, and forms the [2 × 2] square based homoleptic grid $[Zn(II)_4(L6a)_4](PF_6)_8$. The dissymmetric nature of the ligand results in a trapezoidal arrangement of the four Zn ions in the racemic complex (Bark *et al.*, 2001). The related bis-bidentate ligand L6b has a similar *anti*-arrangement of donor groupings, but at present only one self-assembled motif has been reported in the unusual trinuclear oligomeric complex $[Zn(II)_3(L11b)_3Cl_6]$ (**3**), which has three ligands spanning the sides of an equilateral triangle (Fig. 3), with terminal Cl ligands completing six-coordination at the Zn centres (Neels and Stoeckli-Evans, 1999). This indicates that despite the grid based direction provided by the encoded information in the ligand, other competing reactions are taking place, and in this case the lower order oligomer appears to dominate. The ditopic pyrimidine based ligand L7 creates N_3 *mer* tridentate donor groupings in a roughly linear arrangement, and provides an ideal platform for homoleptic square [2 × 2] grid formation with a six-coordinate metal ion. Numerous examples of [2 × 2] grids in this class ($[M_4(L7)_4]X_8$) have been produced (M = Mn(II), Fe(II), Co(II), Zn(II); varying R1, R2, R3). An important property in the case of $[Co(II)_4(L7)_4](BF_4)_4$ (R1 = Ph; R2 = R3 = H), is its redox behaviour, which showed several reversible multistep redox

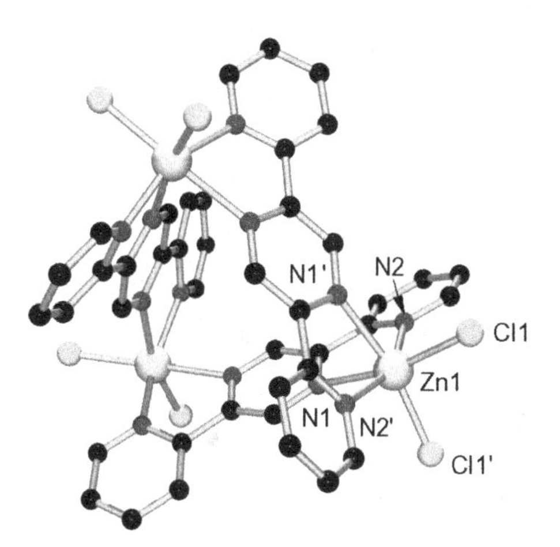

Fig. 3. Structure of Complex 3.

processes in solution associated specifically with ligand reduction, rather than metal based redox (Ruben *et al.*, 2003).

2.1.1.2. Ditopic ligands with more remote coordination pockets

Oxalic bis-hydrazone ligands, e.g. H_2L8–H_2L10, provide a large number of coordination options based on several different conformations which can be adopted, and are created easily by reaction of the parent bis-hydrazone with ketones and nitriles. One such option is based on two N_2O pockets, and in the *cis* form (shown in Chart 1), two metal ions would be separated by a three bond bridging connection. The complexes $[Co(II)_2Co(III)_2(HL8a^-)_4]Br_6\cdot 9H_2O$ (**4**) and $[Fe(II)_2Fe(III)_2(HL8a^-)_4](ClO_4)_6\cdot 7.5H_2O$ (**5**) result from reaction of H_2L8 with $Co(II)Br_2$ and $Fe(II)(ClO_4)_2$ respectively in air, and have homoleptic, remote square [2 × 2] grid structures, with the six-coordinate metals bound just in the N_2O end pockets. The structure of the Co_4 cation in (**4**) is shown in Fig. 4. Two of the ligands adopt *cis* conformations, while the other two adopt *trans* conformations. Co-L distances for Co_3 and Co_4 are short (1.93 $Å_{ave.}$) indicating oxidation of two metal sites to

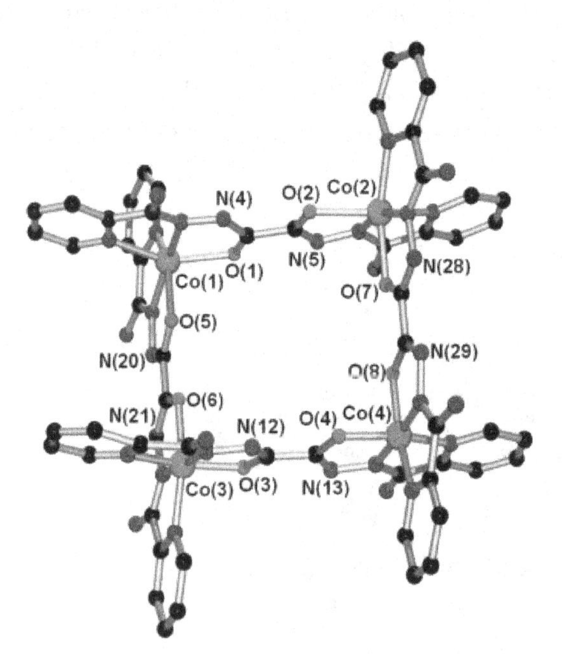

Fig. 4. Structure of Complex 4.

Co(III) in air (Zhao *et al.*, 2004). A similar situation exists for **5**, in which two iron centres oxidized to Fe(III) (Zhao *et al.*, 2004). The remote assembly creates a large internal cavity, and large distance between metal ions (Co-Co distances 6.7–7.3 Å; c.f. Fe_4), resulting in weak magnetic exchange.

2.1.1.3. Other polynuclear oligomers with remote ditopic ligands

Ligands in this class clearly have a great deal of flexibility, and while grid based encoding of coordination is clearly an important feature (*vide supra*), other oligomeric structures might be anticipated. Reaction of H_2L8 with nickel(II) tetrafluoroborate and perchlorate gave a completely different Ni_4 cluster architecture, with six ligands assembled around a tetrahedral core of metal ions, and with each nickel ion bound at the juncture of three N_2 ligand ends, leading to N_6 octahedral coordination spheres. The structure of $[Ni(II)_4(H_2L8)_6]$ $(BF_4)_6F_2$ (**6**) is shown in Fig. 5 (a similar cationic

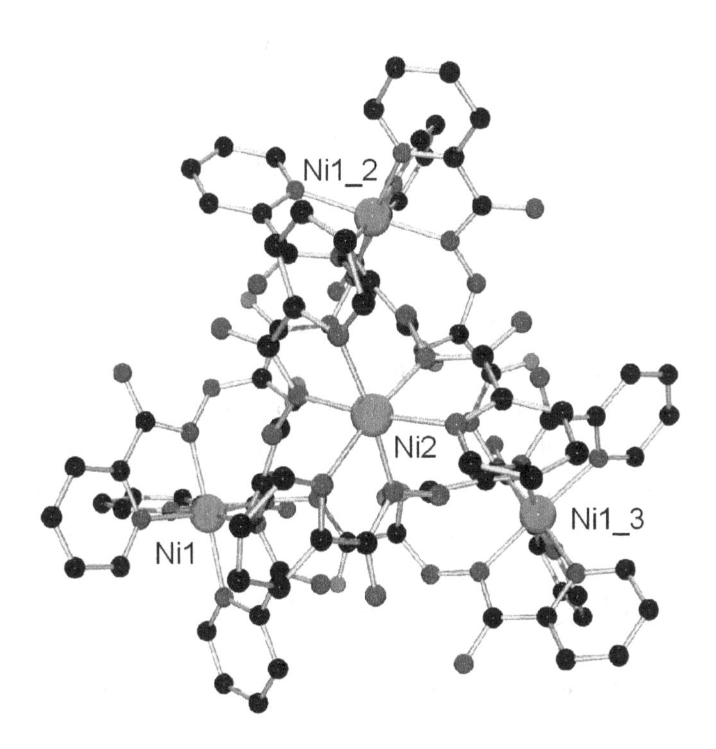

Fig. 5. Structure of Complex 6.

structure is observed for the analogous perchlorate complex). The oxalic hydrazone central part of the ligand is uncoordinated, leading to very large Ni-Ni distances (8.4–9.2 Å). What is significant is that the nickel(II) ions ignore the hydrazone oxygen atoms as potential donors in favour of an all nitrogen coordination sphere from three intersecting N_2 ligand ends. The major reason for this behaviour would reasonably be associated with the enhanced crystal field stabilization energy of Ni(II) in an N_6 crystal field environment, as opposed to N_4O_2, in comparison with Co(II) (*vide infra*) (Zhao *et al.*, 2004). This turns out to be a common feature associated with the self-assembly reactions involving Ni(II).

2.1.1.4. [2 × 2] grids with single atom μ-O and μ-S bridging ditopic ligands

HL11 and HL12 incorporate central secondary alcohol and phenolic subunits respectively, both of which provide a dinucleating focus in a μ-O bridged arrangement (see Chart 2). Examples of [2×2] grids with HL11 and HL12 appear to be limited so far to just $[Mn_4(L11^-)_4](ClO_4)_4$(**7**) (Galasco *et al.*, 1996), and $[Cu_4(L12^-)_4](PF_6)_4$(**8**) (Rojo *et al.*, 1999). The structure of **8** is depicted in Fig. 6, with the four copper ions in a square μ-O bridged arrangement. Magnetic properties are not discussed for **7**, but in **8** very weak antiferromagnetic exchange is associated with significant orthogonality between the metal magnetic orbitals associated with the orientation of the copper Jahn-Teller axes.

Carbo-hydrazone and thio-carbohydrazone ligands (H_2L13^*) are similarly designed and quite amenable to grid formation, with the first example, $[Co(II)_4(HL13a^-)_4](CH_3COO)_4$, reported in 1997 (Duan *et al.*, 1997). Four Co(II) centres are bridged by deprotonated thiolate sulphur atoms in a square [2 × 2] grid arrangement with a large Co-S-Co angle (151°). Very short Co-N distances (1.905–1.983 Å) suggest the possibility of LS (low spin) Co(II), but no magnetic data are reported. $Ni(II)_4$ and $Zn(II)_4$ [2 × 2] grid complexes are reported with H_2L13b, along with a mixed valence iron [2 × 2] grid $[Fe(II)_3Fe(III)(L13b^{2-})_2(HL13b^-)_2](FeCl_4)_3$ (Manoj *et al.*, 2007). Magnetic data for this complex do not clearly indicate the spin state situation and the extent of exchange coupling. Reaction of H_2L13b with $Fe(II)SO_4{\cdot}7H_2O$ and $NaClO_4$, with no apparent exclusion of air, produced

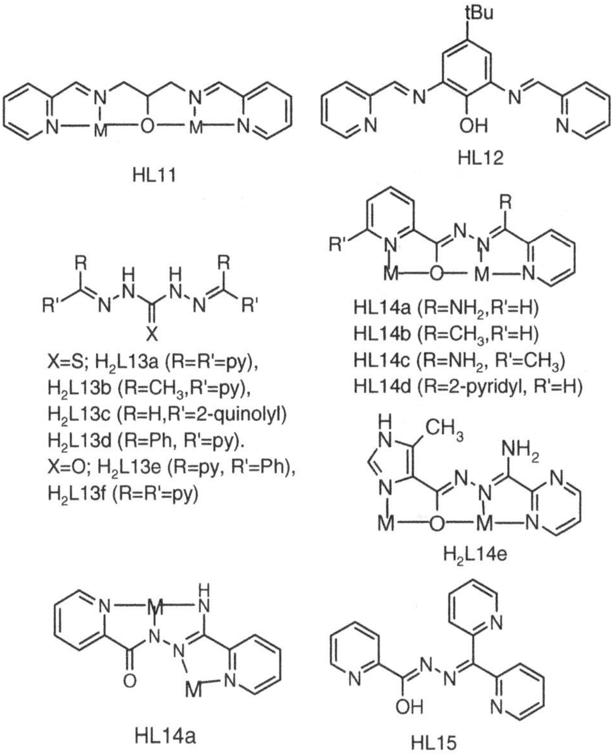

HL11

HL12

HL14a (R=NH_2,R'=H)
HL14b (R=CH_3,R'=H)
HL14c (R=NH_2, R'=CH_3)
HL14d (R=2-pyridyl, R'=H)

X=S; H_2L13a (R=R'=py),
H_2L13b (R=CH_3,R'=py),
H_2L13c (R=H,R'=2-quinolyl)
H_2L13d (R=Ph, R'=py).
X=O; H_2L13e (R=py, R'=Ph),
H_2L13f (R=R'=py)

H_2L14e

HL14a

HL15

Chart 2.

$[Fe(II)_2Fe(III)_2(L13b^{2-})_4](ClO_4)_2$, in which some Fe(II) sites appear to have been oxidized to Fe(III) (Zhao *et al.*, 2008). Electrochemical measurements on this complex showed a suite of well separated one electron redox steps ranging from $Fe(II)_4$ to $Fe(III)_4$ (−0.8 to +0.8 V vs. Ag/AgCl) (Zhao *et al.*, 2008). Molecular based systems of this sort with multiple redox steps have significant potential for device activity based on the possible bistability of the individual redox intermediates. H_2L13a produces the square [2 × 2] grid $[Ni(II)_4(HL13a^-)_4](PF_6)_4$, in which one pyridine ring remains uncoordinated. The magnetic properties indicate strong intramolecular antiferromagnetic exchange between the Ni(II) centres (Cheng *et al.*, 2000). The related ligand H_2L13d reacts with $NiCl_2$ followed by addition of PF_6^- to give $[Ni(II)_4(L13d)_4](PF_6)_4$, in a similar antiferromagnetically coupled

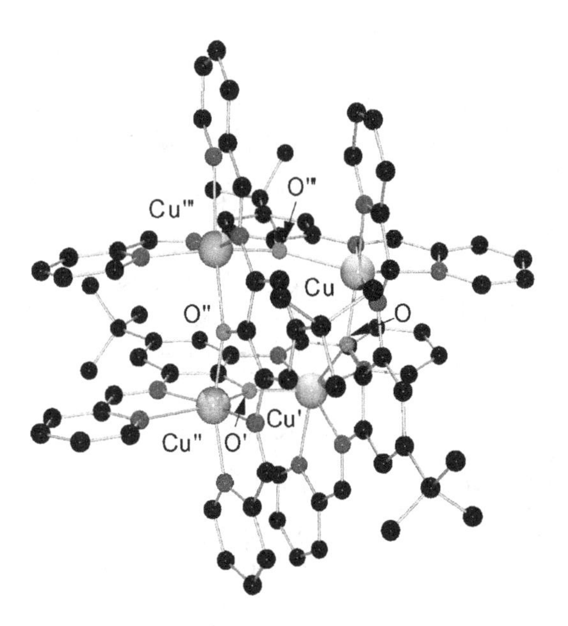

Fig. 6. Structure of Complex 8.

$[Ni_4(\mu\text{-}S)_4]$ square structure (Cheng *et al.*, 2000). Carbohydrazone ligands (H_2L13e,f) have the same ditopic pocket arrangement, and produce $[2 \times 2]$ μ-O bridged antiferromagnetically coupled square grids with Ni(II) (Manoj *et al.*, 2007) and Fe(II) (Wu *et al.*, 2008). In the Fe(II) case an unusual spin crossover (SC) transition is observed at ~130 K associated with two iron centres changing from low spin (LS) to high spin (HS).

2.1.1.5. Ditopic hydrazone ligands with both μ-O or μ-NN bridging modes

Asymmetric ditopic hydrazones (e.g. HL14a–c) are a versatile class of ligands producing self-assembled grids with adjacent, different tridentate *mer*-N_2O and bidentate NO pockets, and in one common conformation can lead to oxygen bridging arrangements (Chart 2). The pockets create five-membered chelate rings, which leads to a contiguous and by default, linear arrangement of the metal ions relative to the oxygen bridge, suitable for grid formation. This group of ligands, and other variants, in general bind two metals in a μ-O fashion, and lead to high yield self-assembly of a large number of square $[2 \times 2]$ grid complexes with Mn(II), Ni(II), Co(II), Cu(II)

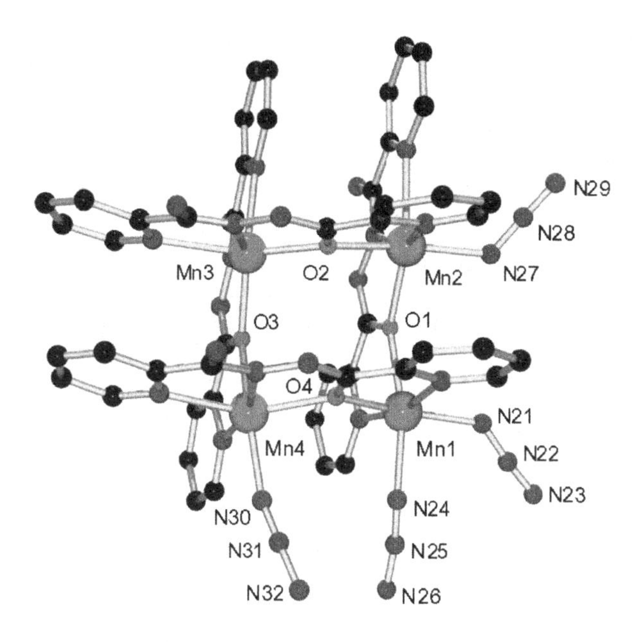

Fig. 7. Structure of Complex 9.

and Zn(II) (Dawe and Thompson, 2008; Dawe *et al.*, 2008; Dawe *et al.*, 2006; Matthews *et al.*, 1999; Thompson *et al.*, 2001; Matthews *et al.*, 1999; Matthews *et al.*, 2001). Figure 7 shows the structure of a typical heteroleptic example, $[Mn(II)_4(L14a^-)_4(N_3)_4]$ (**9**), with four co-ligands (azide) which bind to the otherwise vacant sites left at some of the six-coordinate metal ions resulting from the assembly of the four pentadentate ligands and four metals (Dawe *et al.*, 2006). Some $Cu(II)_4$ complexes are homoleptic, due to the metal centres adopting five-coordinate geometries. Typically anti-ferromagnetic exchange prevails within the grids, due to the large M-O-M angles, except in the case of copper, where strict magnetic orbital orthogonality leads exclusively to intra-molecular ferromagnetic exchange. This occurs as a result of the fact that the copper Jahn-Teller axes arrange themselves in such a way that axial and equatorial orbitals are connected through the oxygen bridges. This is illustrated in Figs. 8(a) and 8(b) for the complex $[Cu(II)_4(HL14e^-)_4](ClO_4)_4{\cdot}3CH_3CN$ (**10**). The hydrazone group in the HL14 type ligands can tautomerize into keto or enolic forms, and so could present either the hydrazone oxygen or the diazine NN group as a possible bridge between metal ions (see Chart 2 for NN bridging alternative for

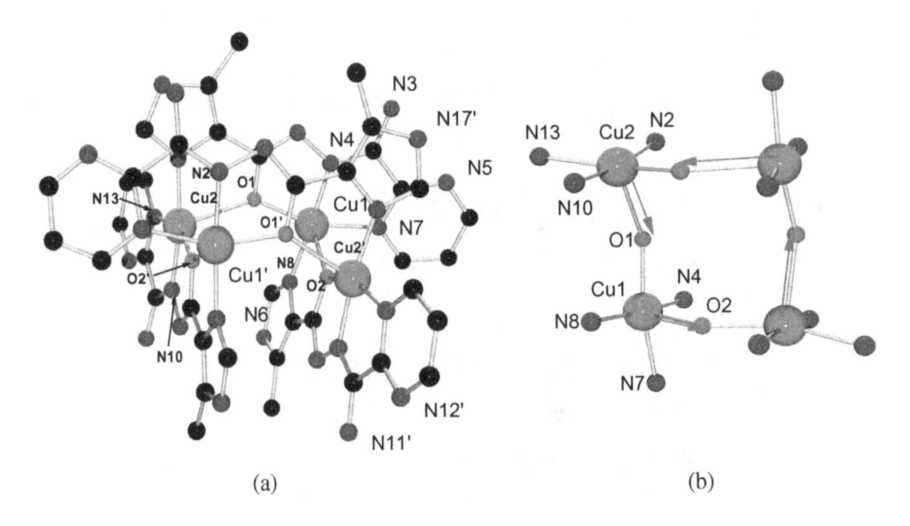

Fig. 8. Structure of Complex 10.

L14e$^-$). In most cases with the ditopic hydrazone ligands, self-assembly results in μ-O bridged square [2 × 2] grids. However, rectangles have been observed occasionally, in which mixtures of both μ-O and μ-NN bridges exist, leading to short and long bridge connections respectively. HL15 (Chart 1) forms the complex $[Mn(II)_4(L15^-)_4(H_2O)_2](ClO_4)_4$ (**11**) in which two ligands bridge via the diazine (NN) and two by the hydrazone oxygen, forming a rectangle (Thompson *et al.*, 2001). Reaction of $Co(CH_3COO)_2$ with HL14a in air, followed by the addition of $NaNO_3$, leads to the complex $[Co(II)_2Co(III)_2(L14a^-)_2(L14a^{2-})_2(H_2O)_2(MeOH)_2](NO_3)_4$ (**12**), which has a similar rectangular structure with both types of bridge (Fig. 9). Co(1) has very short M-L bonds indicative of oxidation of Co(II) to Co(III), while the Co(2) sites remain as Co(II). In this case a second proton loss occurs from the NH_2 group on two ligands, which are coordinated to the Co(III) centres. This unusual ligand behaviour is not usually observed, with a single proton loss per ligand being the normal situation. Additional ligands (H_2O, MeOH) are bound to Co(2) (Thompson *et al.*, 2001). $[Mn(II)_4(L15^-)_4(H_2O)_2](ClO_4)_4$ exhibits weak anti-ferromagnetic exchange ($J_O = -1.9\,cm^{-1}$, $J_{NN} = -0.1\,cm^{-1}$), while **12** is essentially uncoupled due to the large distance of separation of the Co(II) centres (Thompson *et al.*, 2001).

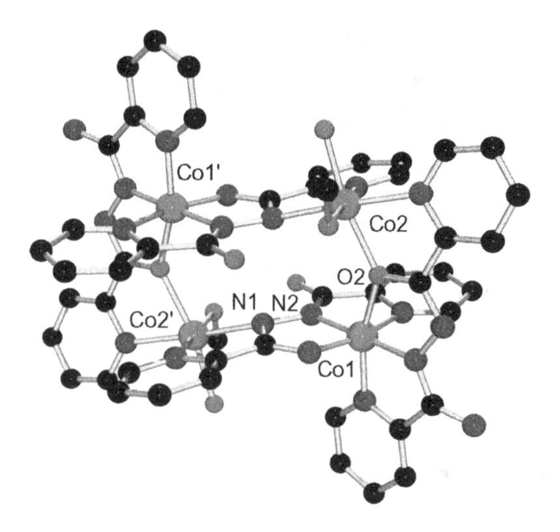

Fig. 9. Structure of Complex 12.

2.1.1.6. Higher order oligomeric clusters based on ditopic ligands

The coordinative unsaturation which occurs with the ditopic hydrazone ligands (HL14), resulting from their nominally pentadentate nature, leaves open the possibility of the formation of other higher order oligomeric self-assembled clusters, in which a homoleptic arrangement could be achieved. The next highest order homologue satisfying this homoleptic situation would be a pentanuclear cluster arrangement (6 × (5) ligand sites for 5 × (6) metal sites). This has been found to occur as an alternative self-assembly product with Mn(II) and Co(II) in the complexes $[Mn(II)_5(L14e^-)_6](ClO_4)_4 \cdot 7H_2O$ (**13**) and $[Co(II)_5(L14e^-)_6](ClO_4)_4$ respectively (Dawe *et al.*, 2008; Dawe *et al.*, 2006; Matthews *et al.*, 1999; Matthews *et al.*, 2001). The structure of the cation in $[Mn(II)_5(L14e^-)_6](ClO_4)_4 \cdot 7H_2O$ is shown in Fig. 10 and the core structure in Fig. 11. The penta-cobalt cluster has a similar structure. Each metal is bridged to its neighbours through μ-O hydrazone oxygen atoms, with the metals adopting a trigonal bipyramidal cluster shape. Each pentadentate ligand binds using five donor sites, such that an exact match exists between the donor complement of the six ligands (30) and the total donor requirement of the five six-coordinate metal ions (30). Large Mn-O-Mn angles in the manganese cluster lead to

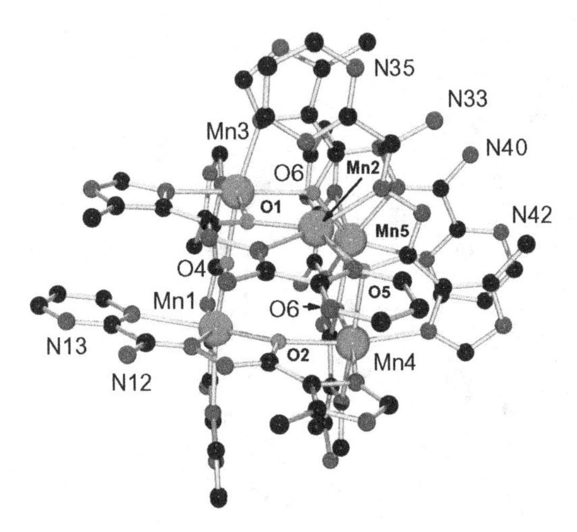

Fig. 10. Structure of Complex 13.

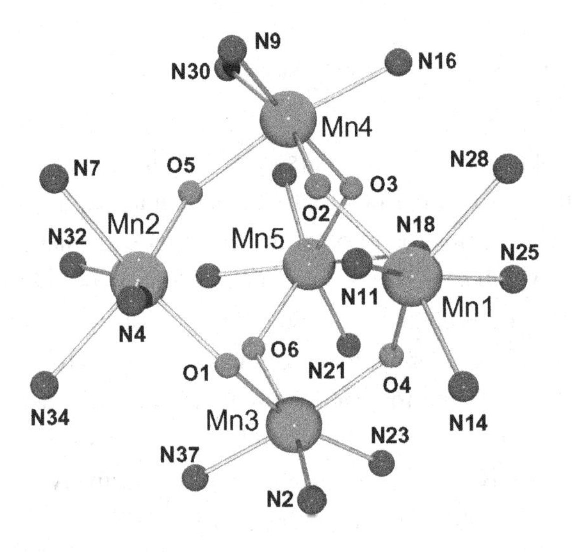

Fig. 11. Core Structure of 13.

intermolecular antiferromagnetic exchange, with an $S = 5/2$ ground state spin, resulting from the odd number of metals present. No other expanded clusters based on ligand coordinative un-saturation with related ligands appear to have been reported.

Electrochemical properties have not generally been considered to be of importance in the context of single molecule 'bistability', despite the fact that such properties are exhibited at room temperature. Due to the dearth of potentially useful magnetic systems (blocking temperatures too low) some attention is now being paid to other features of the electronic nature of clusters and grids, based on the inherent ability of some transition metal ions to display spin fluxionality as a function of applied electrical potential, and the formation of different, and 'bistable' electronic redox states. The pentanuclear cluster complex $[Mn_5(L14a)_6](ClO_4)_4$ (**14**) (same overall structure as **13**) was one of the first examples of such a cluster, and shows remarkable electrochemical response at room temperature, with three well defined reversible one electron CV waves in the range 0.5–1.0 V in acetonitrile (vs SSCE), associated with oxidation of the three equatorial metal centres in the trigonal bipyramidal cluster from Mn(II) to Mn(III) (Matthews *et al.*, 1999). The oxidation process is accompanied by a colour change from orange to brown. Such properties are not normally temperature dependent, and so are attractive at the molecular level in the context of device behavior (*vide infra*). The tetra-cobalt cluster $[Co(II)_4(L7)_4](BF_4)_4$ falls into this category, but its redox behaviour is ligand rather than metal ion based (*vide supra*) (Ruben *et al.*, 2003).

2.1.2. *Heterometallic* [2 × 2] *and mixed spin state grids*

The self assembly of a symmetric ditopic bis-tridentate ligand involving the formation of five-membered chelate rings (e.g. L7, HL11, HL12) with one type of metal ion leads to a homometallic [2 × 2] grid, in which there would be little distinction between the coordination pocket environments, and so no competitive coordinative advantage for one particular site. This would then make it difficult to incorporate different metal ions in the same grid if there was direct competition between the different ions and the ligand in the same reaction mixture. The creation of heterometallic grids is attractive from a magnetic perspective, since this could lead to non-zero and possibly high spin ground states in an antiferromagnetic situation, a desirable magnetic attribute in the search for functionally useful molecular based magnetic properties. Coordinative preferences based for example on different metal ion CFSE (crystal field stabilization energy) requirements would not lead to any thermodynamic selectivity in this case, based on a coordination pocket preference. However, if a mononuclear precursor could be produced, with

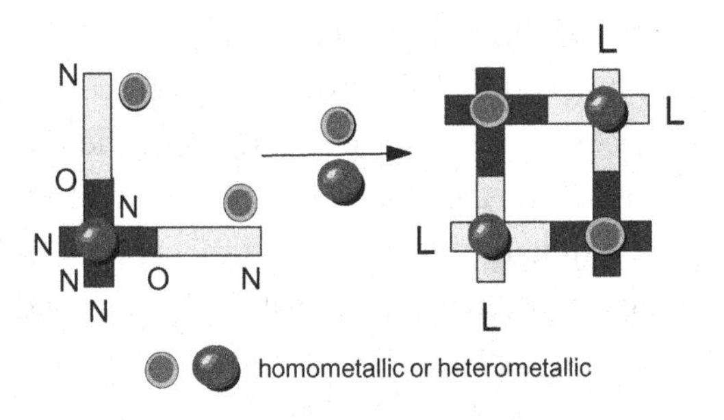

Scheme 3.

an appropriate arrangement of two intersecting ligands, or even one ligand with labile co-ligands, a 'complex as ligand' approach could possibly be used to create a mixed metal grid by reaction with a second metal ion, both in the presence or absence of extra ligand. This assumes that dynamic ligand dissociation effects are minimized.

Topo-selectivity has been achieved with the symmetric ditopic ligand L7 (Chart 1) by first producing a mononuclear complex with one six-coordinate metal ion bound to two *mer* N_3 end groups from two independent ligands (L7; R1 = R3 = S-*n*Pr; R2 = H), followed by reaction with a second six-coordinate metal ion to produce the $M_2M'_2$ mixed metal $[2 \times 2]$ grid. Mixed metal complexes have been produced with Ru(II)/Fe(II), Ru(II)/Co(II), Os(II)/Fe(II) and Os(II)/Ni(II) heterometallic combinations by this route, with success resulting through the formation of mono-nuclear bis-ligand complexes first, with the kinetically less labile metal ion (Bassani *et al.*, 1998). The resulting *mer* arrangement of two N_3 donor pockets oriented mutually at 90° establishes a suitable corner fragment of a grid, which could then effectively double to create the $[2 \times 2]$ grid on reaction with the second metal.

Ligands can be programmed selectively in a number of different ways by constructing compartments with different donor atoms (e.g. soft or hard or soft/hard combinations), and order (e.g. bidentate vs. tridentate), designed to attract metal ions with different donor and coordination geometry preferences; e.g. tetrahedral metals would be accommodated comfortably at the juncture of two bidentate pockets, while an octahedral metal ion would be accommodated comfortably at the juncture of two tridentate pockets (vide

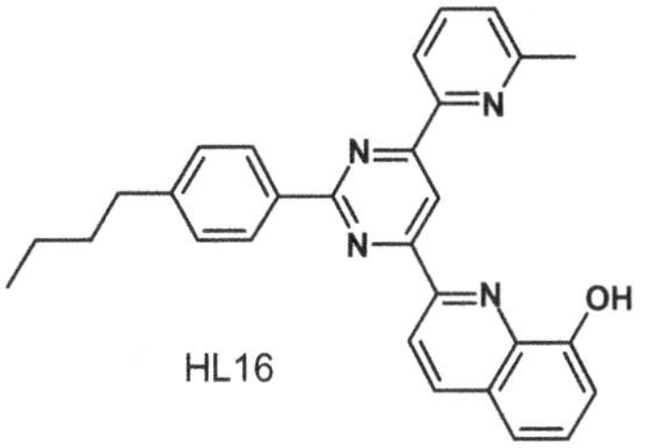

HL16

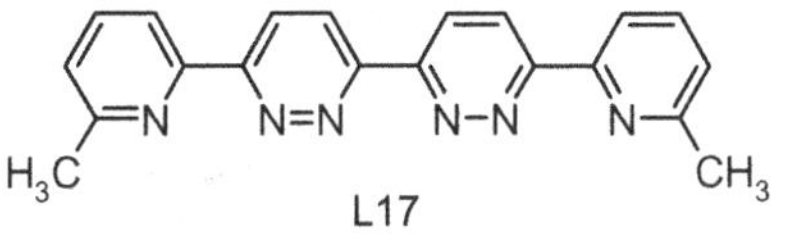

L17

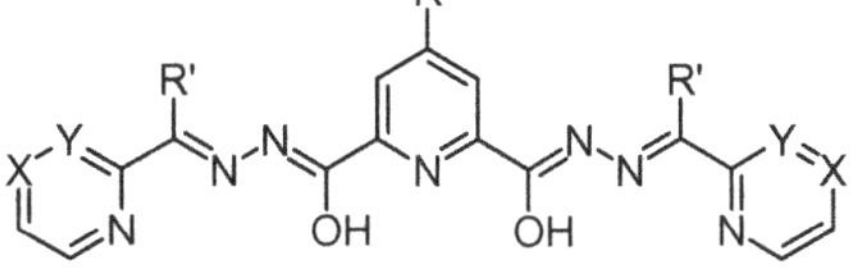

[X=Y=CH]; H_2L18a (R=H,R'=NH_2), H_2L18b (R=Cl,R'=NH_2), H_2L18c (R=H,R'=CH_3),H_2L18d (R=Cl,R'=CH_3),H_2L18e (R=H, R'=Ph), H_2L18f (R=Cl, R'=Ph), H_2L18g (R=OMe, R'=NH_2). [X=N,Y=CH]; H_2L18h (R=H,R'=NH_2). [X=CH,Y=N]; H_2L18j (R=H,R'=NH_2)

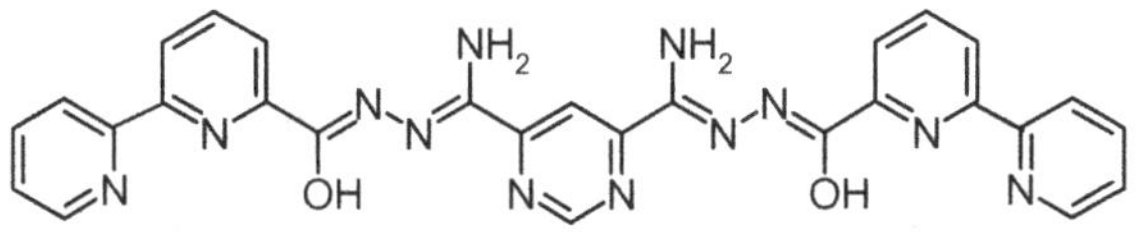

H_2L19

Chart 3.

supra). HL16 (Chart 3) has a potentially bidentate N_2 and a tridentate N_2O pocket, and reaction with a mixture of Cu(I) and Zn(II) ions results in a symmetrical $[Zn(II)_2Cu(I)_2(L17^-)_4]^{2+}$ grid with rhombohedral distortion in which the copper(I) ions occupy the N_4 sites resulting from the congruence of two bidentate N_2 pockets, and the zinc(II) ions the N_4O_2 sites, resulting from the congruence of two tridentate N_2O pockets (Petitjean *et al.*, 2004).

A different approach has been employed using the nominally kinetically stable Fe(III) complex $[Fe(III)(HL14a^-)(NO_3)(H_2O)_2](NO_3)_2$ (**15**), which only has one ditopic ligand bound through a *mer* N_2O arrangement

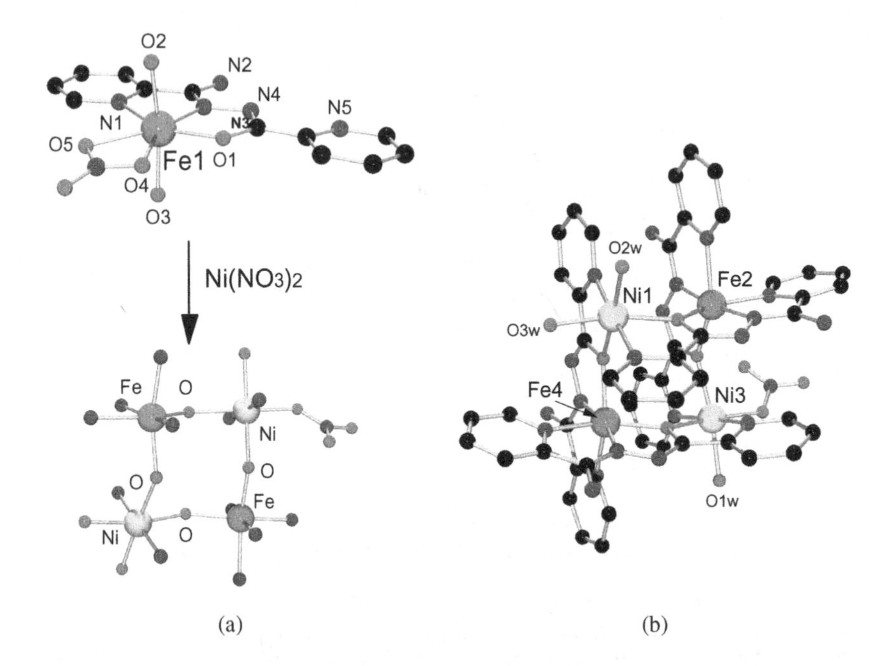

Fig. 12. Formation of Complex 15.

of donors (Fig. 12(a)). Reaction of this starting material with $Ni(NO_3)_2$ produced the symmetric [2 × 2] grid $[Fe(III)_2Ni(II)_2(L14a^-)_4(NO_3)(H_2O)_3](NO_3)_5 \cdot 2.5H_2O \cdot 1.75CH_3OH$ (**16**) (Fig. 12(b)), with two Ni(II) and two Fe(III) centres bridged in the normal μ-O square arrangement. The combination of different antiferromagnetically coupled spin centres ($S = 5/2$, $S = 2/2$; $J = -15.6\,\text{cm}^{-1}$) leads to a non-compensated ground state spin of S = 6/2 in a ferrimagnetic system (Parsons *et al.*, 2006). In another reaction of **15** with $Cu(ClO_4)_2$ a different heterometallic grid $[Cu(II)_3Fe(III)(L14a^-)_4(NO_3)]_2(ClO_4)_4(NO_3)_4 \cdot 12H_2O$ (**17**) was produced with three Cu(II) and one Fe(III) ions bound in a square μ-O bridged arrangement, with two such subunits connected by long μ-O (nitrate)-Cu axial contacts. The two tetranuclear halves are effectively magnetically isolated, and the intragrid exchange results from the sum of antiferromagnetic Cu-Cu and ferromagnetic Cu-Fe exchange terms, giving an $S = 6/2$ ground state (Xu *et al.*, 2001). In this case it is apparent that the system is somewhat dynamic, since some of the mononuclear iron complex must have dissociated in order for free ligand to have been generated.

2.2. *Symmetric tritopic ligands and their complexes*

2.2.1. *Homometallic* [3 × 3] *grids*

M_9 [3 × 3] grids are the next highest homologues in the square [n × n] series, but their synthesis is clearly more of a challenge, since organizing such a large number of metal ions into one small molecular based entity will depend on the appropriate encoding of the required coordination information in the tritopic ligand. The pyridazine core is a convenient multiple in this regard and in L17 two such subunits are joined together in creating a linearly disposed array of three N_2 coordination pockets suitable for binding to three tetrahedral metal ions. The complex $[Ag_9(L17)_6](CF_3SO_3)_9$ (**18**) was successfully synthesized, and represents the first such example in this class with nine tetrahedral Ag(I) ions bound to six ligands arranged in two parallel groups above and below the pseudo planar metal core (Baxter *et al.*, 1994). A structural representation is shown in Fig. 13. The metal ions are bridged by the pyridazine groups. The [3 × 3] grid is not square but has a compressed rhombically distorted shape, presumably in response to a slight mismatch

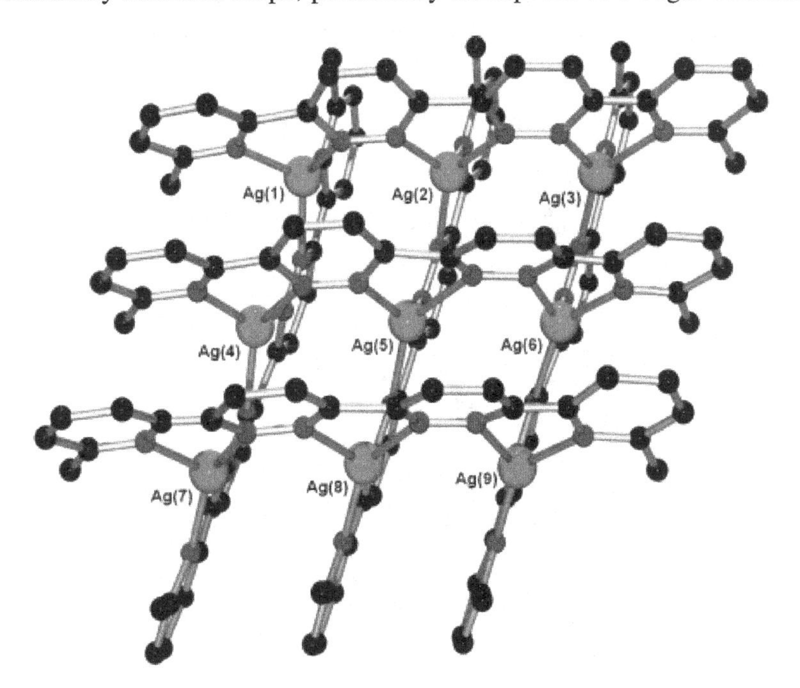

Fig. 13. Structure of Complex 18.

between the ligand dimensions and the coordination sphere requirements of the Ag(I) ion (ionic radius in four-coordinate environment ~1.0 Å), and the resulting long Ag-N distances (e.g. Ag-$N_{pyridazine}$ > 2.3 Å). Examples of related grids with other metals have not been reported.

Hydrazone based ditopic ligands $R(CO)NHNC(NH_2)R'$ are numerous (HL14*, Chart 2), and can be varied in many ways depending on the identity of the end groups (R, R′). However extension to tritopic variants is easy, and can be effected by starting with a pyridine-2,6-bis-hydrazone core (Chart 2, Chart 3; see H_2L18a-j), which generates two symmetrically appended hydrazone fragments, the ends of which can be altered in a variety of ways through simple organic elaboration. Recent review style articles have highlighted numerous examples of complexes of ligands in this class and their important physical properties (Milway *et al.*, 2006; Dawe *et al.*, 2009; Thompson *et al.*, 2003; Thompson *et al.*, 2005). Some typical examples will be discussed here. Tritopic ligands like H_2L18a, and its variants, have three potentially tridentate coordination pockets, which in the conformation shown (Chart 3) can accommodate three six-coordinate metal ions in a μ-O bridged arrangement. [3 × 3] square grids with $[M_9–(\mu–O)_{12}]$ cores can be produced in high yield by self-assembly with Mn(II), Cu(II) and Zn(II) salts. Figure 14 shows the structure of a

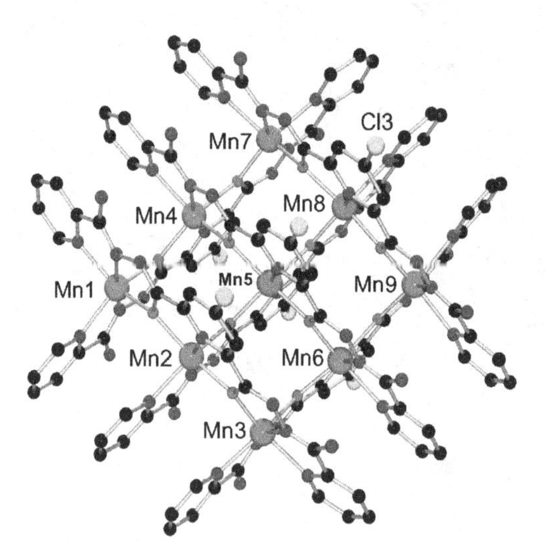

Fig. 14. Structure of Complex 19.

typical example $[Mn(II)_9(L18b^{2-})_6](ClO_4)_6$ (**19**) (Thompson *et al.*, 2003). Six hepta-dentate ligands encompass the nine six-coordinate metal ions in two roughly parallel groups of three above and below the roughly planar $[Mn_9-(\mu-O)_{12}]$ nonanuclear metal core. Mn-Mn separations are typically ~4 Å, with Mn-O-Mn angles in the range 130–140°. The close proximity of the metal centres in this μ–O bridged structure results in dominant intramolecular antiferromagnetic exchange between the Mn(II) ions, and an $S' = 5/2$ ground state, resulting from the combination of all the interconnected exchange pathways, and the odd number of metal ions present (Milway *et al.*, 2006; Dawe *et al.*, 2009; Thompson *et al.*, 2003; Thompson *et al.*, 2005).

$Cu(II)_9$ [3 × 3] grids also form readily, with the same arrangement of six-coordinate metal ions in the μ-O bridged structure. Figure 15 shows a typical example $[Cu(II)_9(HL18j^{-})_4(L18j^{2-})_2](ClO_4)_{10}$ (**20**) (Dawe *et al.*, 2006), with the core structure illustrated in Fig. 16. The copper(II) ions display significant Jahn-Teller distortion, which has important consequences regarding the magnetic properties of **20**. The elongated Jahn-Teller axes in the outer square Cu_8 framework are highlighted in Fig. 16, result in $d_{x^2-y^2}$ magnetic ground states at these centres. The relative orientation of these axes leads to a strictly orthogonal connection of all copper ions in the Cu_8 ring, which would be expected to lead to ferromagnetic exchange within the ring.

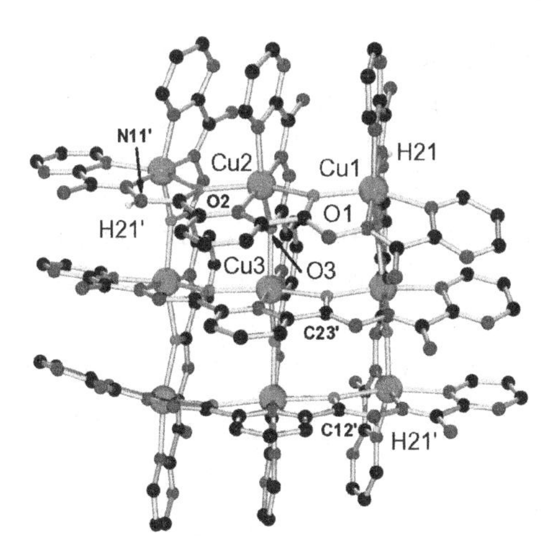

Fig. 15. Structure of Complex 20.

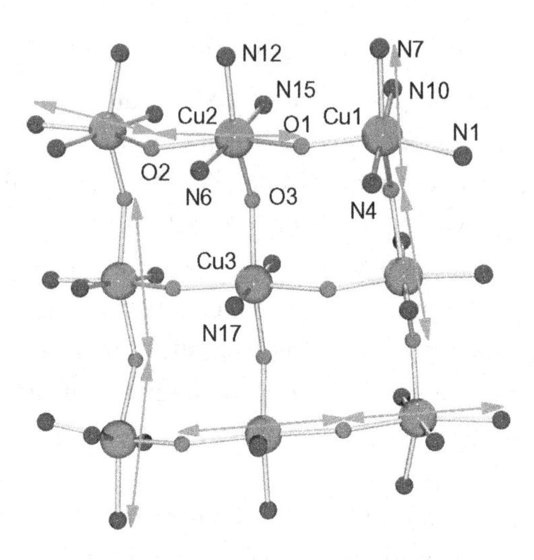

Fig. 16. Core Structure of 20.

The central copper ion has an axially compressed six-coordinate geometry, which results in a d_z2 ground state, and thus a non-orthogonal connection to its immediate neighbours. This would result in anti-ferromagnetism, and so the overall exchange situation would be expected to be the summation of ferromagnetic (Cu_8 ring) and antiferromagnetic (central Cu) exchange components, and effectively create a ferrimagnetic situation (Dawe *et al.*, 2006).

Cobalt(II) [3 × 3] grid complexes of the tritopic hydrazone ligands have been very reluctant to form, in part because of the susceptibility of the Co(II) ion to aerial oxidation to Co(III) in the resulting N,O coordination environments. Also the thermodynamic energy balance associated with grid formation would include crystal field stabilization energy considerations for the Co(II) ions (*vide infra*), which would tend to favor nitrogen rich coordination environments. In one case the complex $[Co(II)_9(L18j^{2-})_6](NO_3)_6 \cdot 24H_2O$ (**21**) was obtained from CH_3CN/MeOH solution as red crystals (Dawe *et al.*, 2009). The structure (Fig. 17) is a typical square [3 × 3] grid, with the normal $[Co(II)_9(\mu\text{-}O)_{12}]$ core structure. Co-L distances and magnetic properties confirm the presence of Co(II) at all sites, and magnetic properties indicate intramolecular anti-ferromagnetic exchange as expected (Dawe *et al.*, 2009).

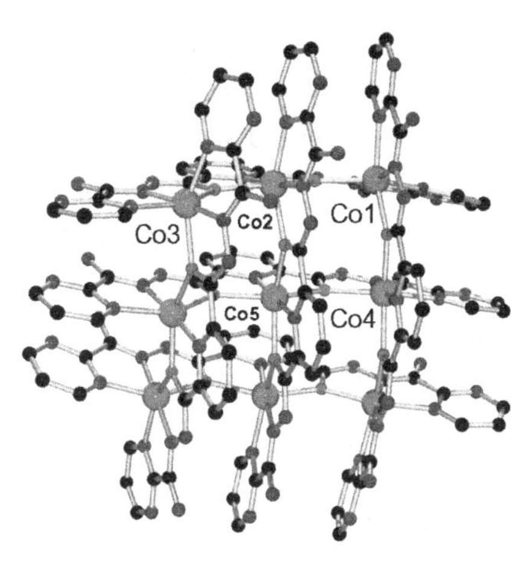

Fig. 17. Structure of Complex 21.

Nickel(II) has been even more reluctant to cooperate in forming [3 × 3] grids, despite numerous attempts with a variety of ligands, different nickel salts, and differing reaction conditions. Instead it prefers to form lower order oligomers (*vide infra*). However in one case the complex $[Ni(II)_9(HL18f^{2-})_5(OH)_2(CH_3CN)_2(H_2O)_3](ClO_4)_6 \cdot 19H_2O$ (**22**) was obtained, which contained nine six-coordinate Ni(II) ions in a grid based framework of five ligands, with one ligand missing from the central segment. Two bridging hydroxide ions complete the $[Ni_9(\mu\text{-}O)_{12}]$ core, with additional water and acetonitrile as co-ligands (Niel *et al.*, 2008). The nickel(II) ion also has a significant CFSE contribution (CFSE Ni(II) > Co(II) > Cu(II)), and so grid formation would be subject to a significant CFSE bias regarding the ultimate choice of donors in any product. The products obtained with Ni(II) and this class of tritopic ligand have been, almost exclusively, lower order oligomers, with dinuclear, and hexanuclear complexes as common examples (Niel *et al.*, 2008).

2.2.2. *Heterometallic and mixed spin state* [3 × 3] *grids*

The systematic construction of hetero-metallic grids using the 'complex as ligand' approach described for **16** and **17** with a ditopic ligand (*vide supra*) is difficult, and requires that a stable mononuclear fragment of one metal ion can be created prior to reaction with the second metal ion. This approach is complicated further in the case of the tritopic ligands by the fact that within the [3 × 3] grids there are three different types of coordination spheres, based on the differing arrangements of N and O donors, associated with the two different coordination pockets. However, the μ-O bridged [3 × 3] grid structure leads to a symmetric or pseudo-symmetric arrangement of the six ligands and nine metals, which effectively creates equivalent coordination groupings at all the side and all the corner positions. The most labile sites in terms of metal substitution are likely to be the corner sites, where ligand dissociation could occur more readily, since these sites only have two μ-O bridging connections, as opposed to three and four for the side and center sites respectively. The synthesis of symmetric heterometallic complexes may therefore be possible by starting with e.g. a $Mn(II)_9$ [3 × 3] grid and reacting it with a different metal ion directly. Kinetic considerations aside, by starting with a Mn(II) grid the advantage of not involving any initial CFSE barrier in terms of metal substitution, particularly if the second metal had a significant CFSE contribution to the overall thermodynamic energy balance, would tend to favour substitution. The differing coordinating nature of the metal ions, and their differing ionic radii, might also allow a site preferential occupancy in a reaction where the ligand competes with two different metal ions in a reaction mixture.

Reaction of H_2L18j with a mixture of $Mn(NO_3)_2{\cdot}6H_2O$ and $Zn(CH_3COO)_2{\cdot}2H_2O$ in methanol/acetonitrile led to the formation of $[Mn(II)_5Zn(II)_4(L18j^{2-})_6]$ $(NO_3)_6{\cdot}14H_2O$ (**23**). The structure (Fig. 18) revealed that the Zn(II) ions occur at the side sites, with Mn(II) ions occupying the remaining sites (Dawe *et al.*, 2009). The structure has tetragonal crystallographic symmetry, which is manifested as effective four-fold symmetry within the grid cation (point symmetry S_4). Since neither Mn(II) or Zn(II) gain any crystal field stabilization energy in the substitution process, there is no apparent site selectivity based on donor preferences. However, the site choice may simply be the result of a better coordination sphere match for the slightly smaller Zn(II) ion (ionic radii: Mn(II), 0.80 Å; Zn(II), 0.74 Å). Therefore from a site-charge perspective, and metal ion charge/radius ratio considerations, the side sites (nominally −3 resulting from the three deprotonated oxygen donors) would create a more attractive coordination environment for the Zn(II) ion. The choice of Mn(II) at the center site is

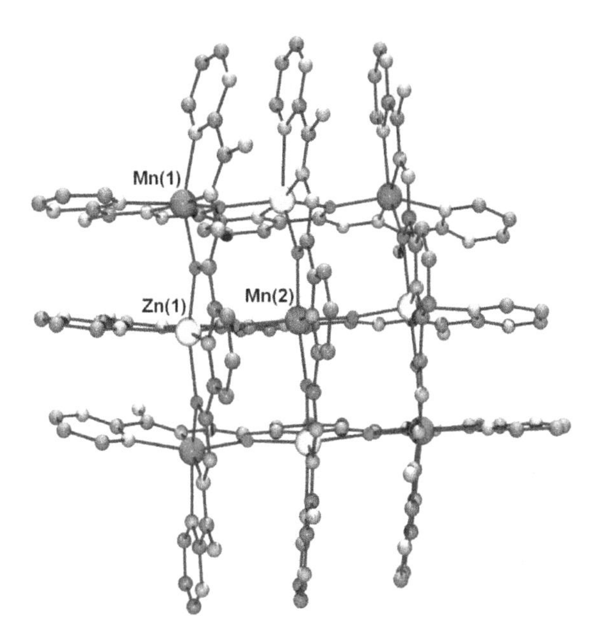

Fig. 18. Structure of Complex 23.

perhaps unexpected, but may be a consequence of kinetic factors associated with the nominally more labile Mn(II) ion as it undergoes solvent (ligand) substitution by the two intersecting deprotonated ligands, which reasonably constitute a primary grid construction fragment. The magnetic properties of this heterometallic grid clearly support the remote location of the five Mn(II) ions, exhibiting very weak antiferromagnetic exchange, in agreement with the presence of the intervening Zn(II) ions between the Mn(II) centres (Dawe *et al.*, 2009). The remote arrangement of the five $S = 5/2$ spin centres is directly analogous to the remote grid arrangement of five high spin Fe(III) ($S = 5/2$) in the incomplete grid based complex $[Fe(III)_5(HL18b^{1-})_6](ClO_4)_9$ (**24**) (Thompson *et al.*, 2003). The magnetic properties of the two complexes are almost identical.

Using the metal substitution approach reaction of the $Mn(II)_9$ $[3 \times 3]$ grid complex $[Mn(II)_9(L18j^{2-})_6](NO_3)_6 \cdot 13H_2O$[25] with $Cu(NO_3)_2 \cdot 6H_2O$ in methanol/acetonitrile under mild (room temperature) conditions gave the symmetrically arranged heterometallic grid $[Mn(II)_5Cu(II)_4(L18h^{2-})_6]$ $(NO_3)_6 \cdot 24H_2O$ (**25**), in which copper(II) substitution occurs just at the

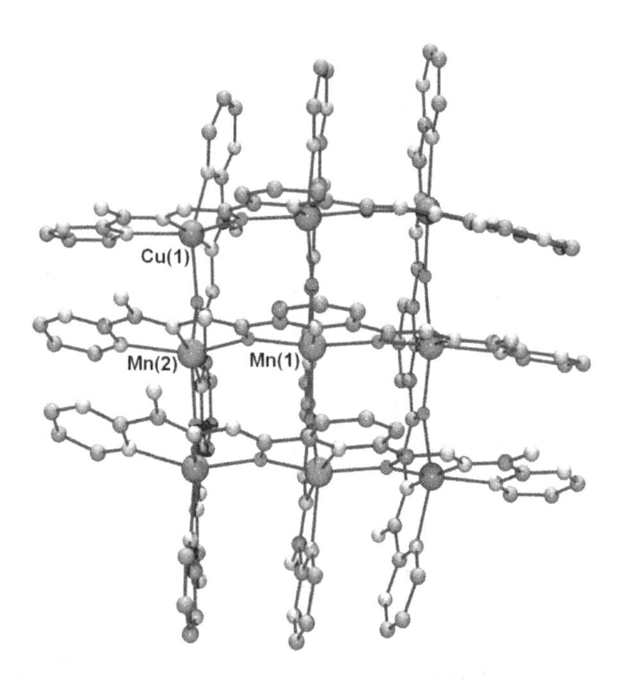

Fig. 19. Structure of Complex 25.

corner sites. The structure of the grid cation is shown in Fig. 19 and the core structure in Fig. 20, with the cation exhibiting four-fold symmetry (Dawe *et al.*, 2009). It is clear that in this case there is a substitution site preference, illustrated by the symmetric substitution arrangement. Cu(II) does gain a small CFSE advantage over Mn(II), and so could reasonably take advantage of the nitrogen rich corner sites to maximize this effect. Also, the rather distorted corner sites which occur in the grids as a whole, frequently with some very long metal-ligand bonds (~2.4 Å), would be a more comfortable environment for six-coordinate Cu(II), which is typically highly distorted, exhibiting Jahn-Teller elongation or compression. In this case the corner Cu(1) sites have axially compressed geometries, resulting in a d_z2 magnetic ground state (arrows in Fig. 20), with long contacts to N(7′), N(17), O(2′) and O(3) (2.134–2.244 Å), and short bonds to N(10) and N(14) (1.970(5) and 1.931(5) Å respectively). This arrangement leads to antiferromagnetic exchange coupling throughout the grid, as a result of the effective alignment of metal based 'd' orbitals (M-O-M angles in the range 130.1–134.4°) (Dawe *et al.*, 2009). The presence of the different spin

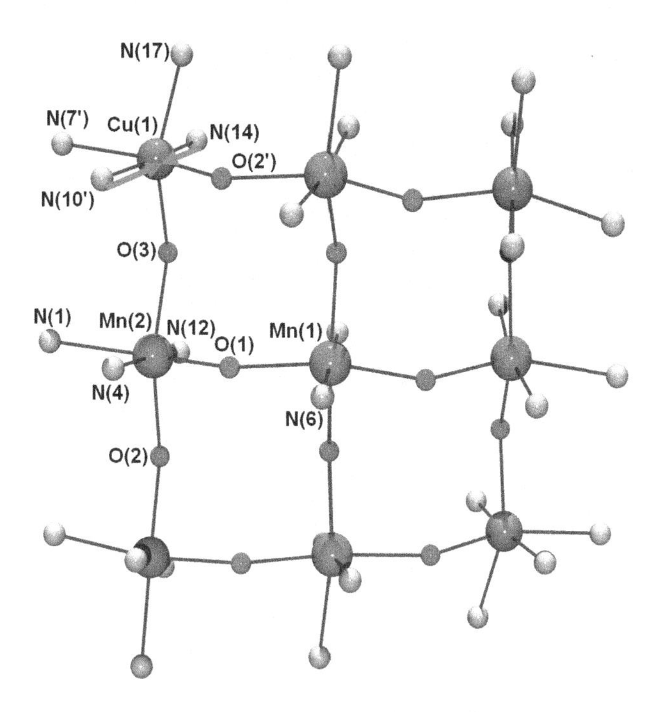

Fig. 20. Core Structure of 25.

centres (S = 5/2, 1/2) at these [3 × 3] grid positions in an antiferromagnetic system would be expected to lead to non-compensation of spins in the ground state and ferrimagnetic behavior. The magnetic moment drops from 12.8 μ_B at 300 K to about 9.1 μ_B at approximately 20 K, followed by a rise to 10.5 μ_B at 2 K, indicative of a high spin ground state. The expected ground state for this situation should be $S' = 11/2$, but magnetization data as a function of field at 2 K suggest that the ground state is closer to 9/2. Theoretical calculations for this spin model show that, depending on the choice of coupling constants, there is a low lying $S' = 9/2$ spin state, which is energetically close to the $S' = 11/2$ ground state, suggesting that the predicted ground state is not populated significantly at 2 K (Dawe *et al.*, 2009).

Reaction of $[Mn(II)_9(L18j^{2-})_6](NO_3)_6 \cdot 13H_2O^{25}$ with $Cu(NO_3)_2 \cdot 6H_2O$ in methanol/acetonitrile under more vigorous extended reflux conditions, led to the more copper rich heterometallic complex $[Mn(II)Cu(II)_8(L18j^{2-})_6](NO_3)_6 \cdot 23H_2O$ (**26**). The structure of the cationic fragment of **26** and its core structural representation are shown in Figs. 21 and 22,

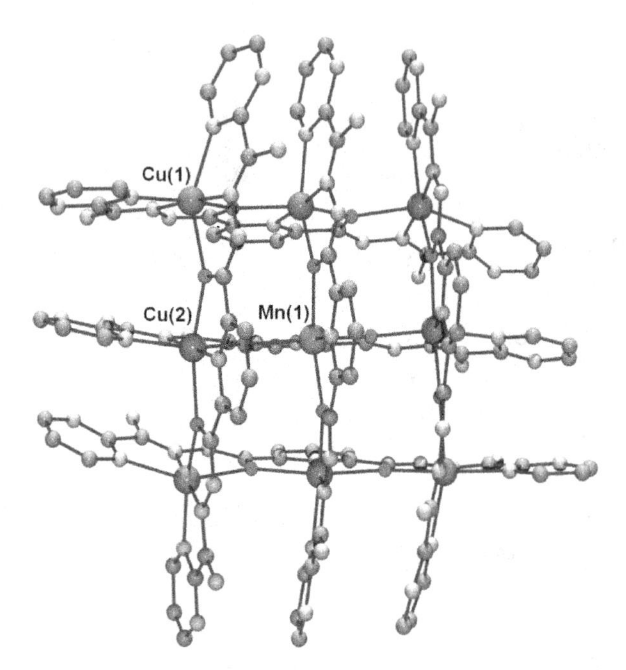

Fig. 21. Structure of Complex 26.

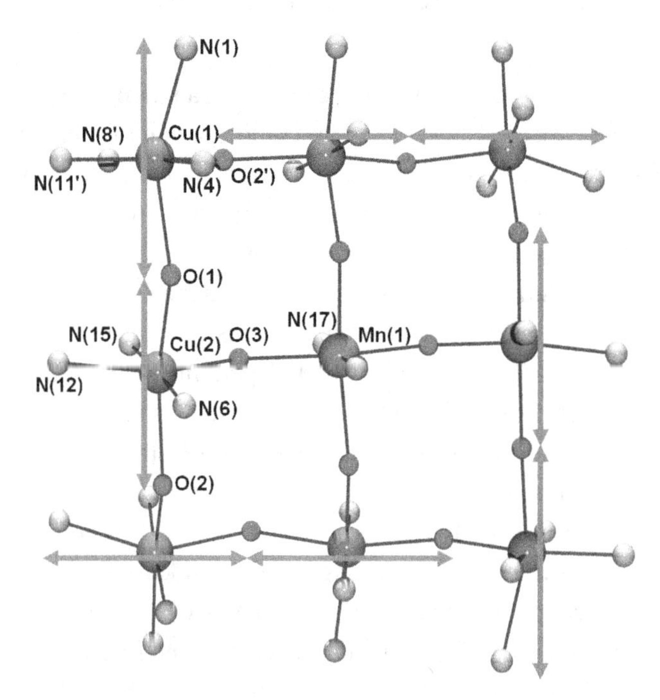

Fig. 22. Core Structure of 26.

respectively (Dawe *et al.*, 2009). Compound **26** also exhibits four-fold symmetry in the grid cation (Fig. 22), and each copper ion has a Jahn-Teller distorted, axially elongated octahedral geometry (arrows in Fig. 22). Metal-metal distances fall in the range 4.01–4.25 Å, with M-O-M angles in the range 135.7–138.3°. The Cu(1) and Cu(2) sites have $d_{x^2-y^2}$ magnetic ground states due to the axial elongation, and the Jahn-Teller axes (Fig. 22) are arranged so that there is strict orthogonality between the copper ions in the outer ring of eight copper ions. However the connection between the central Mn(II) ion and its copper nearest neighbours is non-orthogonal, and so one would expect a combination of both ferromagnetic and antiferromagnetic exchange overall, and as a result a non-compensation of spins due to the combination of different spin centres. The magnetic moment drops from 8.7 μ_B at 300 K to a value of approximately 5.2 μ_B at 18 K followed by a slight rise to 2 K.

The magnetic data for **26** were fitted to a model incorporating both ferromagnetic and antiferromagnetic terms to give $g_{av.} = 2.3(1)$, $J1 = 0.45(2)\ cm^{-1}$, $J2 = -22.5(2)\ cm^{-1}$, where J1 refers to the Cu-Cu coupling in the outer ring and J2 to the inner Mn-Cu coupling, in agreement with the structural details (Dawe *et al.*, 2009).

Mixed metal grids result in sites of mixed spin state by default in most cases, but this can also be achieved by having different oxidation state centres of the same metal ion within the grid. Mn(II) and Co(II) are redox active metal centres, with oxidation potentials which could vary depending on their specific coordination environment. This could lead to specific site oxidation in a typical [3 × 3] grid, given the differing nature of the ligand donor environments at the three unique grid sites. $Mn(II)_9$ [3 × 3] grids have shown remarkable versatility in this regard, with site selective redox behavior involving Mn(II) to Mn(III) oxidation. $[Mn_9(L18a^{2-})_6](ClO_4)_6$ (**27**) (Zhao *et al.*, 2000; Thompson *et al.*, 2004; Waldmann *et al.*, 2006) is the prototypical example of the [3 × 3] Mn_9 grids, with a similar structure to **19** (vide supra), and like other similar grids (Zhao *et al.*, 2000; Thompson *et al.*, 2004; Waldmann *et al.*, 2006) with related tritopic ligands and exhibits remarkable electrochemical properties, with a well defined suite of redox waves in a narrow potential window (0.5–1.6 V vs Ag/AgCl) (Zhao *et al.*, 2000; Thompson *et al.*, 2004; Zhao *et al.*, 2004; Dey *et al.*, 2007). Five clearly defined redox events are observed for several iso-structural Mn_9

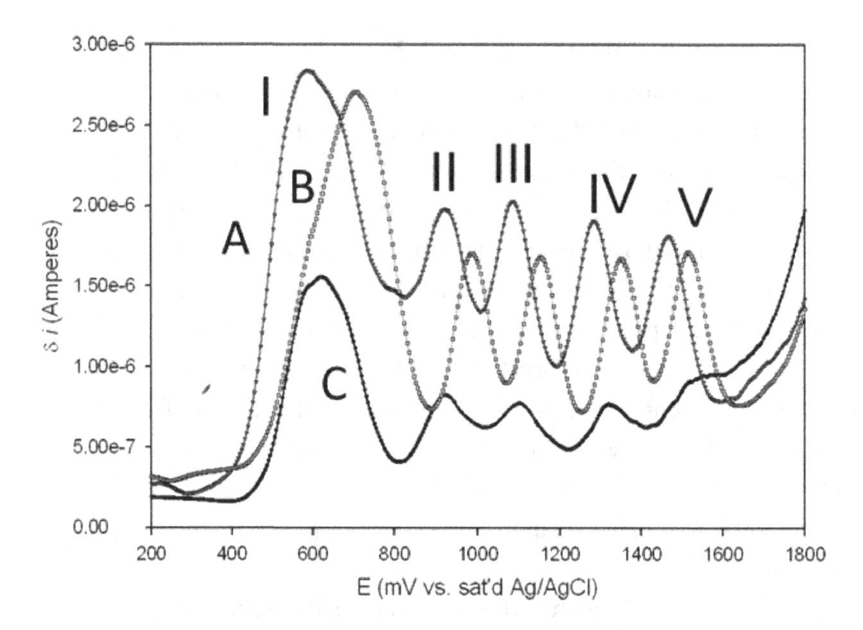

Fig. 23. DPVs for $Mn(II)_9$ [3 × 3] grids.

grids (Fig. 23: B(27); A (H2L18g grid); C(H2L18b grid)) corresponding to a single quasi-reversible four electron step at ~0.6 V (I), which is assigned specifically to the oxidation of the four corner Mn(II) sites to Mn(III), and then a sequence of four reversible one electron steps (II–V), in the range 0.9–1.6 V, which are assigned to oxidation of the side sites to Mn(III). The remote but identical nature of the four N_4O_2 corner sites would reasonably lead to four equivalent redox events, which would occur at the 'same' or very close potentials, because of the effective lack of communication between these sites due to their large distances of separation. All of the side N_3O_3 sites are effectively connected via the central Mn(II) site, and so communication between metals would be more likely, with the result that oxidation at one site would affect the subsequent oxidation of the other nominally equivalent sites. Also since positive charge on the grid would increase with increased oxidation level one would expect to see separate one electron events at successively increasing potentials, which is shown clearly in Fig. 23. The preferential oxidation of the corner sites first may well be associated with the stronger crystal field environment (N_4O_2), and their more open and flexible nature (Zhao *et al.*, 2000; Thompson *et al.*, 2004; Waldmann *et al.*, 2006).

The isolation of oxidized species would provide the structural proof of the location of the oxidized sites, and by using controlled oxidation via electrochemical and chemical means it has been possible to selectively oxidize just the corner Mn(II) ions in several cases. However, isolation of oxidized species beyond the four electron level proved to be very difficult, because the oxidized species themselves had high enough oxidation potentials to oxidize water or solvent. Interestingly, unstable products with a higher oxidation level could be isolated as black solids, but reverted to the four electron oxidized species on recrystallization from solvents containing even small amounts of water, presumably by oxidation of water, and perhaps even the solvent. The complex $[Mn(III)_4Mn(II)_5(L18a^{2-})_6](ClO_4)_{10}\cdot 10H_2O$ (**28**) can be isolated as dark brown, air stable crystals by bulk electrochemical synthesis, revealing clearly the presence of Mn(III) just at the corner sites, identified unequivocally by the fourfold symmetry of the grid and the very short Mn–L distances (Thompson *et al.*, 2004).

The $Mn(II)_9$ grids exhibit antiferromagnetic exchange with an $S = 5/2$ ground state, expected on the basis of the odd number of spin centres (*vide supra*). **28** also exhibits antiferromagnetic exchange, with a comparable coupling constant, but ends up with an $S' = 1/2$ ground state, in what is essentially a ferrimagnetic system, resulting from the non-compensation of the different spin sites in the grid. Such species are rare, and represent an important class of mesoscopic $S = 1/2$ 'qubit' analogues, which are considered as potentially important nanoscale components in future quantum computing (Waldmann *et al.*, 2006; Meier *et al.*, 2003). Other important systems in this rare class include the ferrimagnetic wheels, e.g. Cr_7Ni, which also have $S = 1/2$ ground states (*vide infra*) (Corradini *et al.*, 2007).

The [3 × 3] grid cations have footprints of $\sim 2.5 \times 2.5$ nm, which represents a very small subunit dimension, and in a close packed 2D arrangement would represent a high density spatial medium for data storage, if the individual grids displayed some functionally useful property, which could lead to molecular bistability. A convenient 2D spatial medium would be a surface application, if the grids could be attached in an ordered array. Surface studies have been carried out and monolayer assemblies of [3 × 3] $Mn(II)_9$ grids have been produced successfully on Au(III), and probed using STM techniques, revealing a close packed 2D surface arrangement (Zhao *et al.*, 2004). Similar studies on HOPG using STM/CITS reveal surface adhesion and spatial resolution at the molecular and metallic level, clearly indicating that the probing of individual molecules and also individual metal atoms can be achieved by tuning tunneling current energies (Dey *et al.*, 2007). Figure 24 shows the STM image of a single molecular cation of **19** adsorbed on HOPG, represented as an unresolved 'blob' of electron density, with poorly defined size limits (Dey *et al.*, 2007). However dramatic structural resolution can be achieved by using CITS imagery (Fig. 25), where tuning of the tunneling current to selectively image the metal 'd' orbitals (ligand atom orbital energies are different and so do not interfere) reveals clearly image

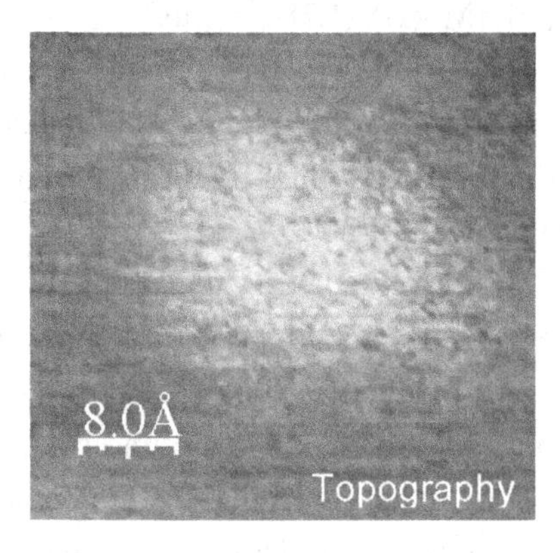

Fig. 24. STM image of 19.

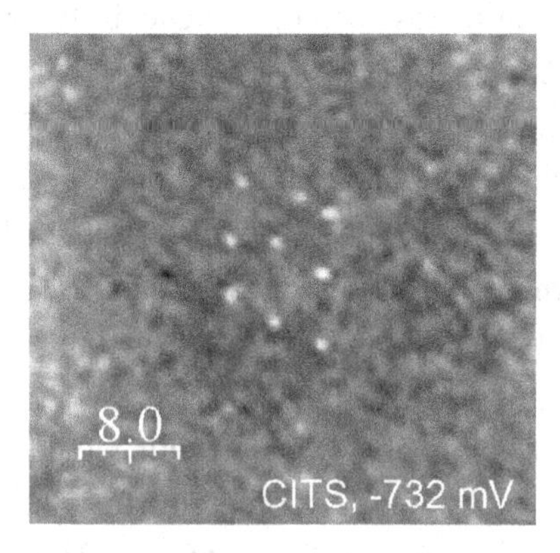

Fig. 25. CITS image of 19.

spots, which represent the positions of the Mn(II) centres. Their spacing and arrangement match that obtained through single crystal structural measurements, clearly showing that a single grid molecular cation sits on the HOPG surface. DFT calculations closely match the image based on an analysis of the metal 'd' orbital composition in the grid (Dey *et al.*, 2007).

These results present the realistic possibility of creating a surface bound molecular based medium, and the selective probing of individual molecular components, and also selective metal site redox control, based on the reversible nature of the redox chemistry demonstrated by the [3 × 3] $Mn(II)_9$ grids in solution. This could allow for a variety of different, stable redox intermediates to be created in a fixed spatial state on the surface, with appropriate tuning of the tunneling current energies. Given that the individual redox steps are reversible, then from the practical perspective a single molecule could be encoded with binary information (1,0 in terms of an oxidized or a reduced state), not only as a single molecular entity, but also through the individual changes associated with the energy selective redox steps. This would potentially lead to an enormous capacity for data storage, particularly if a surface monolayer assembly could be created and probed appropriately. Detection of an organic radical at the molecular level on a surface has been achieved at room temperature (Durkan and Welland, 2002), and so in principle the paramagnetic grid species could well be probed in this way also by using their EPR signatures. Stable redox intermediates would be key to the success of such an approach, and the stability of **28** augers well for its viability. Clearly this approach would be difficult using magnetic measurements (*vide supra*).

2.3. *Tetratopic ligands and complexes*

2.3.1. *Homometallic* [4 × 4] *grids*

Even numbered square [n × n] grids ($n > 3$) would logically be based on symmetric ligands with two ditopic fragments linked appropriately by a symmetric central dinucleating core. Pyrimidine and pyridazine groups are suitable core dinucleating elements, which would bring two metal centres within reasonable distances for spin communication, and can be modified with various end groups to produce ligands capable of forming M_{16}[4 × 4] grid examples. H_2L19 (Chart 3) is derived from 2,4-dicyanopyrimidine, and provides two pairs of adjacent N_2O pockets capable of *mer* coordination to six-coordinate metal ions. Reaction of H_2L19 with $Pb(CF_3SO_3)_2$ gave the grid based complex $[Pb(II)_{16}(L19^{2-})_8](CF_3SO_3)_{16}$, with lead ions arranged in four square [2 × 2] groups, with μ-O bridges between metal ions, partitioned by the bridging pyrimidines. 6, 7 and 8 coordinate Pb(II) ions were observed with some coordination of triflate and solvent (Onions *et al.*, 2003). L20 (R = $SCH_2CH_2CH_3$) (Chart 4) has a

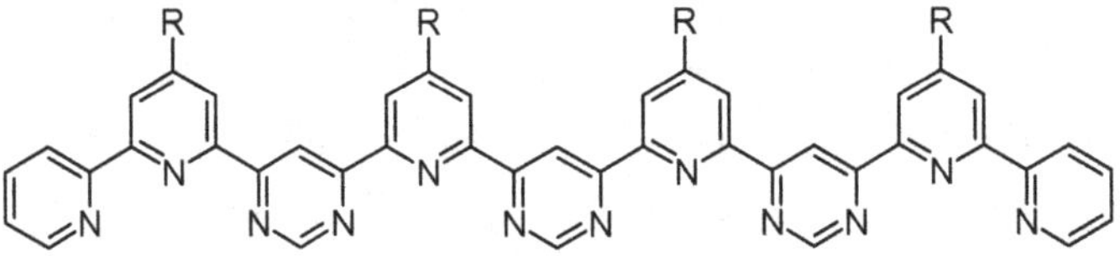

L20 (R=$SCH_2CH_2CH_3$)

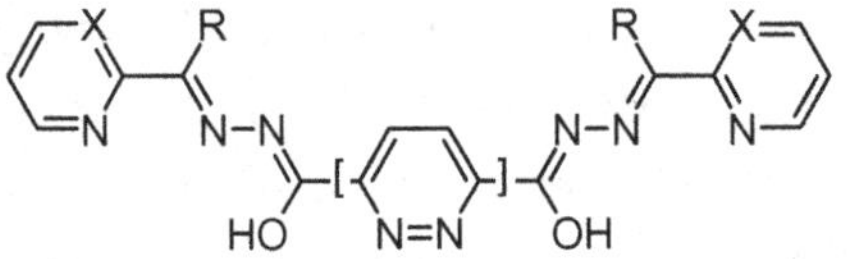

H_2L21a (X=CH,R=NH_2), H_2L21b (X=CH,R=H), H_2L21c (X=N,R=NH_2)

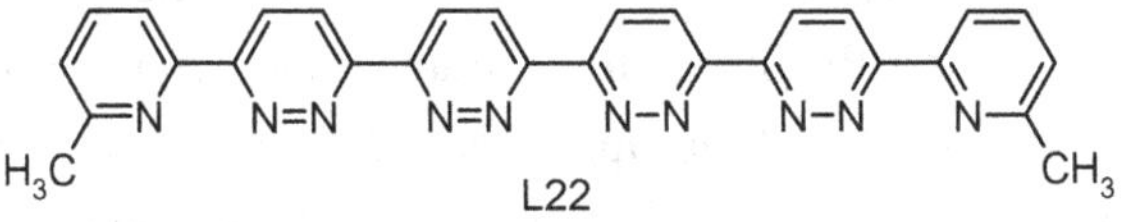

L22

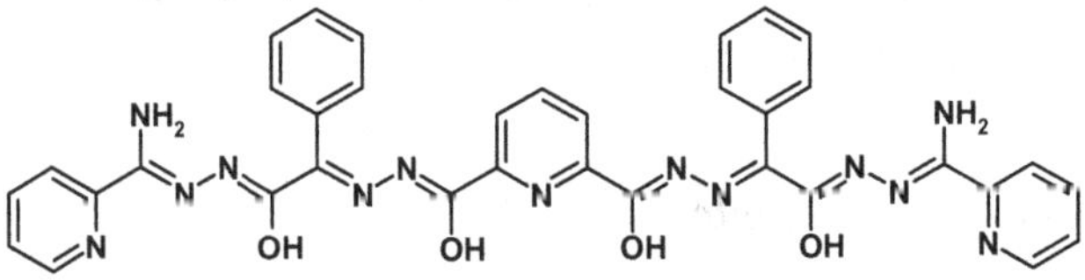

H_4L23

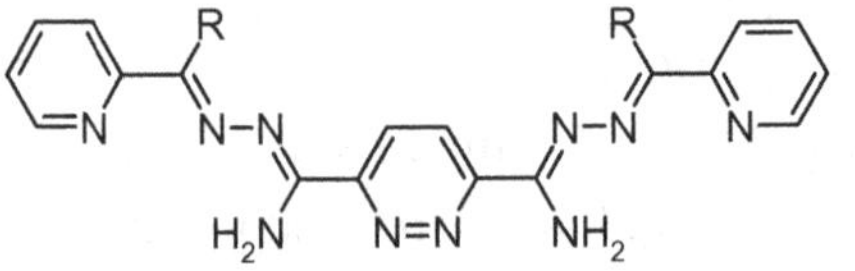

L24 (R=NH_2), H_2L25 (R=OH)

Chart 4.

similar arrangement of four contiguous ligand pockets, but with three linked pyrimidine groups providing the bridging connections. The Pb(II) complex $[Pb(II)_{16}(L20)_8](CF_3SO_3)_{32}$ has a grid based array of eight ligands arranged in two roughly parallel groups of four above and below the Pb_{16} core, with the Pb ions bridged by the pyrimidine groups (Barboiu *et al.*, 2003). The critical design feature of this type of ligand rests with the formation of five-membered chelate rings, and the essentially linear disposition of the tridentate pockets. This creates optimal conditions for the appropriate alignment of the metal ions bound to each ligand, which would lead to a minimal organizing of these subunits into the [4 × 4] grid. The intersection of the *mer*-N_2O and N_3 coordination groups would be optimized with a metal ion which preferred six-coordinate geometries (*vide supra*).

Pyridazine based tetratopic dihydrazone ligands H_2L21a-c (Chart 4) also create ditopic fragments capable of forming μ-O bridges, connected by the potentially bridging pyridazine central fragment. However each half ligand is comprised of a potentially bidentate and a tridentate pocket, and cannot provide sufficient internal donors to satisfy the full coordination requirements of sixteen six-coordinate metal ions in a putative [4 × 4] grid structure. Co-ligands are required, and are usually scavenged from the solvent environment. This mirrors the situation for the ditopic hydrazones (*vide supra*).

The reaction between H_2L21a and $Cu(CF_3SO_3)_2$, with added $NaBF_4$ led to the self-assembled partial grid $[NaCu(II)_{12}(L21a^{2-})_8](BF_4)_9$, which has copper ions coordinated just in the outer square framework of the intersecting arrangement of eight ligands, with no copper coordination at the four central sites (Matthews *et al.*, 2003). Partial coordination of a Na^+ ion at one central site indicated that a fully occupied grid arrangement with sixteen metal ions is possible, but it was not immediately obvious in this case as to why incomplete metallation occurred. [4 × 4] square M_{16} grids have now been successfully produced with Mn(II), Co(II) and Cu(II) salts and this type of ligand. The complex $[Mn(II)_{16}(L21b^{2-})_8(OH)_8](NO_3)_8$ (**29**) is shown in Fig. 26, and consists of eight ligands arranged in two roughly parallel groups of four above and below the sixteen metal based core (Fig. 27) (Dey *et al.*, 2006). The structure can be envisaged as four [2 × 2] $[Mn_4(\mu\text{–}O)_4]$ square subunits, similar to those observed with the

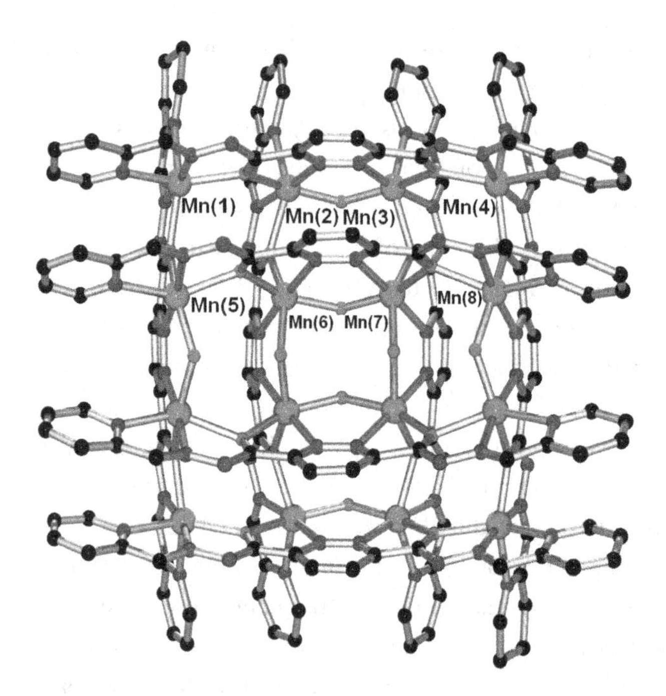

Fig. 26. Structure of Complex 29.

simpler ditopic ligands, connected by the eight pyridazine (N-N) bridges. In the *syn*-μ–O bridging conformation each ligand can only fill a maximum of ten metal coordination sites (80 sites per grid of eight ligands), and so extra co-ligands are required to complete the coordination spheres of the sixteen six-coordinate metal ions ($6 \times 16 = 96$). The solvent medium acts as a source of extra donors, and in this case they are hydroxide ions, which fill the remaining sites by forming second bridges in combination with each pyridazine bridge. The combination of pyridazine and hydroxide bridges leads to antiferromagnetic exchange throughout the grid, and an $S' = 0$ ground state (Dey *et al.*, 2006).

A similar Cu(II)$_{16}$ grid [Cu(II)$_{16}$(L21c^{2-})$_6$(L22c^{3-})$_2$(O)$_2$(OH)$_4$(H$_2$O)$_2$] (CF$_3$SO$_3$)$_6$ (**30**) is produced on reaction of the ligand with Cu(CF$_3$SO$_3$)$_2$ in the presence of base (Dawe and Thompson, 2007). In this case the exogenous bridges include an unusual combination of oxide, hydroxide and water to complete the internal bridging connections to the six-coordinate Cu(II) ions. Figure 28 shows the structure of the $4 \times [2 \times 2]$ grid cation and part of the core

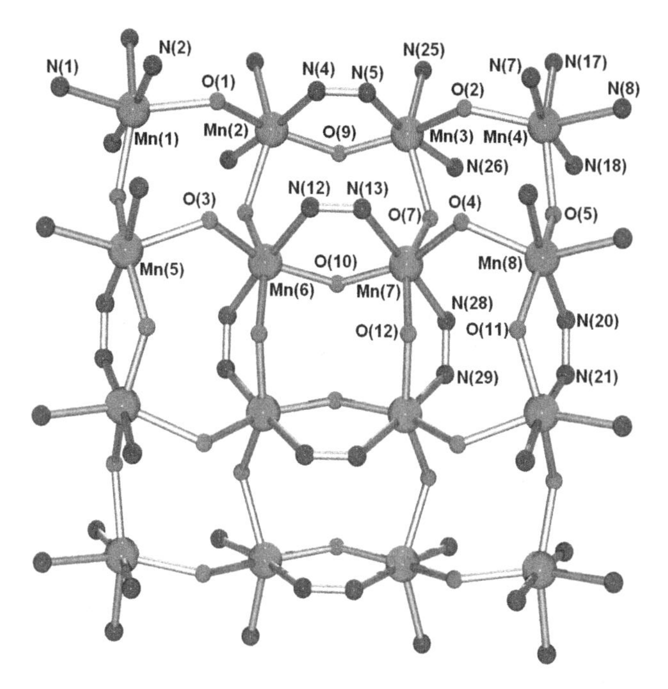

Fig. 27. Core Structure of 29.

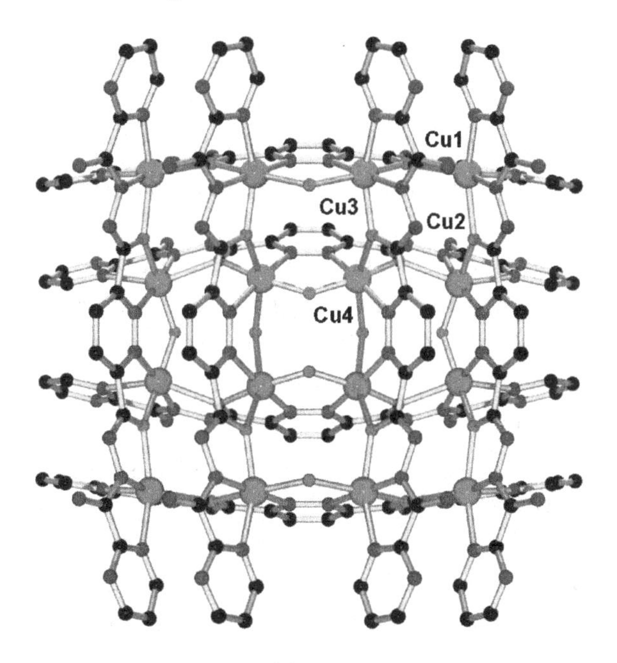

Fig. 28. Structure of Complex 29.

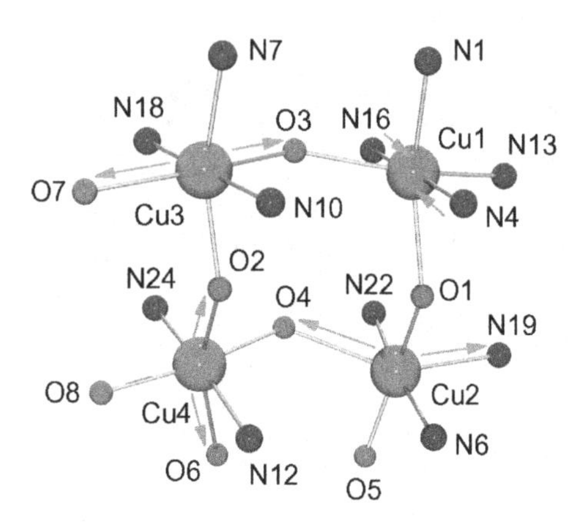

Fig. 29. Core Fragment of 29.

structure is highlighted in Fig. 29. Magnetic exchange coupling within the grid is complicated by the presence of both orthogonal and non-orthogonal magnetic connections between copper(II) ions (Fig. 29 shows the orientation of the Jahn-Teller axes in the distorted copper coordination spheres at a corner Cu_4 subunit), but the high symmetry has allowed an evaluation of the exchange picture, and overall, unlike the $Cu(II)_9$ [3 × 3] grids, the [4 × 4] system is dominated by anti-ferromagnetic coupling associated with both $\mu_{pyridazine}$ and μ_O bridges (Dawe and Thompson, 2007).

Reaction of the same ligand with $Cu(CF_3SO_3)_2$ in the presence of $KAg(CN)_2$ gave the unusual mixed oxidation state homoleptic grid $[Cu(II)_{12}Cu(I)_4(L21c^{2-})_8](CF_3SO_3)_{12}$ (**31**). The structure (Fig. 30) shows the familiar 4 × [2 × 2] grid arrangement, with the core structure is illustrated in Fig. 31. There are no adventitious oxygen bridges, which results from the fact that the central group of four copper sites are tetrahedral Cu(I) centres, while the remaining sites are Cu(II), but have five and six-coordinate geometries, in response to the available donors from just the H_2L21c ligands. Magnetic exchange within the grid only concerns just the outer framework of twelve Cu(II) centres, since the inner grouping of Cu(I) centres is diamagnetic. The bridging connections in the outer ring of Cu(II) centers are all orthogonal (Fig. 31) with the result that the magnetic properties of the grid are dominated by ferromagnetic coupling (Dawe *et al.*, 2009). The redox reaction producing Cu(I) is associated with the use

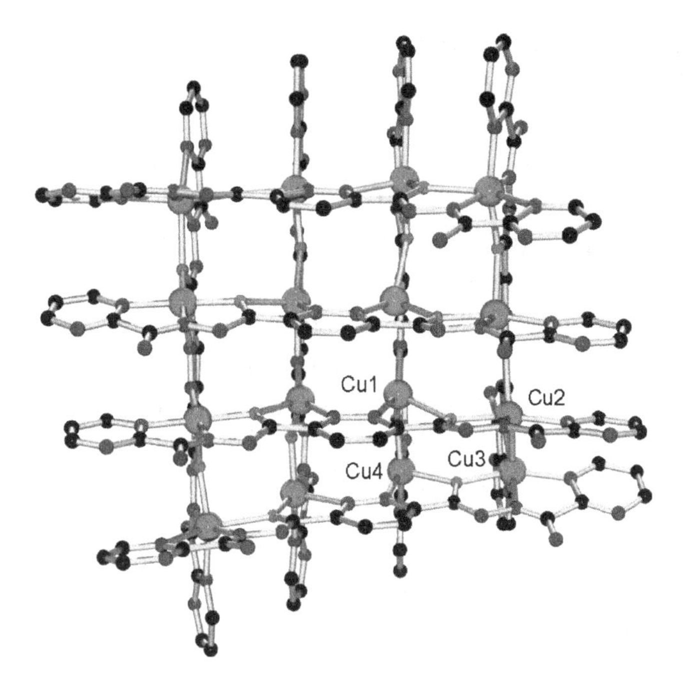

Fig. 30. Structure of Complex 31.

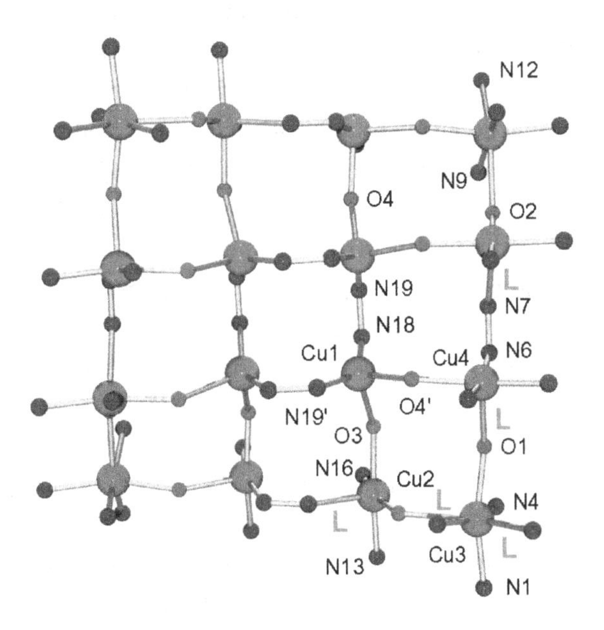

Fig. 31. Core Structure of 31.

of acetonitrile (helps to stabilize Cu(I)) as a solvent, and the presence of $[Ag(CN)_2]^-$, which appears to be the reductant (Dawe *et al.*, 2009).

Co(II) and Ni(II) have been reluctant to react with this type of ligand, and this may be associated as before with the increased CFSE associated with these ions in comparison with Cu(II) and Mn(II). However, in one case, in the reaction of $Co(BF_4)_2$ with H_2L21c, in the presence of air (CH_3CN/MeOH), a small quantity of red crystals of $[Co(II)_{16}(L21c^{2-})_8(OH)_8](BF_4)_8$ (**32**) was obtained and shown to have the same structure as **29**, with hydroxide bridges completing the internal cobalt(II) coordination spheres (Dawe, 2008). The bulk product in this reaction was a mixed oxidation state Co(II)/Co(III) grid, resulting from exposure of the reaction to air. No $Ni(II)_{16}$ grids have so far been identified.

2.4. *Pentatopic ligands and their complexes*

2.4.1. *Homometallic* [5 × 5] *grids*

The construction principles for the convergent synthesis of [n × n] grids are in principle not limited by size, providing that an appropriate polytopic ligand, with the appropriate donor complement and pocket arrangement, can be created. The poly-pyridazine ligand L22 (Chart 4) has five bidentate pockets, with the potential to coordinate five four-coordinate metal ions in a linear array, assuming that no ligand rotation about the C-C bonds linking the pyridazine rings occurs. Reaction of $Ag(CF_3SO_3)$ with L22 produced two oligomers, which were characterized structurally; a quadruple helicate structure involving four ligands folded around 10 four-coordinate Ag(I) ions in $[Ag(I)_{10}(L22)_4]^{10+}$, and a 2 × [2 × 5] grid like arrangement in which a twist around the central C-C bond occured and prevented full [5 × 5] grid formation, and led to two connected [2 × 5] rectangular arrays, each with ten tetrahedral Ag(I) centres, built on the end ditopic ligand fragments in $[Ag(I)_{20}(L22)_{10}]^{20+}$ (Baxter *et al.*, 2000). In this case the flexibility of the ligand allowed for the formation of a linear based helical structure, in which the encoded potential [5 × 5] grid coordination information was not interpreted appropriately by the four-coordinate Ag(I) ions, in favour of a preferred, alternate and stable structural arrangement. However a partial grid also formed, which indicates that in the reaction many complex formation equilibria are probably involved. This situation may also be complicated by the low charge on the cation, and its low coordination number, both factors which would tend to detract from full grid formation.

H_4L23 (Chart 4) is also a pentatopic ligand, based on a 2,6-bis-hydrazone core, with effectively two connected ditopic hydrazone fragments on each side (Dey *et al.*, 2007). It has five contiguous coordination pockets, which could bind five metal ions in *mer*-N_2O and *mer*-NO_2 groupings in a linear fashion, thus creating the correct primary coordination subunit for grid formation, such that other ligands could then intersect at the metals with the appropriate 90° twist. The mutual intersection of ten ligands in this way in two intersecting groups of five would then provide the requisite 25 six-coordinate pockets for the construction of a $[5 \times 5]$ grid. The advantage in this case would rest with the tridentate nature of the coordination pockets, and the creation of a six-coordinate metal environment, which would lead to a homoleptic grid arrangement. H_4L23 can theoretically exist in a large number of different tautomeric forms, and can also undergo conformational twisting like L22, which clearly would complicate the coordination process and possibly lead to oligomer formation. To illustrate this point a neutral mononuclear complex $[Mn(II)(H_2L23^{2-})]$ (**33**) was obtained as light orange crystals in one reaction with $Mn(NO_3)_2$, in which the two deprotonated tridentate end pockets join to form a N_4O_2 coordination sphere around a single six-coordinate Mn(II) centre (Dey *et al.*, 2007). The middle part of the ligand remains uncoordinated, and is testimony to the large number of different coordination possibilities which such large ligands could take part in, and the overall ligand flexibility, which increases with its length. Reaction of the ligand with $Mn(ClO_4)_2$ produced deep red-orange crystals of **34**, which, despite numerous attempts, would not diffract strongly enough to allow for a structural study. Elemental analysis, suggested the formation of a $Mn(II)_{25}[5 \times 5]$ grid, but in such a complex system, these data alone were not convincing evidence for grid formation. Mass spectral and magnetic data were also somewhat inconclusive (Dey *et al.*, 2007).

STM and CITS studies have been successful in characterizing the Mn_9 $[3 \times 3]$ grid complex $[Mn(II)_9(L18b^{2-})_6](ClO_4)_6$ (**19**) on a HOPG surface (Figs. 24 and 25) (Dey *et al.*, 2007), and in cases of other grids and clusters (Petukhov *et al.*, 2009) (*vide supra*), including an image of the sixteen

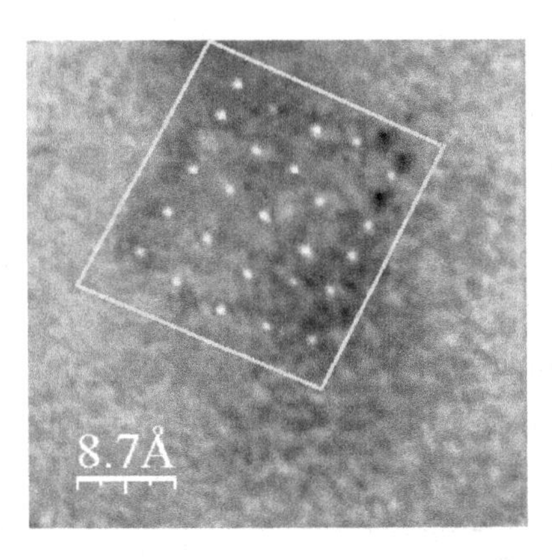

Fig. 32. CITS image of 34 on HOPG.

Mn(II) ions in $[Mn(II)_{16}(L21b^{2-})_8(OH)_8](ClO_4)_8$ (**29**) in the expected [4 × 4] arrangement (Dey *et al.*, 2007). Given the success of this unprecedented approach to molecular identity on the basis of the metal ion structural arrangement in the $Mn(II)_9$ and $Mn(II)_{16}$ cases, the CITS imagery of the putative $Mn(II)_{25}$[5 × 5] grid complex (**34**) was also examined as a surface application on HOPG. The remarkable image of **34** (Fig. 32) shows very clearly the positions of the 25 metal ions in a well defined [5 × 5] square array, with Mn-Mn spacings which exactly match the expected dimensions for a $[Mn_{25}(\mu\text{–}O)_{40}]$ core structure. The logical implication based on this structure would be the presence of the ten ligands with the expected μ–O bridging arrangement (Dey *et al.*, 2007).

3. Other Oligomers in the Assembly Process

3.1. *Incomplete grids, clusters and chains*

Since the types of n-topic ligands used for the formation of [n × n] grids are flexible, in the sense that different conformations are possible, these would inevitably become more numerous with increase in size, and so there is a real possibility of non-grid oligomers forming as well or even in preference to the expected grid (*vide supra*). The hydrazone based ligands represent a very flexible class in this context, and have been extended systematically within the n-polytopic range n = 2–5 easily (*vide supra*), but could be extended further using similar synthetic methodologies. This is less likely with some other classes of polytopic ligands, where synthetic difficulties appear to have limited their scope and study.

Oligomeric lower order ($n = 1, 2$) complexes have been observed with the ditopic hydrazone ligands, in addition to tetranuclear complexes with different bridging arrangements (e.g. μ–NN vs. μ-O), but the tritopic class has produced the widest variety of examples with $n = 1–3, 5–8$, in addition to the [3×3] grids. A recent review, and other references summarizes the current state of this situation (Dawe *et al.*, 2009; Niel *et al.*, 2008; Dawe *et al.*, 2009; Milway *et al.*, 2004). In the case of tetratopic hydrazone ligands, while there are fewer grid examples at present, due to their more limited study, dinuclear, trinuclear and tetranuclear oligomers have been observed with Cu(II) and Ni(II) (Dey *et al.*, 2007; Thompson *et al.*, 2001; Shuvaev *et al.*, 2009). The limited studies with H_4L23 have so far only revealed mononuclear and [5×5] grid examples (*vide supra*).

Ligand L24 is a closely related ligand to the H_2L21 series, but does not have an OH group capable of acting in a bridging capacity, and so despite its sequential arrangement of four potential donor pockets, it would not be expected to form a [4×4] grid with six-coordinate metal ions. It would, however, have a great deal of rotational flexibility. In a reaction with $Mn(ClO_4)_2$ the linear tetranuclear complex $[(L24)_3Mn(II)_4]\ (ClO_4)_8 \cdot 21H_2O$ (**34**) was obtained as orange crystals (Dey *et al.*, 2007). The structure of the tetranuclear cation in **34** is shown in Fig. 33 (Dey *et al.*, 2007), revealing an unusual linear spiral arrangement of four six-coordinate Mn(II)

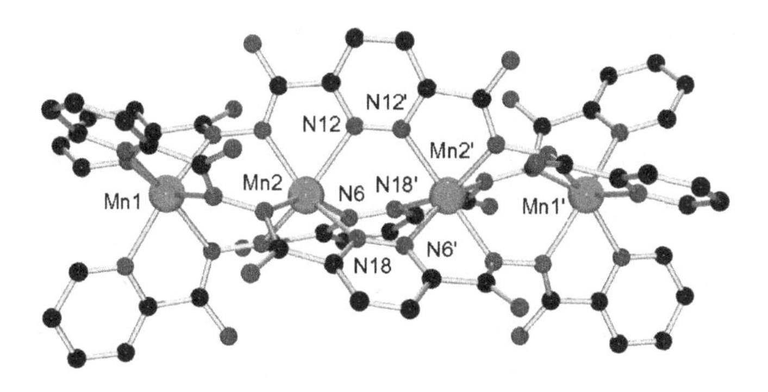

Fig. 33. Structure of Complex 34.

ions encompassed by three ligands, which all have pronounced twists around the flexible bridging diazine N-N linkages. The metal centers are bridged alternately by open chain diazine and pyridazine NN groups. Mn-Mn distances along the chain are 3.802 Å (Mn(1)–Mn(2)) involving the open chain diazine bridge, and 3.850 Å involving the pyridazine bridge. The spiral twisting can be expressed through the Mn–N–N–Mn torsional angles, which fall in the range 26.9–34.1° for Mn(1) and Mn(2), and in the range 21.7–32.9° for Mn(2) and Mn(2)′. There appears to be only one other related tetranuclear example of a spiral chain of this sort with Cu(II) and the same ligand $[Cu_4(L25)_3](ClO_4)_8$, which has a similar linear spiral structure (Onions *et al.*, 2004). However smaller dinuclear spiral complexes with the simpler, closely related ligand PAHAP (picolinamide azine) and Mn(II), Fe(II), Co(III) and Ni(II) are well documented (Xu *et al.*, 1998). These involve just open chain diazine bridges.

Variable temperature magnetic data for the Mn_4 spiral strand (**34**) are dominated by intramolecular antiferromagnetic exchange, but the acute torsional angles around the single N-N diazine bridges might signal a ferromagnetic contribution. Fitting of the data to a linear model gave $g = 2.015(10)$, $J1 = +1.6(1)\,cm^{-1}$, $J2 = -3.2(1)\,cm^{-1}$, $TIP = 0\,cm^3mol^{-1}$, where J1 represents the open chain diazine contribution (Dey *et al.*, 2007). Pyridazine bridges typically lead to antiferromagnetic exchange. Magnetic data for the closely related Cu_4 spiral chain complex $[Cu_4(L5)_3](ClO_4)_8$ were not analyzed, but show a very slight drop in moment from 280 to 5 K, associated with a possible combination of ferromagnetic and antiferromagnetic exchange terms (Onions *et al.*, 2004).

The closely related tetratopic ligand H_2L21a has a similar backbone structure, but the presence of the hydrazone oxygen groups on the central pyridazine leads to grid formation with Mn(II) and Cu(II) (*vide supra*). However with Ni(II) comparable [4 × 4] grids appear not to form, but instead the linear trinuclear complex ion $[Ni(II)_3(Na)_{0.15}(H_2L21a)_3]^{6.15+}$ (**35**) is formed, in which three spirally arranged ligands encapsulate three six-coordinate Ni(II) ions in a linear arrangement with one inner site nickel free (Fig. 34) (Shuvaev *et al.*, 2009). The presence of partial occupancy Na^+ in the fourth site is a result of residual sodium carried through from

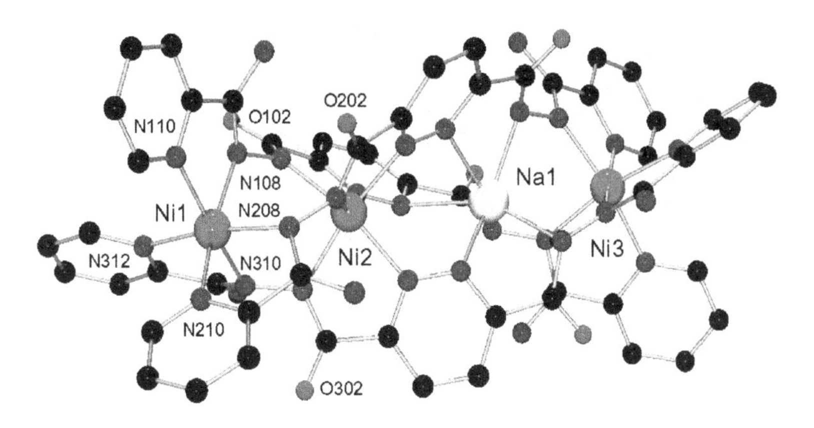

Fig. 34. Structure of Complex 35.

the preparation of the ligand in the complex synthesis. All attempts to form the $Ni(II)_4$ fully metallated chain complex have failed, which appears to be a consequence of the unfavourable geometry created at the fourth site (Shuvaev *et al.*, 2009). A major factor in this case appears to be the strong preference of Ni(II) for a strong field ligand environment (larger CFSE than Cu(II), Mn(II), Co(II)) (*vide supra*), and so it ignores the oxygen atoms, despite the potential for the formation of a grid, where large entropy contributions would perhaps be considered to favour grid assembly.

The isomeric ligand H_2L25 is potentially tetratopic and is built on the same pyridazine core, but has the hydrazone function at the ligand ends, which are similar topically to the simple ditopic hydrazones. It can be compared with the ligand H_2L20, but lacks the third terminal *mer*-donor nitrogen atom. However, it again effectively presents a metal with the opportunity to form a self-assembled grid, which would be based on four connected $[2 \times 2]$ square subunits, which would of necessity be heteroleptic in combination with six-coordinate metal ions. Recent studies with Ni(II) and H_2L25b show a consistent pattern with nickel, in that once again it rejects the option of forming a grid by combining N and O donors, but instead forms a linear tetranuclear chain, comparable with the $Mn(II)_4$ and $Cu(II)_4$ chains with L24 (*vide supra*). The structure of $[Ni(II)_4(L25)_3](ClO_4)_2$ (**36**) is shown in Fig. 35, clearly indicating the nickel preference for an optimized crystal field environment based just on nitrogen donors (Shuvaev *et al.*).

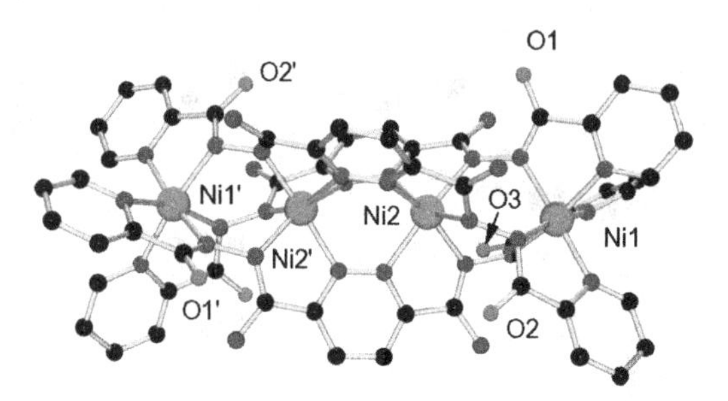

Fig. 35. Structure of Complex 36.

4. Nano-scale Molecular-Based Devices?

The quest for nano-scale relevancy in science has driven chemists and physicists to delve deeper into the 'nano-realm', while at the same time public 'hype' is triggered and manipulated by industrial advertising tactics, which promote small as better, and the key to the future. If there is a pressing necessity for using molecules as fundamental subunits, which can store information, or become components in extended molecular electronic circuitry, then chemists perhaps have the best chances of success in designing and constructing the fundamental subunits. From the simpler molecular perspective the first step would be to generate a molecular system which is small, but yet has a manipulable footprint, and also has some inherent property which can lead to 'bistability' in a 'convenient' temperature regime (ideally close to room temperature).

Single molecule magnets, exhibiting quantum tunneling of magnetization, have been a primary focus in the synthetic chemical community in this context for at least twenty years, but have yet to match the simple metal oxides (e.g. Fe_2O_3) in terms of readily accessible and functional magnetic properties. The most important limitation with such SMM systems is the low temperature at which these properties manifest themselves (typically <5 K). Therefore at present they have little practical utility. Spin transition or spin crossover (SC) materials, however, do show some promise for application purposes, with examples exhibiting bistable behaviour close to room temperature. The now classic Fe(II)-triazole SC chain compounds of Kahn and co-workers represent examples which have demonstrated some serious practical application in a display device (Kahn and Martinez, 1998). Despite working at room temperature practical devices based on this system have not been commercially developed.

Another property directly related to the presence of unpaired spins located in metal ions includes epr (electron paramagnetic resonance), which in principle is more sensitive than magnetic response, and also measurable at low concentration even at room temperature. Early experiments on an oxidized silicon (Si-(111)) surface showed that at constant magnetic field the spin magnetic moments precess around the field direction in the classical way, and induce a modulation of the tunneling current at the same frequency under STM conditions (Manassen *et al.*, 1989). The successful detection of an organic radical on HOPG using STM/EPR further supports the concept (Durkan and Welland, 2002), but success in this area has been limited, and experimental conditions are such that at this time practical use is not in the immediate future.

The creation of species with regularly organized centres of specific charge (e.g. transition metal ions in a specific oxidation state) arranged predictably in a rigid geometric fashion can be considered as quantum dot cellular automata (QDCA), which could be used to create 'logic gates' in a close packed 2D arrangement. These would have significant advantages over traditional field effect transistors (FETs), which act in a current switching capacity, but suffer from scaling problems, when considering dimensions approaching the nano-regime (Amlani *et al.*, 1999). Grids are ideal systems for nano-scale approaches to functional molecular based species. They can be created to order within limits, and have metal separations and donor environments which allow for site charge stabilization, despite the possibilities of spin exchange, and involve transition metals with accessible redox properties. The $Mn(II)_9$ [3 × 3] grids have been shown to lead to predictable and stable mixed valence species by simple controlled oxidation, within a four electron redox window (*vide supra*), and specifically oxidized species can be isolated. In principle a series of mixed valence intermediates could be created (Fig. 36), which would have different net spins based on the arrangement of the Mn(II) and Mn(III) centres (Dawe *et al.*, 2008; Milway *et al.*, 2006). In principle, each would have a different magnetic and epr signature. The demonstrated surface applicability of these grids to

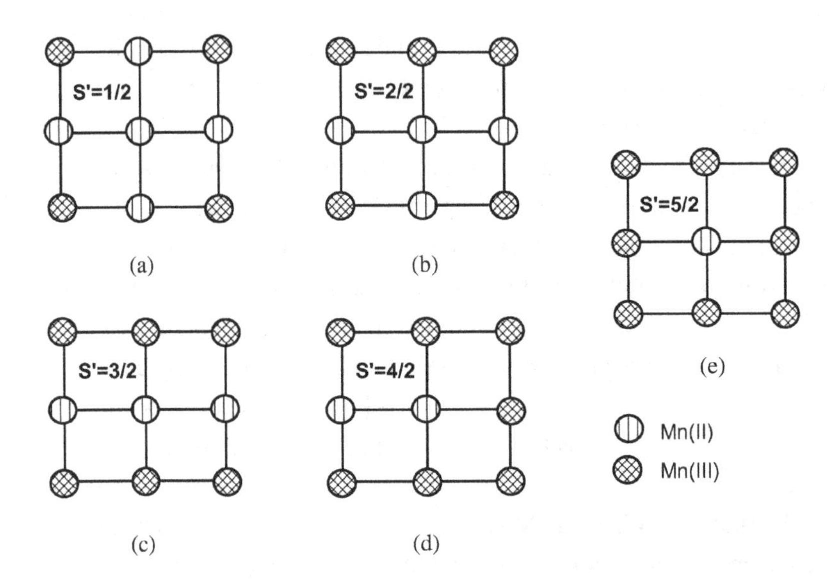

Fig. 36. Mixed valence Mn_9 intermediates.

Au(III) and HOPG would allow a 2D based medium to be created, and by selective tuning of tunneling currents in STM the creation of surface bound grids with specific spin signatures may be possible, paving the way for possible QDCA applications.

Quantum computing is seen as another way forward in the quest for bigger and faster systems, and can be approached by substituting single electron equivalent species for isolated electrons (Thompson *et al.*, 2004; Waldmann *et al.*, 2006). The creation of such species can be achieved by mixing sites of different spin such that in an antiferromagnetic system, a non-compensation of the spins leads to a ground state spin of $S' = 1/2$. The mixed oxidation grid $[Mn(III)_4Mn(II)_5(L18a^{2-})_6](ClO_4)_{10}\cdot 10H_2O$ (**28**) has an $S' = 1/2$ ground state, in what is essentially a ferrimagnetic system, resulting from the non-compensation of the different spin sites in the grid (*vide supra*) (Thompson *et al.*, 2004; Waldmann *et al.*, 2006). The ferrimagnetic wheels, e.g. Cr_7Ni, also have $S = 1/2$ ground states (*vide supra*) (Corradini *et al.*, 2007), and these mesoscopic $S = 1/2$ 'qubit' analogues are considered to be of potential importance as nanoscale components in future quantum computing.

5. Conclusions and Future Perspectives

Large grids form relatively easily, and the entropic advantage clearly helps, and serves to promote grid formation as size increases. However, ligand rotational flexibility, and tautomeric variability, can lead to alternative coordination modes, despite appropriate ligand programming. Of the classes of ligands discussed the hydrazones have been shown to be the most versatile, and have led to the largest range of grids than any other group. They also have synthetic advantages in that their construction is based on simple organic procedures, which can be extended to larger members with relative ease. Competing tendencies based on metal based properties, e.g. CFSE considerations, may limit the choice of metals used, but thus far with Mn(II) and Co(II) mixed valence species can be created, and with Cu(II) ferromagnetic species can be created. Applicable properties, e.g. magnetic and electronic, have been identified, which have significant potential for molecular device creation, and organized surface arrangements have been examined, which indicate that suitable media for information storage can be created. The challenge of taking such systems forward to a stage where applicability is more realistic will have to rest with a combination of efforts from chemists on the synthetic front, physicists on the theoretical and experimental fronts, and also micro-electronics experts, who will ultimately face the problem of device implementation, and electronic component development. Ultimately, the question of whether it is deemed necessary to carry this technology forward into molecular size regimes, will rest with pushing current limits to the extreme, and also, more importantly, whether society and industry are satisfied with the technological status quo, or are still hungry for 'faster and bigger' approaches to future problems.

Acknowledgments

NSERC (the Natural Sciences and Engineering Research Council of Canada) has provided generous research support over the years, and is gratefully acknowledged. Memorial University is thanked for support during a sabbatical year, in which this chapter was written.

References

Amlani, I., Orlov, A.O., Toth, G., Bernstein, G.H., Lent, C.S. and Snider, G.L. (1999) *Science* **284**, 289–291.

Barboiu, M., Vaughan, G., Graff, R. and Lehn, J.-M. (2003). *J. Amer. Chem. Soc.* **125**, 10257–10265.

Bark, T., Düggeli, M., Stoeckli-Evans, H. and Von Zelewsky, A. (2001). *Ang. Chem. Int. Ed.* **40**, 2848–2851.

Bassani, D.M., Lehn, J.-M., Fromm, K. and Fenske, D. (1998). *Angew. Chem. Int. Ed.* **37**, 2364–2367.

Baxter, P.N.W., Lehn, J.-M., Fischer, J. and Youinou, M.-T. (1994). *Angew. Chem. Int. Ed.* **33**, 2284–2287.

Baxter, P.N.W., Lehn, J.-M., Baum, G. and Fenske, D. (2000). *Chem. Eur. J.* **6**, 4510–4517.

Bu, X.-H., Kentaro, T., Mitsuhiko, S., Biradha, K., Morishita, H. and Furusho, S. (2002). *Chem. Commun.* 971–972.

Campos-Fernández, C.S., Clérac, R. and Dunbar, K.R. (1999). *Angew. Chem. Int. Ed.* **38**, 3477–3479.

Campos-Fernández, C.S., Schottel, B.L., Chifotides, H.T., Bera, J.K., Basca, J., Koomen, J.M., Russell, D.H. and Dunbar, K.R., *J. Am. Chem. Soc.* 2005, **127**, 12909–12923.

Cati, D.S., Ribas, J., Ribas-Ariño, J. and Stoeckli-Evans, H. (2004). *Inorg. Chem.* **43**, 1021–1030.

Cheng, H., Duan, C.-Y., Fang, C.-J., Liu, Y.-J. and Meng, Q.-J. (2000). *J. Chem. Soc. Dalton Trans.* 1207–1212.

Cornia, A., Fabretti, A.C., Sessoli, R., Sorace, L., Gatteschi, D., Barra, A.L., Daiguebonne, C. and Roisnel, T. (2002). *Acta Crystallogr.* **C 58**, 371–373.

Corradini, V., Biagi, R., del Pennino, U., De Renzi, V., Gambardella, A., Affronte, M., Muryn, C.A., Timco, G.A. and Winpenny, R.E.P. (2007). *Inorg. Chem.* **46**, 4937–4943 and references therein.

Dawe, L.N. and Thompson, L.K. (2007). *Angew. Chem. Int. Ed.* **46**, 7440–7444.

Dawe, L.N. and Thompson, L.K. (2008). *Dalton Trans.* 3610–3618.

Dawe, L.N., Abedin, T.S.M., Kelly, T.L., Thompson, L.K., Miller, D.O., Zhao, L., Wilson, C., Leech, M.A. and Howard, J.A.K. (2006). *J. Mater. Chem.* **16**, 2645–2659.

Dawe, L.N., Abedin, T.S.M. and Thompson, L.K. (2008). *Dalton Trans.* 1661–1675.

Dawe, L.N., Shuvaev, K.V. and Thompson, L.K. (2009). *Inorg. Chem.* **48**, 3323–3341.

Dawe, L.N., Shuvaev, K.V. and Thompson, L.K. (2009). *Chem. Soc. Rev.* DOI:10.1039/b807219c.

Dawe, L.N. (2008). Ph.D. Thesis, Memorial University of Newfoundland.

Dey, S.K., Thompson, L.K. and Dawe, L.N. (2006). *Chem. Commun.* 4967–4969.

Dey, S.K., Abedin, T.S.M., Dawe, L.N., Tandon, S.S., Collins, J.L., Thompson, L.K., Postnikov, A.V., Alam, M.S. and Müller, P. (2007). *Inorg. Chem.* **46**, 7767–7781.

Duan, C.-Y., Liu, Z.-H., You, X.-Z., Xue, F. and Mak, T.C.W. (1997). *Chem. Commun.* 381–382.

Durkan, C. and Welland, M.E. (2002). *App. Phys. Lett.* **80**, 458–460.

Galasco, A., Askenas, A. and Pecoraro, V.L. (1996). *Inorg. Chem.* **35**, 1419–1420.

Hausmann, J., Jameson, G.B. and Brooker, S. (2003). *Chem. Commun.* 2992–2993.

Kahn, O. and Martinez, C.J. (1998). *Science* **279**, 44–48.

Klingele (née Hausmann), J., Prikhod'ko, A.I., Leiberling, G., Demeshko, S., Dechert, S. and Meyer, F. (2007). *Dalton Trans.* 2003–2013.

Lis, T. (1980). *Acta Crystallogr.* **B 36**, 2042–2046.

Manassen, Y., Hamers, R.J., Demuth, J.E. and Costellano, A.J., Jr. (1989). *Phys. Rev. Lett.* **62**, 2531–2534.

Manoj, E., Prathapachandra Kurup, M.R., Fun, H.-K. and Punnoose, A. (2007). *Polyhedron.* **26**, 4451–4462.

Matthews, C.J., Avery, K., Xu, Z., Thompson, L.K., Zhao, L., Miller, D.O., Biradha, K., Poirier, K., Zaworotko, M.J., Wilson, C., Goeta, A.E. and Howard, J.A.K. (1999). *Inorg. Chem.* **38**, 5266–5276.

Matthews, C.J., Xu, Z., Mandal, S.K., Thompson, L.K., Biradha, K., Poirier, K. and Zaworotko, M.J. (1999). *Chem. Commun.* 347–348.

Matthews, C.J., Thompson, L.K., Parsons, S.R., Xu, Z., Miller, D.O. and Heath, S.L. (2001). *Inorg. Chem.* **40**, 4448.

Matthews, C.J., Onions, S.T., Morata, G., Bosch Salvia, M., Elsegood, M.R. and Price, D.J. (2003). *Chem. Commun.* 320–321.

Meier, F., Levy, J. and Loss, D. (2003). *Phys. Rev.* **B68**, 134417-1–134417-15.

Milway, V.A., Niel, V., Abedin, T.S.M., Xu, Z., Thompson, L.K., Grove, H., Miller, D.O. and Parsons, S.R. (2004). *Inorg. Chem.* **43**, 1874–1884.

Milway, V.A., Abedin, S.M.T., Niel, V., Kelly, T.L., Dawe, L.N., Dey, S.K., Thompson, D.W., Miller, D.O., Alam, M.S., Müller, P. and Thompson, L.K. (2006). *Dalton Trans.* 2835–2851.

Neels, A., and Stoeckli-Evans, H. (1999). *Inorg. Chem.* **38**, 6164–6170.

Niel, V., Milway, V.A., Dawe, L.N., Grove, H., Tandon, S.S., Abedin, T.S.M., Kelly, T.L., Spencer, E.C., Howard, J.A.K., Collins, J.L., Miller, D.O. and Thompson, L.K. (2008). *Inorg. Chem.* **47**, 176–189.

Onions, S.T., Frankin, A.M., Horton, P.N., Hursthouse, M.B. and Matthews, C.J. (2003). *Chem. Commun.* 2864–2865.

Onions, S.T., Heath, S.L., Price, D.J., Harrington, R.W., Clegg, W. and Matthews, C.J. (2004). *Angew. Chem. Int. Ed.* **43**, 1814–1817.

Parsons, S.R., Thompson, L.K., Dey, S.K., Wilson, C. and Howard, J.A.K. (2006). *Inorg. Chem.* **45**, 8832–8834.

Petitjean, A., Kyritsakas, N. and Lehn, J.-M. (2004). *Chem. Commun.* 1168–1169.

Petukhov, K., Alam, M.S., Rupp, H., Strömsdörfer, S., Müller, P., Scheurer, A., Saalfrank, R.W., Kortus, J., Postnikov, A., Ruben, M., Thompson, L.K. and Lehn, J.-M. (2009). *Coord. Chem. Rev.* DOI:10.1016/J.ccr.2009.01.024.

Rojo, J., Lehn, J.-M., Baum, G., Fenske, D., Waldmann, O. and Müller, P. (1999). *Eur. J. Inorg. Chem.* 517–522.

Ruben, M., Breuning, E., Barboiu, M., Gisselbrecht, J.-P. and Lehn, J.-M. (2003). *Chem. Eur. J.* **9**, 291–299.

Ruben, M., Rojo, J., Romero-Salguero, F.J., Uppadine, L.H. and Lehn, J.-M. (2004). *Angew. Chem. Int. Ed.* **43**, 3644–3662.

Sessoli, R., Gatteschi, D., Caneschi, A. and Novak, M.A. (1993). *Nature(London)*, **365**, 141–143.

Sessoli, R., Tsai, H.L., Schake, A.R., Wang, S.Y., Vincent, J.B., Folting, K., Gatteschi, D., Christou, G. and Hendrickson, D.N. (1993). *J. Am. Chem. Soc.* **115**, 1804–1816.

Shuvaev, K.V., Abedin, T.S.M., McClary, C.A., Dawe, L.N., Collins, J.L. and Thompson, L.K. (2009). *Dalton Trans.* In press, DOI:10.1039/B818939k.

Shuvaev, K., Dawe, L.N. and Thompson, L.K., Unpublished results.

Thompson, L.K., Matthews, C.J., Zhao, L., Xu, Z., Miller, D.O., Wilson, C., Leech, M.A., Howard, J.A.K., Heath, S.L., Whittaker, A.G. and Winpenny, R.E.P. (2001). *J. Solid State Chem.* **159**, 308–320.

Thompson, L.K., Matthews, C.J., Zhao, L., Wilson, C., Leech, M.A. and Howard, J.A.K. (2001). *Dalton Trans.* 2258–2262.

Thompson, L.K., Zhao, L., Xu, Z., Miller, D.O. and Reiff, W.M. (2003). *Inorg. Chem.* **42**, 128–139.

Thompson, L.K., Kelly, T.L., Dawe, L.N., Grove, H., Lemaire, M.T., Howard, J.A.K., Spencer, E.C., Matthews, C.J., Onions, S.T., Coles, S.J., Horton, P.N., Hursthouse, M.B. and Light, M.E. (2004). *Inorg. Chem.* **43**, 7605–7616.

Thompson, L.K., Waldmann, O. and Xu, Z. (2005). *Coord. Chem. Rev.* **249**, 2677–2690.

van der Vlugt, J.I., Demeshko, S., Dechert, S., Meyer, F. (2008). *Inorg. Chem.* **47**, 1576–1585.

Waldmann, O., Güdel, H.U., Kelly, T.L. and Thompson, L.K. (2006). *Inorg. Chem.* **45**, 3295–3300.

Wu, D.-Y., Sato, O., Einaga, Y. and Duan, C.-Y. (2008). *Angew. Chem.* **47**, 1–5.

Xu, Z., Thompson, L.K., Miller, D.O., Clase, H.J., Howard, J.A.K. and Goeta, A.E. (1998). *Inorg. Chem.* **37**, 3620–3627.

Xu, Z., Thompson, L.K., Matthews, C.J., Miller, D.O., Goeta, A.E. and Howard, J.A.K. (2001). *Inorg. Chem.* **40**, 2446–2449.

Youinou, M.-T., Rahmouni, N., Fischer, J. and Osborn, J.A. (1993). *Angew. Chem. Int. Ed.* **31**, 733–735.

Zhao, L., Matthews, C.J., Thompson, L.K. and Heath, S.L. (2000). *Chem. Commun.* 265–266.

Zhao, L., Niel, V., Thompson, L.K., Xu, Z., Milway, V.A., Harvey, R.G., Miller, D.O., Wilson, C., Leech, M., Howard, J.A.K. and Heath, S.L. (2004). *Dalton Trans.* 1446–1455.

Zhao, L., Xu, Z., Grove, H., Milway, V.A., Dawe, L.N., Abedin, T.S.M., Thompson, L.K., Kelly, T.L., Harvey, R.G., Miller, D.O., Weeks, L., Shapter, J.G. and Pope, K.J. (2004). *Inorg. Chem.* **43**, 3812–3824.

Zhao, Y., Guo, D., Liu, Y., He, C. and Duan, C. (2008). *Chem. Commun.* 5725–5727.

Chapter 2

RECENT SYNTHETIC RESULTS INVOLVING SINGLE MOLECULE MAGNETS

GUILLEM AROMÍ
Departament de Química Inorgànica
Universitat de Barcelona
08028 Barcelona, Spain

ERIC J L McINNES and RICHARD E P WINPENNY
School of Chemistry,
The University of Manchester,
Manchester M13 9PL, UK

1. Introduction

Some unusual molecular species show slow relaxation of magnetization, similar to the slow relaxation of magnetization shown by magnetic nanoparticles. This was first observed in the early 1990s, and the term single molecule magnet (SMM) was coined to describe such molecules. The term is widely-used although it is slightly misleading, and the physics involved is very different to a ferromagnet.

After approximately 20 years, there is a great deal of literature on SMMs, and it is not the intention to review all this work. Here we will briefly introduce the physics of SMMs, not going into huge depth as that is covered in an excellent book by Gatteschi, Sessoli and Villain (Gatteschi *et al.*, 2006).

A comprehensive review of the literature of SMMs to May 2005 was published in 2006 (Aromí and Brechin, 2006), and here we concentrate on work published since that date. Where necessary we include earlier work, otherwise much discussion would be incomprehensible, but the aim is to show where this science is at the end of 2009.

In the earlier review (Aromí and Brechin, 2006) 91 SMMs were described; 49 of those SMMs were homometallic manganese cages, some homovalent Mn(III) cages but the majority mixed-valent. Nine iron(III) SMMs were included and two iron(II) cages. There were eight nickel(II) SMMs reported, two cobalt(II) SMMs and two vanadium(III) examples. Ten further examples involve cyanide bridges, normally creating heterometallic SMMs; there were also three heterometallic SMMs featuring 3d-metals but not containing cyanide. Two monometallic 4f-complexes had shown slow relaxation of magnetization, and three heterometallic 3d-4f complexes.

The growth since 2005 is incredible. By the end of 2008 there were 309 SMMs. Over 170 of these are homometallic manganese cages. For iron(III) and nickel(II) the number of new examples is very limited. Cobalt(II) SMMs have always been controversial, and new examples have been reported but still no consensus exists on the behavior displayed. The major growth in new compounds involves heterometallic 3d-4f complexes; this is now the second largest class of SMMs, after the homometallic manganese cages.

2. A Brief Introduction to the Physics of SMMs

The physics underlying SMMs has been discussed at great length elsewhere (e.g. Gatteschi *et al.*, 2006) and we restrict ourselves to describing the features that have informed the chemical synthesis that has been pursued in the attempt to design new SMMs, and ideally SMMs with higher operational temperatures.

SMMs arise from the quantum mechanical fact that for any molecule with a spin S the projection of that spin on a specific direction is quantized, given the quantum number M_s, with the condition that M_s can vary from $+S$ to $-S$ in integer steps. Therefore for any S state contains $2S + 1$ M_s states, and these M_s states need not be degenerate in zero external magnetic field. The loss of degeneracy is termed the zero-field splitting (ZFS) and

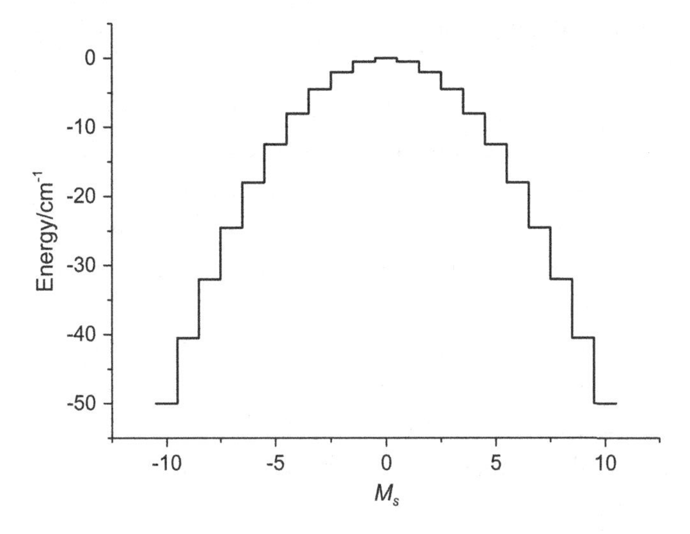

Fig. 1. The zero-field energies of M_s levels for $S = 10$ and $D = -0.5\,\text{cm}^{-1}$.

is parameterized with a series of ZFS parameters, the most significant of which is the axial ZFS, normally given the symbol D.

It is easiest to see with an example: for $S = 10$ there are 21 M_s levels (Fig. 1). The energies of each M_s level is given by DM_s^2, so if D is zero, the 21 levels would be degenerate. If D is positive, the lowest energy M_s level would be $M_s = 0$, and if D is negative the lowest energy levels would be $M_s = \pm S = \pm 10$. The latter is the case shown in Fig. 1.

The physical meaning of this picture is that if you imagine the spin projected against a specific direction, it is saying that having the spin lying perpendicular to the direction ($M_s = 0$) is higher in energy than having the spin lie parallel to the specified direction ($M_s = \pm S$). This leads to the term easy axis to describe such systems, as it is easier for the spin to lie near the axis. In zero-field the energy will lead to equal occupation of $M_s = +S$ and $M_s = -S$, as these two levels are degenerate, however in the presence of an external field one of these two levels will fall in energy and the other rise; the rate of change is given by the Zeeman term, $M_s g\beta H_z$, where β is the Bohr magneton, H_z is the external field and g is the electronic g-value. Hence the population will all fall into one of the two energy levels, say $M_s = -S$; the sample becomes magnetized. If the field is then switched off, the energy level picture returns to that given in Fig. 1, however to return to equilibrium the population which is entirely in the $M_s = -S$ needs to return to the $+S$ state. In principle it can do this in two ways. It can rotate, passing through the position where S is perpendicular to the field direction ($M_s = 0$), before moving to the $+S$ level. This is a thermal process, and absorption of energy needs to occur. The amount of energy is the gap between the $M_s = -S$ and $M_s = 0$ levels, which is given by DS^2 (for integer spins). The second way by which the system can return to equilibrium is to tunnel through the energy barrier. This is not a thermal process.

Therefore the slow relaxation seen for SMMs takes place when both the thermal and quantum tunneling processes are sufficiently slow. In the earliest SMMs the thermal process was the more important, and the key consideration was that the thermal energy barrier, $U_{\text{eff}} = DS^2$ was larger than the thermal energy of the system. For the first example observed the barrier is *ca.* $50\,\text{cm}^{-1}$ and slow relaxation was then seen at around 4 K. As lower temperatures were studied it was found that for many SMMs quantum tunneling can become important (Thomas *et al.*, 1996).

The thermal energy barrier for integer spins is given by DS^2; for non-integer spins it is given by $D|(S^2 - 1/4)$. These two parameters remain the most significant for SMMs: the size of the spin, given by S, and the axial anisotropy, given by D. However for many of the recent SMMs reported with 4f-elements quantum tunneling is also important, and here other ZFS parameters become important. This is because the tunneling between energy levels is promoted by mixing of states, which is more efficient if other ZFS parameters are needed to describe the system. These additional terms arise when the symmetry of the system falls. So a third consideration for slow relaxation becomes axial symmetry — ideally the presence of a three- or four-fold rotation axis.

D can appear to be a mysterious parameter. The major contribution to the D-value for a polymetallic cage compound comes from the anisotropy of the single ions present in the cage compound. The anisotropy of these ions arise from the electronic structure of the ions. The first consideration has to be the number of unpaired electrons on the ion in question: as D is the parameter to describe the energy splitting in zero-field, it can only arise for $S > 1/2$. Therefore for a Cu(II) ion, for example, the single ion D is necessarily zero.

For the other common ions present in SMMs we can classify them depending on whether the *electronic structure* is anisotropic in the ground state or introduced by mixing in of excited states. For example, for Mn(III), which is present in over half the known SMMs, the ground state in an octahedral crystal field would be $t_{2g}^3 e_g^1$ — which is a ^{4}E state and hence anisotropic because the two e_g orbitals (d_{z^2} and $d_{x^2-y^2}$) must be unequally occupied. In almost all cases this leads to a Jahn-Teller distortion, which makes the *atomic structure* anisotropic, i.e. ligands attached to the z-axis will be further away than ligands attached to the x- and y-axes. The ground state anisotropy of Mn(III) explains why it is the ideal 3d-ion for building SMMs.

For Ni(II) or Cr(III) ions, by contrast, the ground state electronic structures are isotropic. For Ni(II) this is $t_{2g}^6 e_g^2$ in an octahedral field — giving a ^{3}A state. For Cr(III) this is $t_{2g}^3 e_g^0$ — giving a ^{4}A state. Both ground states are isotropic, however promoting a single electron in either case will give unequal occupation of the e_g levels and hence an anisotropic excited state. Therefore, there can be single ion anisotropy but due to mixing in of excited states. Spin-orbit coupling is also important in this mixing.

For Fe(III) the ground state is $t_{2g}^3 e_g^2$, which is a ^{6}A state — which is again isotropic. However here there is no excited state with the same spin, and so there is much less possibility of mixing in an excited state to generate single ion anisotropy. Hence Fe(III) tends to have much smaller single ion anisotropies than ions such as Ni(II) or Cr(III).

Ions such as Co(II) or V(III) generate a different type of complexity. For both these ions the electronic ground state — $t_{2g}^5 e_g^2$ for Co(II) or t_{2g}^2 for V(III) — are T states. Spin orbit coupling leads to anisotropy even in the ground state. Therefore the single ion anisotropy should be very large. However in these cases, where the orbital contribution to the magnetic moment is very high, even the description using S and D becomes problematic, i.e. S is not a good quantum number and therefore D, which is a parameter based on the projection of S, is not a very meaningful parameter. This means that SMMs featuring V(III) and Co(II) are very different to SMMs featuring Mn(III). For example, for most if not all reports of Co(II) SMMs there is strong quantum tunneling at zero-field and very fast relaxation.

As further metal ions — either 4d-, 5d- or 4f-ions — have been studied the behavior becomes steadily more complicated due to spin-orbit coupling. For the 4f-SMMs described below, spin-orbit coupling means that the total angular momentum has to be considered rather than S. The other up-shot is that while higher *thermal* barriers to relaxation of magnetization have been seen for 4f-SMMs, quantum tunneling becomes dominant and relaxation remains fast because of the second mechanism. This creates a further challenge: the means for raising the thermal barrier seems to increase the tunneling rate. Therefore as well as controlling D and S, control of the symmetry of SMMs is also a vital consideration.

3. Further SMMs Based on Mn(III)

Manganese has been the most featured metal among the hundreds of SMMs prepared since this property was discovered on the cluster $[Mn_{12}O_{12}(O_2CMe)_{16}(H_2O)_4]$ (Lis 1980, Sessoli *et al.*, 1993). In fact, Mn clusters currently represent more than 50% of this category of compounds, the vast majority of which having some of the metal ions, if not all, in the oxidation state 3+. The reason for this strong presence is a combination of factors; (i) the large number of unpaired electrons (u.e.) exhibited by high spin Mn ions in various oxidation states (Mn^{II}, 5 u.e.; Mn^{II}, 4 u.e.; Mn^{IV}, 3 u.e.), which facilitates the access to large total spin ground states, S_T, (ii) the presence of Ising type magnetoanisotropy in Mn^{III}, favoring a significant negative zero field splitting (ZFS) parameter, D, of the cluster, (iii) a good probability of establishing ferro- or ferrimagnetic interactions, also increasing the chances for a large S_T. It is therefore not surprising that many of the highlights occurred in SMM research have involved manganese compounds, let alone the fact that this property was discovered on a complex of this metal. Some of these landmarks are briefly described in the following section.

3.1. *The largest SMM; a* $[Mn_{84}]$ *torus*

The largest molecule behaving as a SMM is a manganese cluster; $[Mn_{84}O_{72}(O_2CMe)_{78}(OMe)_{24}(OH)_6(MeOH)_{12}(H_2O)_{42}]$ (Tasiopoulous *et al.*,

2004). This large molecule has the shape of a wheel composed of a succession of alternative linear [Mn_3O_4] and cubic [$Mn_4O_2(OMe)_2$] fragments, that stay together also with help of AcO^- bridges. An extensive study of the magnetic properties of the Mn_{84} has revealed that the combination of intramolecular Mn$\cdots$Mn interactions in this system lead to a total spin ground state of $S_T = 6$. The magnetization of the molecule displays temperature dependent hysteresis loops and several experiments demonstrate that it also undergoes quantum tunneling. The wheel has an external diameter of 4.2 nm, which is larger than some of the smallest Co nanoparticles reported (Jamet *et al.*, 2001) ($\approx$3 nm) and therefore it is the meeting point of the bottom-up and top-down approaches to nanoscale magnetic materials.

3.2. *Record spin number, S_T = 83/2, but no slow relaxation*

The mixed valence manganese cluster [$Mn_{19}O_8(N_3)_8(bhmp)_{12}(MeCN)_6$] Cl_2 (H_3bhmp = 2,6-bis(hydroxymethyl)-4-methylphenol), (Ako *et al.*, 2006) exhibits the highest ground state spin number of any molecule ever prepared, with a value of $S_T = 83/2$. The latter arises from the total alignment of the spin magnetic moments of twelve Mn^{III} ($S = 2$) and seven Mn^{II} ($S = 5/2$) ions. Despite this very large spin ground state, this complex does not exhibit SMM behavior since the overall D parameter turns out to be positive, as found by Q-band EPR spectroscopy (Waldmann *et al.*, 2008). The SMM with the highest reported spin ground state, $S_T = 37$, is also a cluster of manganese (Fig. 2), [$Mn_{17}O_8(N_3)_4(O_2CMe)_2(pd)_{10}(py)_{10}$ $(MeCN)_2(H_2O)_2$]$(ClO_4)_3$ (Moushi *et al.*, 2009). The magnetic moment of the ground state in this complex is the highest possible for this combination of ions ($11Mn^{III}$ and $6Mn^{II}$) as a result of exclusively ferromagnetic interactions present within the molecule.

As in the Mn_{19}, these interactions are facilitated by the presence of end-on N_3^- bridges, which have become a common tool for the preparation of high-spin transition metal clusters (Escuer and Aromí, 2006, Stamatatos and Christou, 2009). Interestingly, the value of the D parameter turns out to be rather small ($D = -0.009\,cm^{-1}$). Theoretical calculations demonstrate that large spin ground states favor a smaller anisotropy (Ruiz *et al.*, 2008).

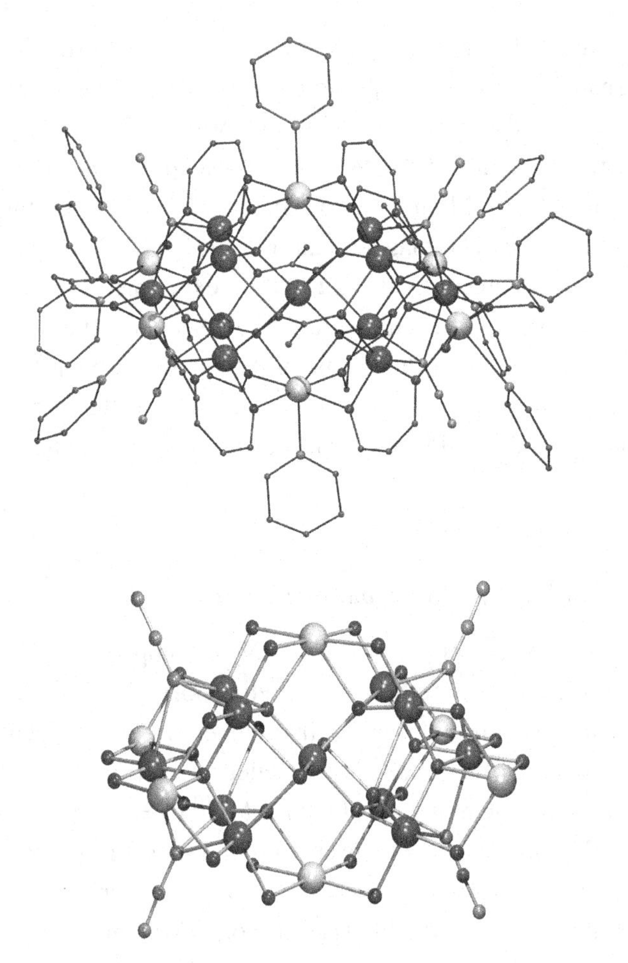

Fig. 2. PovRay representation of $[Mn_{17}O_8(N_3)_4(O_2CMe)_2(pd)_{10}(py)_{10}(MeCN)_2(H_2O)_2]$ $(ClO_4)_3$ with hydrogen atoms omitted for clarity (top) and its core (bottom). Code for atoms: *largest grey*, Mn(III); *largest light grey*, Mn(II); *second largest grey*, O; *second largest light grey*, N; *rest*, C.

3.3. *Record magnetic anisotropy barrier; a Mn_6 cluster*

A series of Mn^{III} clusters with a hexanuclear core (see below) have emerged, exhibiting high spin ground states (S_T from 4 to 12) and SMM properties. One member of this family, $[Mn_6O_2(Etsao)_6(O_2CPh(Me)_2)_2(EtOH)_6]$ ($EtsaoH_2$ = Ethylsalicyl-aldoxime), (Milios *et al.*, 2007) with $S_T = 12$, shows the highest energy barrier ever observed ($U_{eff} = 86.4$ K) for the magnetization reversal in an SMM. This complex is the first polymetallic SMM with slower relaxation of the magnetization (by one order of magnitude) than observed for the pioneering complex $[Mn_{12}O_{12}(O_2CMe)_{16}(H_2O)_4]$. High resolution inelastic neutron scattering (INS) studies and frequency domain magnetic resonance spectroscopy (FDMRS) show that the giant spin model is not appropriate to explain the relaxation dynamics of this molecule, since the levels from several excited spin multiplets are found at lower energy than some of the levels of the spin ground state. On the other hand, recent micro-SQUID measurements reveal that the mechanism of magnetic relaxation is best described using a model that considers two weakly coupled $S = 6$ halves of the molecule. This provides for a large number of tunneling pathways between spin multiplets that explain well the pattern of steps observed in the magnetization curves (Bahr *et al.*, 2008).

3.4. *Quantum entanglement between SMMs; first discovered in a pair of Mn_4 clusters*

Entanglement of the spin wave functions of two magnetic clusters is a required property if such an assembly is to be considered as a two qubit quantum gate for quantum information processing (Affronte, 2009). The first pair of SMMs showing evidence of mutual quantum superposition of their spin wave functions is formed by two tetranuclear Mn clusters with formula $[Mn_4O_3Cl_4(O_2CEt)_3(py)_3]$ (Hill *et al.*, 2003). This complex is found in the crystal lattice in form of dimers, weakly bound through an array of $[Cl\cdots H\text{–}C]$ supramolecular interactions, which provide the pathway for their feeble magnetic exchange. Such coupling causes the entanglement of the spin ground states of the clusters, as evidenced by multifrequency high field EPR spectroscopy (Hill *et al.*, 2003). This interaction also causes a bias to the magnetic fields at which quantum tunneling of the magnetization (QTM) takes place within the individual clusters (Wernsdorfer *et al.*, 2002). Such effects are of particular relevance, since they suggest the capacity of tuning the process of magnetic relaxation through chemical control at the supramolecular level.

Similar effects have been studied more recently between two differentiated parts of the same molecule, treated as weakly coupled individual nanomagnets (i.e. as a pair of SMMs). Two most relevant examples are the cluster

$[Mn_6O_2(Etsao)_6(O_2CPh(Me)_2)_2(EtOH)_6]$ (see above) studied as a pair of exchange-coupled $S = 6$ Mn_3 moieties (Bahr *et al.*, 2008) and the wheel $[Mn_{12}(O_2CMe)_{14}(adea)_8]$ ($AdeaH_2 = N$−allyldietanolamione, Shah *et al.*, 2008) which behaves as two $S = 7/2$ halves, each with SMM properties, that exhibit quantum interference of their quantum tunneling because of the weak magnetic interactions between both parts of the molecule (Ramsey *et al.*, 2008).

The ubiquitousness of manganese among SMMs has allowed the creation of several families of closely related compounds with this behavior. This has granted the possibility of studying properties inherent to the phenomenon in systematic ways, by following the effects of small or gradual changes or by establishing magnetostructural correlations. In the following paragraphs we review some of the most relevant families of SMMs of manganese. Clearly, the best known is that of clusters derived from the first described SMM; $[Mn_{12}O_{12}(O_2CMe)_{16}(H_2O)_4]$ (Lis, 1980, Sessoli *et al.*, 1993). This family is the most numerous and has been extensively investigated since the beginning of the field. A comprehensive account on the properties of this category of SMMs and the studies performed on them has been recently featured in a review (Bagai and Christou, 2009). For these reasons, the focus here will be on less known families.

3.5. *[$Mn^{III}_3Mn^{IV}$] clusters with an S = 9/2 ground state*

This group of compounds was one of the first well established families of SMMs, and their study in depth has produced important contributions to the fundamental understanding of this phenomenon. This work has been partially reviewed on previous occasions (Aromí and Brechin, 2006, Feng *et al.*, 2008) and only a brief account will be given here. The title group is formed by tetranuclear Mn aggregates possessing a mixed-valence distorted cubane core described as $[Mn^{III}_3Mn^{IV}O_3X]$ (Fig. 3). In all these complexes, the Mn^{IV} center is linked to the other three Mn^{III} ions through the O^{2-} ligands of the core, and by three carboxylate groups. The two other positions on each Mn^{III} ion remain as terminal sites. All the molecules of this type exhibit a spin ground state of $S_T = 9/2$, as a result of the ferromagnetic interactions (of the order of $J = -20\,cm^{-1}$) between the three high-spin

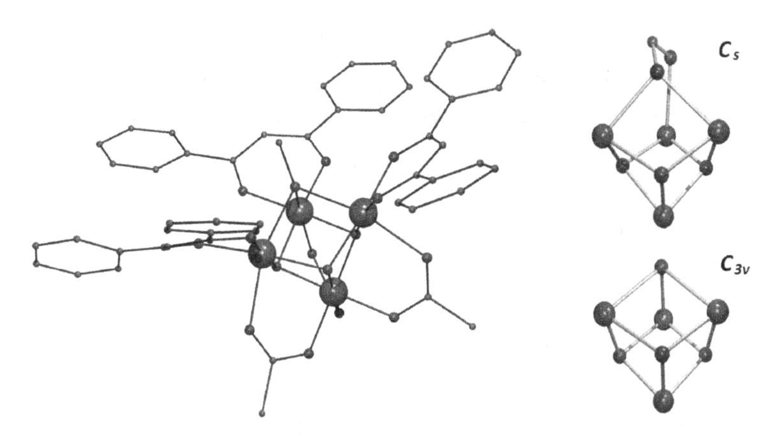

Fig. 3. (Left) PovRay representation of $[Mn_4O_3(OMe)(O_2CMe)_3(dbm)_3]$ with hydrogen atoms omitted for clarity. (Right) and the core for this family of compounds, emphasizing the two types of idealized symmetries (C_s and C_{3v}) that they exhibit. Code for atoms as in Fig. 2.

Mn^{3+} ions ($S = 2$) together with the antiferromagnetic coupling (near $J = +10\,cm^{-1}$) between these three centers and the Mn^{4+} ion ($S = 3/2$).

With the advent of SMMs in the early nineties, these $[Mn_4]$ molecules, some of which had been synthesized long before, were investigated under a new perspective and were found to exhibit slow relaxation of their molecular magnetic moment (Aubin *et al.*, 1996). Their case was at the time very special, since unlike the $[Mn_{12}]$ and $[Fe_8]$ molecules that pioneered the research on SMMs (both types with $S_T = 10$), the $[Mn_4]$ clusters exhibit a non-integer spin ground state, which in theory precludes the observation of QTM. However, the later was clearly observed on the complex $[Mn_4O_3Cl(O_2CMe)_3(dbm)_3]$ and was believed to result from transverse terms of the anisotropy or small internal magnetic fields (Aubin *et al.*, 1998). Precise values for the parameters of the Hamiltonian describing the magnetic anisotropy of the spin ground state in this compound could be obtained through High Field EPR (HFEPR) spectroscopy (Aubin *et al.*, 1998).

This group of molecules is amenable to extensive chemical variations: (i) A rich disparity of synthetic methodologies has brought about the possibility of changing the nature of the group ‘X’ within the core, which can be Cl^-, Br^-, F^-, OH^-, MeO^-, EtO^-, PhO^-, AcO^-, N_3^-, NCO^- or NO_3^-,

(Aromí *et al.*, 1998, 2002, Wemple *et al.*, 1995) all acting as μ_3, monodentate donors, or be a carboxylate bound in the [3.21] fashion (Harris notation); (ii) the carboxylate group bridging the Mn^{IV} ion to the other metals can vary, either during the cluster formation or by ligand substitution; (iii) the terminal positions of the Mn^{III} metals are either occupied by chelating 1,3-diketonates derived from dibenzoylmethane (Hdbm) or monodentate ligands. This versatility has offered the chance to investigate properties of SMMs through controlled chemical changes.

One such outstanding study is the INS investigation of the series $[Mn_4O_3X(O_2CMe)_3(dbm)_3]$ with $X = Cl^-$, Br^-, F^-, OAc^- (Andres *et al.*, 2000). This method allows for the observation of transitions between the various levels of a particular spin state, in the absence of an external magnetic field. This work served to indentify the influence of the 'X' group on the rhombic ZFS parameter E, which in turn, has an important impact on the QTM of the cluster. In particular, it was found that $|E|$ for the 'Cl' derivative is larger than for the 'Br' compound (0.022 and 0.017 cm^{-1}, respectively), explaining a faster rate of relaxation through quantum tunneling for the former cluster. Such a systematic study on a family of SMMs was unprecedented.

A subset of the $[Mn_4]$ family possesses idealized C_s symmetry; lower than C_{3v}, as exhibited by the majority of compounds in the series (Fig. 3, right). This is because the 'X' group of the core is a didentate carboxylate rather than a monodentate ligand. The consequence of this symmetry reduction was studied by HF-EPR and was found to be an increase of the rhombicity, as manifested by higher E values (Aliaga-Alcalde *et al.*, 2004). This leads to an increase of the rate of the magnetic relaxation through the mechanism of quantum tunneling, which could be observed experimentally by means of ultra-low temperature (<1 K) specific heat experiments (Evangelisti *et al.*, 2002). Indeed, the relaxation rate of the magnetization for the cluster $[Mn_4O_3Cl(O_2CMe)_3(dbm)_3]$ (with idealized C_{3v} symmetry) was found to be ten times longer than for $[Mn_4O_3(O_2CPh\text{-}p\text{-}Me)_4(dbm)_3]$ (possessing C_s symmetry).

The chemical manipulation of the peripheral ligation of this family of clusters enabled the control of their supramolecular organization and the extent of their intermolecular interactions. One outstanding example is the crystallization of the SMM $[Mn_4O_3Cl_4(O_2CEt)_3(py)_3]$ (Hill *et al.*, 2003) as pairs of weakly coupled dimers thanks to weak inter-cluster hydrogen bonds, leading to very important consequences (see above). A theoretical DFT calculation could reproduce reasonably well the magnetic anisotropy and the exchange interaction between both clusters in this dimer (Park *et al.*, 2003). More recently, a very extensive theoretical study included calculations on model complexes related to this family of $[Mn_4]$ clusters, changing the three categories of peripheral ligation; the 'X' group, the bridging carboxylates and the terminal ligands. Several properties were investigated, however, the study focused on the importance of the ligands on the nature and strength of the magnetic exchange between the Mn ions of the cluster. Interestingly, it was concluded that besides the decisive impact that ligands can have on the exchange by influencing on the structure of the cluster, the use of bridging ligands with sp^2 carbon atoms enhances the coupling between the Mn ions through a spin polarization mechanism (Tuan *et al.*, 2009).

3.6. *The $[Mn_2^{III}Mn_2^{II}]$ family of "rhombic" SMMs*

This group of compounds possess a core of the type $[Mn_2^{III}Mn_2^{II}(\mu_3{-}O)_2(\mu{-}O)_4]$ (Fig. 4) that has been termed in several different ways; 'diamond-shaped', 'butterfly', 'rhombic', 'defective dicubane', etc. The first published member of this family, $[Mn_4(O_2CMe)_2(pdmH)_6](ClO_4)$ (Brechin *et al.*, 1999, Yoo *et al.*, 2000) joined then the group of molecules that had been identified as SMMs, which was at that time very small. The robustness of this assembly explains why a large family of structurally related complexes with similar properties followed that first report. In all cases, the Mn(II) ions occupy the external positions of the core, whereas the Mn(III) metals are located at the central positions. In this core, the oxygen atoms are provided by alkoxide moieties from polydentate ligands (most often 2-hydroxymethylpyridine, hmp^-, Fig. 4, left), which also chelate the metals, thereby preventing the growth of the system into larger aggregates. With very few exceptions, the Mn(II) ions always exhibit labile terminal ligands to complete hepta- or hexacoordination (the latter occurs if steric reasons impede the coordination number seven). These ligands are

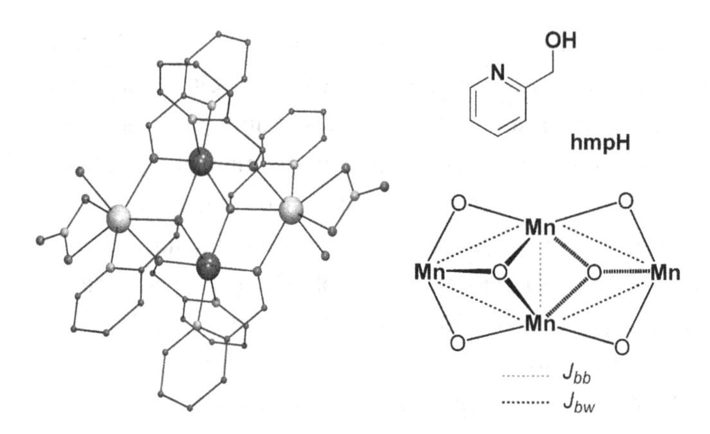

Fig. 4. (Left) PovRay representation of the cation $[Mn_4(hmp)_6(H_2O)_2(NO_3)_2](NO_3)_2]^{2+}$ with hydrogen atoms omitted for clarity and code for atoms as in Fig. 2. (Right) The ligand hmpH and a representation of the intramolecular spin coupling scheme used for this family of molecules.

usually neutral molecules from reaction solvents (MeCN, MeOH, H_2O) or weakly coordinating anions (NO_3^-, Cl^-, Br^-, RCO_2^-, etc.), which can be conveniently manipulated to favor the assembly of the [Mn_4] SMMs into 1D, 2D or 2D polymeric networks. A recent review shows, with great structural detail, an extensive compilation of tetranuclear complexes of this type, together with the polymeric networks that have been constructed with some of them (Roubeau and Clérac, 2008).

The interesting magnetic properties of the rhombic [Mn_4] clusters arise from the combination of a large spin ground state ($S_T = 9$) and the presence of negative axial magnetoanisotropy, D. The S_T value stems from the occurrence of intramolecular ferromagnetic interactions, as reflected by positive J_{bb} ($Mn^{III} \cdots Mn^{III}$) and J_{bw} ($Mn^{III} \cdots Mn^{II}$) coupling constants (Fig. 4, right), leading to a total spin ground state of $S_T = 9$. The intensity of this coupling is only moderate to very weak (J_{bb} not exceeding 6.3 cm^{-1} and being generally one order of magnitude larger than J_{bw}), which causes the presence of excited spin states very close in energy to the ground state (Yoo *et al.*, 2000). In very exceptional occasions, the J_{bw} coupling constant falls over to the antiferromagnetic side, causing a dramatic decrease of the spin ground state (from $S_T = 9$ to $S_T = 1$) (Lecren *et al.*, 2005) and thus, a very significant change in magnetic behavior, such as the loss of the SMM behavior. This is caused by a slight increase of one Mn–O–Mn angle within the $Mn^{III}\cdots Mn^{II}$ bridge, which highlights the fact that small structural variations induced at the chemical level can serve to control the functional properties of a molecular system. The D value in the [Mn_4] complexes (found to be in the -0.15 to $-0.26\,cm^{-1}$ range) results from the presence of two *quasi* parallel Jahn-Teller axes, located on the Mn^{III} ions.

The quantum properties of the spin ground state have been studied in great detail for some members of this family. A few compounds have been examined by HF-EPR, from where axial, rhombic and quartic longitudinal (D, E, B_4°, respectively) ZFS parameters could be obtained (Yoo *et al.*, 2000, 2001). This has thrown light onto the mechanism of relaxation of the magnetization in these molecules. From the observed D parameters, activation energies for the reversal of the spin angular momentum are predicted (using the $U=|DS_z^2|$ relationship) to be larger than observed studying the relaxation dynamics. This underscores the existence of other mechanisms of relaxation, other than thermal equilibration, such as QTM, as suggested by the other ZFS parameters provided by EPR. Very low temperature isothermal magnetization *vs* magnetic field hysteresis loops obtained with a micro-SQUID proof the existence of quantum tunneling in form of well defined steps (Lecren *et al.*, 2005, 2008). The spacing between these steps, ΔH, have served to calculate very good estimates of the D parameter in some compounds, according to the equation $\Delta H=n|D|/g\mu_B$ (n indicates the nth step), consistent with the values obtained with other techniques. These curves exhibit larger coercive fields upon cooling, becoming thermally independent below a certain temperature. In this regime, the relaxation occurs only from the lowest energy m_s level ($m_s=-9$) through a mechanism of quantum tunneling. Relaxation time measurements also show that at higher temperatures the dynamics of the process follows an Arrhenius curve while below certain temperature ($\approx$0.4 K) the relaxation occurs through a tunneling pathway that is temperature independent. Micro-SQUID investigations on single crystals of the derivative $[Mn_4(hmp)_6(H_2O)_2(NO_3)_2](NO_3)_2]$ (Lecren *et al.*, 2008) allowed to extract estimates of the D and E parameters by using the phenomenon of quantum phase interference.

A remarkable aspect of the family of rhombic [Mn_4] SMMs is that their chemical and structural features have been used to organize these molecular entities by means of coordination bonds, into polymeric arrangements of various dimensionalities and, in some cases, exhibiting novel magnetic properties (Roubeau and Clérac, 2008). This has been achieved mainly thanks to the presence of labile ligands at the external Mn^{II} ions that can be easily removed and replaced by difunctional groups, capable to link the clusters to each other. In this manner, [Mn_4] clusters have been linked to form: (i) polymeric chains (1D); (ii) coordination polymers in form of sheets (2D); and (iii) networks of clusters in the three directions (3D). Chains of rhombic [Mn_4] SMMs have been constructed using various types of bridging ligands, including chloride (Yoo *et al.*, 2005), end-to-end double N_3^- (Lecren *et al.*, 2005) and *syn, syn* carboxylates ($CH_3CO_2^-$, $CClH_2CO_2^-$) (Lecren *et al.*, 2008). In all cases, the magnetic coupling achieved is weak and antiferromagnetic. This precludes the observation of single chain magnet (SCM) behavior (Coulon *et al.*, 2006), which is one of the sought properties behind this designed synthesis. Nevertheless, the uniqueness of such type of arrangement is at the root of the unprecedented observation of interesting physical properties, such as the first experimental manifestation of finite size effects in antiferromagnetic Ising-type chains in the compound $[Mn_4(hmp)_6(N_3)_2](ClO_4)_2$ (Lecren *et al.*, 2005). The flexibility and small volume of the dicyanamide anion (dcn^-) allows to link [Mn_4] clusters into three different 2D assemblies, depending on the identity of the other terminal ligands, which exhibit canted antiferromagnetic ground states (Miyasaka *et al.*, 2006). The clusters in these three polymers are mutually oriented in three different angles, leading to three ordering temperatures. The consequence of this gradual change is the observation of an efficient ordered canted system (for the smallest angle), one that behaves as an SMM (since an inefficient ordering is masked by the slow relaxation of the building blocks), and a third system that exhibits an intermediate situation. A 3D assembly using these versatile [Mn_4] units has also been prepared. The variety of extended architectures just summarized, attained through rationally designed synthetic strategies, makes this family of compounds perhaps the most versatile to date, for the supramolecular organization of SMMs.

3.7. *Oxime bridged SMMs with the core* $[Mn_3^{III}O]$ *and* $S_T = 6$

Oxime-based ligands have served as bridges along the three sides of oxo-centered Mn^{III} triangles, yielding an extensive family of $[Mn_3^{III}O]$ SMMs. The bridging occurs through both atoms of the oxamato groups in the [Mn–N–O–Mn] form. The oximes are generally of two types, pyridylketoximes or salicylaldo- or ketoximes, which also use their pyridyl or phenoxy donor atom, respectively, to bind to manganese (Fig. 5, left). Other bridging or terminal ligands complete the two positions remaining on each Mn ion not taken by the central oxide and the oxime ligands (see below). This group of small clusters are remarkable in the fact that most of them exhibit $Mn^{III} \cdots Mn^{III}$ ferromagnetic interactions, thus having an $S_T = 6$ ground state. This was surprising when seen originally, since such triangles are related to the well known family of $[Mn_3O(RCO_2)(L)_3]^{+/0}$ triangles (where RCO_2^- can be almost any carboxylate and L are neutral monodentate ligands), also known for many other transition metals, which always exhibit intramolecular antiferromagnetic interactions. The complex $[Mn_3O(O_2CMe)_3(mpko)_3](ClO_4)$ (Hmpko = methyl 2-pyridyl ketone oxime, Stamatatos *et al.*, 2005) belongs to one of the early series of oxime bridged triangles. In this subgroup, the oxime also includes a pyridyl

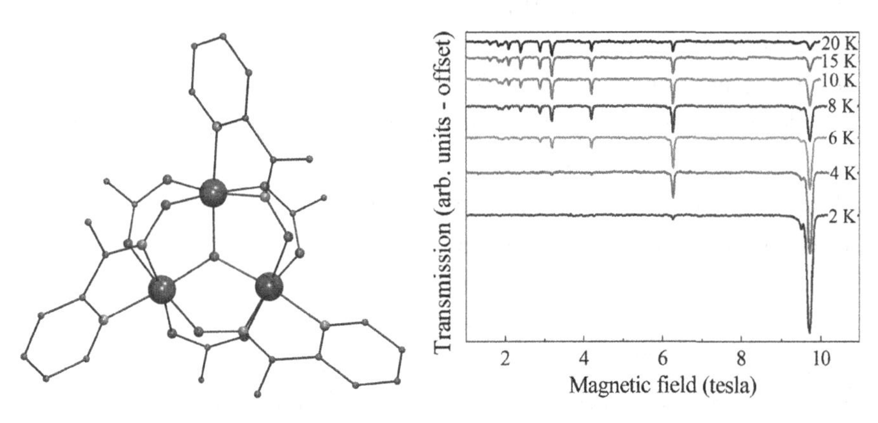

Fig. 5. (Left) PovRay representation of the cation $[Mn_3O(O_2CMe)_3(mpko)_3]^{3+}$ with hydrogen atoms omitted for clarity and code for atoms as in Fig. 2. (Right) Variable temperature HF EPR (104.1 GHz) on an oriented sigle crystal of $(NEt_4)_3[Mn_3Zn_2(salox)_3O(N_3)_6Cl_2]$ (Reprinted with permission, from Feng *et al.*, 2008).

moiety, whereas the additional ligands are *syn,syn* bridging carboxylates. Their $S_T = 6$ ground state, together with significant axial ZFS (D was determined to be $\approx -0.3\,\text{cm}^{-1}$ by simulating reduced magnetization and EPR measurements) (Stamatatos *et al.*, 2005, 2007) confer to these complexes SMM behavior. This finding was portrayed as an example of 'switching-on' the properties of SMM of triangular $[Mn_3^{III}O]$ clusters by introducing a ligand induced distortion. In this case, the distortion consisted on removing the central μ_3–O^{2-} ligand out of the Mn_3 plane, thereby causing a change from anti- to ferromagnetic coupling between metals. Interestingly, the related complex $[Mn_3O(bamen)_3](ClO_4)$ (Sreerama and Pal, 2002), which exhibits three $[Mn–(NO)_2–Mn]$ bridges on the triangle sides, also features an $S_T = 6$ ground state, even if the central oxide atom is only 0.015 Å away from the plane. This complex, however, unlike the rest of the oxamato $[Mn_3]$ triangles, exhibits heptacoordinated Mn^{III} ions. Another possible origin of the switch from anti- to ferromagnetic interactions in most of the oxamato triangles is the distortion of the Mn-NO-Mn bridge, gauged by the Mn-N-O-Mn torsion angle.

A group of $[Mn_3]$ clusters were prepared using derivatives of salicylaldoxime that included groups of different bulkiness in α to the oxime function. Examples of complexes of this series, exhibiting increasing bulkiness of the aldoxime ligand are $[Mn_3O(Mesao)_3(AcO)(py)_4]$, $[Mn_3O(Etsao)_3(O_2CPh(CF_3)_2)(EtOH)(H_2O)_3]$ and $[Mn_3O(Etsao)_3(O_2CPh(Cl)_2)(MeOH)_3(H_2O)]$ (Inglis *et al.*, 2008). The increased distortion has been correlated with the magnetic response; Indeed a change from anti- to ferromagnetic coupling (AF to F) was observed as the distortion induced by the bulkiness of the aldoxime was strong enough (the average torsion angle for the above three complexes are 10.7° (AF), 41.9° (F) and 40.6° (F), respectively). For the smallest Mn-N-O-Mn torsion angles, spin ground states of $S_T = 2$ where observed, even if the central O^{2-} atom was more than 0.35 Å above the Mn_3 plane. This distortion has also been induced following other synthetic strategies. One of these consists of introducing tripodal ligands capable of binding simultaneously to the three metals, causing a severe structural strain to the whole molecule, included an increase of the "twisting" of the oxamato bridge. This was brought about using perrhenate or perchlorate, respectively, as tripodal ligands in the complexes $[Mn_3O(Etsao)_3(ReO_4)(EtOH)(H_2O)_2]$ and $[Mn_3O(Etsao)_3(ClO_4)(MeOH)_3]$ (Inglis *et al.*, 2008). These two compounds were found to be SMMs, with a particularly high effective energy barrier, U_{eff}, for the reversal of the molecular magnetic moment in the spin ground state. Another way of tuning the distortion of the oxamato-bridge in $[Mn_3]$ triangles is by influencing the crystal packing of the molecules in the solid state. This is the case in the comparison between the clusters $(NEt_4)_3[Mn_3Zn_2(salox)_3O(N_3)_6Cl_2]$ and $(AsPh_4)_3[Mn_3Zn_2(salox)_3O(N_3)_6Cl_2]$ (Feng *et al.*, 2008). The difference in size of the cations causes these two complexes to crystallize in different systems (non-centrosymmetric $R3c$ and centrosymmetric $R\overline{3}c$, respectively) which results in significant disparities in the Mn-N-O-Mn torsion angles. Thus, the complex with NEt_4^+ exhibits larger angles and is an SMM with $S_T = 6$, whereas the complex with the other cation, with smaller torsion, exhibits antiferromagnetic $Mn\cdots Mn$ couplings and therefore an $S_T = 2$ ground state.

Besides the chemical control of the spin ground state and magnetic properties described above, the oxamato bridged $[Mn_3]$ clusters have been the object of other advanced physical studies. For example, single crystal HF EPR spectroscopy was performed on complex $[Mn_3O(O_2CMe)_3(mpko)_3](ClO_4)$ (Stamatatos *et al.*, 2007) at different angles. This experiment turned out to be particularly challenging, since the molecules of this trinuclear cluster are found in two different orientations within the crystal, rendering the assignment and simulation of the spectra very complicated. Exceptionally sharp and clear HF EPR spectra (Fig. 5, right) were also recorded for oriented single crystals of $(NEt_4)_3[Mn_3Zn_2(salox)_3O(N_3)_6Cl_2]$, which allowed a highly accurate determination of parameters g_x, g_x, g_z, D and B_4^0 (Feng *et al.*, 2008). Another interesting observation is the shift of external magnetic field at which QTM in the complex takes place. This phenomenon, originally discovered in a supramolecular dimer of $[Mn_4]$ SMMs (see above), called at the time 'exchange-biased quantum tunneling', has now been observed within this family of $[Mn_3]$ clusters in $[Mn_3O(Etsao)_3(O_2CPh(CF_3)_2)(EtOH)(H_2O)_3]$ (Inglis *et al.*, 2008), as a result of weak intermolecular interactions in the crystal lattice *via* hydrogen bonds, which enable the establishment of a feeble magnetic exchange between clusters. This exchange is significant enough to modulate the QTM of the clusters but not sufficiently strong to cause the antiferromagnetic ordering of the molecular magnetic moments.

3.8. *Magnetostructural correlations within a family of* $[Mn_6^{III}]$ *SMMs*

One of the largest families of SMMs is formed by hexanuclear complexes of Mn^{III} with general formula $[Mn_6O_2(salox)_6(X)_2(S)_{4-6}]$, where $salox^{2-}$ is a salicylaldo- or ketoxime, X^- is a carboxylate or a halide and S is a solvent that can be MeOH, EtOH or H_2O. The interest of this extensive series of analogues not only lies on their number (35 hits in the CCDC, version 5.30, May 2009 update), but mainly because this has served to establish the first magnetostructural correlation involving clusters of more than two metals (see below) in addition to allowing several other advanced physical studies.

The structure of all these compounds consists of two bridged parallel $[Mn_3^{III}O]$ oxamato triangles of the kind discussed in the above section. Each triangle binds to its counterpart through the oxygen atom of one of its oxamato groups, which then turns from double to triply bridging. In addition of the two bonds formed in this way, many complexes exhibit two additional links between both triangles *via* the phenolate group of one salicyloxime ligand, which then adopts a bridging mode instead of being just terminal. In the complexes from the latter class, all Mn ions are hexacoordinated, whereas in the absence of the phenolate bridges, two Mn^{III} remain as square-based pyramids. The remaining coordination sites are occupied by two terminal or bridging carboxylates or two halide ligands, and the corresponding solvate ligands.

The first two members of this family were the complex $[Mn_6O_2(O_2CMe)_2(sao)_6(EtOH)_4]$ and its benzoate analogue (Milios *et al.*, 2004). Both compounds were found to have an $S_T = 4$ spin ground state as a result of dominant antiferromagnetic interactions within each triangle (some of which frustrated to give two $S = 2$ units) together with the ferromagnetic interaction between both triangles. A remarkable feature of these clusters is the value of the ZFS parameter; $D = -1.22\,cm^{-1}$ for the acetate Mn_6, which is very large in comparison with other SMMs and originates as a consequence of the alignment of the Jahn-Teller axes of the Mn^{III} ions. The combination of S_T and D furnishes the conditions for the observation of SMM behavior, with a measured energy barrier to the relaxation of $U_{eff} = 28.0\,K$. These properties, together with suggestions that increasing the torsion Mn-N-O-Mn angle in the related $[Mn_3^{III}O]$ clusters favored ferromagnetic interactions *via* the oxamato bridge, encouraged some authors to seek enhancement of the SMM behavior in the $[Mn_6]$ complexes by deliberately manipulating structural distortions through chemical methods. The approach proved successful; the use of a sterically hindered salicyloxime and carboxylate led to the derivative $[Mn_6O_2(O_2CPh)_2(Etsao)_6(EtOH)_4(H_2O)_2]$ (Milios *et al.*, 2007), which featured significantly twisted Mn-(NO)-Mn bridges (as gauged by large Mn-N-O-Mn torsion angles, averaging 36.5°). This induced a switch of the coupling through these bridges from antiferromagnetic to ferromagnetic (AF-to-F), resulting on a dramatic jump in spin ground state of the cluster from $S_T = 4$ to $S_T = 12$, with a raise of the effective barrier up to $U_{eff} = 53.1\,K$, even if the parameter D is smaller ($D = -0.43\,cm^{-1}$) than in the original Mn_6. Once demonstrated, this principle was employed to cause a breakthrough in SMMs research; for the first time after almost fifteen years, the effective energy barrier to the reversal of the magnetization was raised above the values found originally for the $[Mn_{12}]$ family of SMMs. Thus, the bulky carboxylate 3,5-dimethylbenzoate, afforded the complex $[Mn_6O_2(Etsao)_6(O_2CPh(Me)_2)_2(EtOH)_4(H_2O)_2]$, which exhibits the highest effective anisotropy barrier found so far for any SMM ($U_{eff} = 86.4\,K$) (Milios *et al.*, 2007). This is due to the stronger twisting of the Mn-(NO)-Mn bridges as caused by a bulkier carboxylate, which increases the magnitude (J value) of the ferromagnetic interactions occurring through these bridges. Such increase does not bring the spin ground state beyond the value of 12, but it causes this ground state to be more isolated in energy with respect to the excited states. A larger energy separation reduces dramatically the possibilities of QTM through pathways involving higher energy spin states, which leads to an effective energy barrier considerably higher than seen for the benzoate analogue, even if both clusters exhibit almost exactly the same value of D at the ground state. In addition to this achievement, the structure-property relation observed in these compounds stimulated an extensive research program aimed at establishing a possibly underlying magnetostructural correlation.

Such a correlation could be demonstrated, at least qualitatively, first with a series of twelve analogues (Milios *et al.*, 2007), and subsequently including detailed studies on as many as twenty four members of this family (Inglis *et al.*, 2009). A plot of the Mn-N-O-Mn torsion angle, α, and the J value of the coupling mediated by that bridge shows a clear relation between these two variables. No apparent correlation could be established between the magnetic properties and any other structural parameter other than α. From the relationship shown in this plot, the AF-to-F switch should occur for $\alpha \approx 31°$. Thus, among the series of clusters studied, a progressive distortion of the Mn$\cdots$Mn oxamato bridge is observed with the consequent increase in J, which leads to an array of cluster spin ground states, ranging from $S_T = 4$ to $S_T = 12$, but also including values of 5, 6, 7, 9 and 11 (these intermediates being, however, subject to a ± 1 incertitude). Several factors must be born in mind before any attempt to interpret this correlation; (i) the coupling constants involved are small (ranging -2.20 to $+1.83\,\text{cm}^{-1}$), therefore, subject to a large relative error, (ii) the strength of the coupling also must have a certain dependence on other variables not taken into account in this correlation, such as terminal ligands, the exact nature of the oxime and carboxylate or other structural parameters, (iii) the determination of J is subject to several approximations, and the model used to fit the magnetic susceptibility data varied from complex to complex.

Despite these shortcomings, the trends shown by these studies represent a remarkable contribution to the area, not only for its uniqueness, but for the achievements that it has been possible to reach by combining these predictions with synthetic deftness. More recently, it has been possible to provoke the above shown changes to the magnetic properties of [Mn_6] complexes by structural distortion, using hydrostatic pressure instead of chemical methods. This was demonstrated with complex [$Mn_6O_2(Etsao)_6(O_2CPh(Me)_2)_2(EtOH)_4(H_2O)_2$], which was subject to pressures up to 17.5 kbar (Prescimone *et al.*, 2008). The effects of increasing pressure on the structure were monitored through single-crystal X-ray crystallography, whereas the magnetic properties under these conditions could be followed by means of a specially designed SQUID set up, in form of variable temperature magnetic susceptibility and variable field isothermal magnetization measurements. Increasing the pressure causes a compression of the Jahn-Teller axes of the Mn^{III} ions (up 2%), the shortening of some intermolecular interactions and a remarkable flattening of the Mn-N-O-Mn torsion angles, of up to 7° (≈20%). The changes observed in some of the torsion angles are expected to cause a decrease of the magnetic exchanged occurring through these bridges, as was confirmed through simulation of $\chi_{m}T$ *vs* T curves recorded at various pressures. These changes were not large enough to result in a variation of the spin ground state of the complex ($S_T = 12$), but caused a dramatic reduction of the separation between the latter and the excited states, which has an important impact on the relaxation properties of the SMM. This was clearly observed on the corresponding hysteresis curves.

The intense research work generated by this rich family of $[Mn_6]$ SMMs produces some conclusions while opening other intriguing questions. Deliberate structural manipulations permit control, not only the spin ground state, but also its isolation in energy with respect to the other spin states. We have discussed that both effects contribute to an increase of the energy barrier to the magnetic relaxation. This, however, does not occur to the extent that one may have desired. This is because the increase on ground state S_T has been accompanied by a concomitant reduction of D. Such effect has been studied in detail at the theoretical level, thanks to the availability of these objects of study. Ligand field analysis carried out on complexes $[Mn_6O_2(O_2CH)_2(sao)_6(MeOH)_4]$ (Milios *et al.*, 2006) and $[Mn_6O_2(Etsao)_6(O_2CPh(Me)_2)_2(EtOH)_4(H_2O)_2]$ (S_T of 4 and 12, respectively) (Piligkos *et al.*, 2008), indicate that the disparities in ground spin state magnetic anisotropy between both complexes seems to be related to differences in the projection coefficients of the single-ion anisotropies to the total anisotropy of the cluster. DFT calculations performed also on these two clusters lead to similar conclusions (Ruiz *et al.*, 2008); the contributions of the individual metal ions to the total anisotropy (conforming the so-called *spin channels*) depend on the nature of the spin ground state, in ways that, indeed, cause the final D value to be smaller if S_T is larger. These studies underscore the challenges still ahead if the properties of SMMs are to be observed at higher temperatures than are currently operative; is not sufficient with increasing the spin ground state in a controlled manner, or greatly enhancing the magnetic anisotropy. Both avenues need to be advanced in ways that are not detrimental to each other.

4. MMs Based on Fe(III) Ions

While there has been little progress in making new Fe(III) SMMs, there has been considerable advances towards understanding and making hybrid materials using Fe(III) cages. In particular, much work has been published on the "star-shaped" tetra-iron cages. Some of this involves a thorough analysis, through EPR spectroscopy, of the spin Hamiltonian parameters within these compounds (Accorsi *et al.*, 2006). The observations are that the zero-field splitting parameter can vary from $-0.21\ cm^{-1}$ to $-0.45\ cm^{-1}$, depending on the precise derivative used. The report ties this variation to the helical pitch of the coordination sphere around the iron(III) centers, making it one of the earliest example of a magneto-structural correlation in SMMs.

Also of interest is the ability to substitute the coordination sphere of these $\{Fe_4\}$ cages, introducing groups that can be used to attach these SMMs to surfaces (Barra *et al.*, 2007). While $\{Mn_{12}\}$ cages had been previously attached to gold surfaces, most of the studies of such materials suggested a change in the chemical composition of the cage once bound (Cornia *et al.*, 2006). The work on $\{Fe_4\}$ cages shows that not only chemical composition but also magnetic properties can be retained on binding to surfaces (Mannini *et al.*, 2009). If SMMs are ever to have a technological application this could prove to be very important work.

5. New SMMs Based on Divalent 3d-Ions

The earliest SMMs featuring divalent 3d-ions involved nickel(II). One was the cyclic dodecametallic cage $[Ni_{12}(chp)_{12}(O_2CMe)_{12}(THF)_6(H_2O)]$ (chp = the anion of 2-chloro-6-hydroxypyridine, THF – tetrahydrofuran) (Cadiou *et al.*, 2001), and the other a tetrametallic heterocubane $[Ni(hmp)(ROH)Cl]_4$ (hmp = the anion of hydroxymethylpyridine) (Edwards *et al.*, 2003). In both cases the exchange interactions present are ferromagnetic, leading to $S = 12$ and $S = 4$ ground states respectively.

The tetrametallic cage is one of a family of similar $\{Ni_4\}$ SMMs, and the magnetism shows inter-molecular exchange interactions are important leading to exchange-biasing of the quantum tunneling, i.e. a nearby local field due to neighbouring molecules moves the main tunneling step from $H_{ext} = 0$ (del Barco *et al.*, 2004). EPR studies of this phenomena are discussed in detail by McInnes in Chap. 5.

An isostructural $\{Co_4\}$ cage is the first reported cobalt(II) compound claimed to be an SMM (Yang *et al.*, 2002). Since these early reports, the number of nickel(II) SMMs has only increased slowly, and most are covered by the earlier review (Aromí and Brechin, 2007). The lack of progress in nickel SMMs is surprising, given the very large single ion anisotropies that have been reported (Rebilly *et al.*, 2008) and the regular observation of ferromagnetic exchange between Ni centers.

Two recent reports of slow relaxation in nickel cages should be mentioned. Firstly, a $\{Ni_5\}$ cage has been reported (Boudalis *et al.*, 2008); the structure contains a Ni_5 helix with each Ni...Ni edge bridged by an azide but also by other ligands. Detailed studies of this compound show that the ground state in field is $S = 5$, however the magnetization behavior of the compound does not follow a Brillioun function for an $S = 5$ state and there is an inflection at around 1 T which suggests that the zero-field ground state might be lower than $S = 5$. The dynamic behavior of the magnetization is complicated, and at least two relaxation mechanisms seem to be in operation. The authors also observe that only a small fraction of the sample seems to relax slowly. The variable temperature d.c. susceptibility data measured with external fields of 2.5 T and above are fitted, but data recorded with smaller external fields do not fit. The fits reported, with a D value of $-3\,\text{cm}^{-1}$, suggest that the giant spin approximation is breaking down with this system, as for the $\{Ni_4\}$ heterocubanes (Ferguson *et al.*, 2008).

This second new example also illustrates the need to go beyond a simple spin Hamiltonian description of the magnetism of a single magnetic molecule (Ferguson *et al.*, 2008); $[NiCl(HN(CH_2CH_2O)(CH_2CH_2OH)]_4$ contains a heterocubane core, with the O-atoms provided by a mono-deprotonated 2,2'-iminodiethanol molecule. Here magnetic measurements and high frequency EPR spectroscopy confirm an $S = 4$ spin ground state,

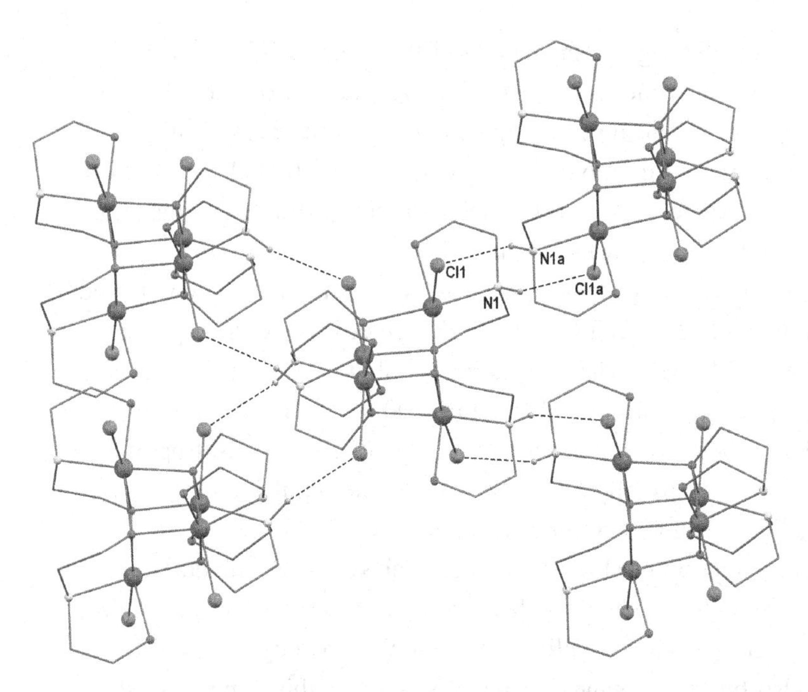

Fig. 6. The intermolecular H-bonding in the $\{Ni_4\}$ SMM reported in Ferguson *et al.*, 2008. Only hydrogen atoms involved in intermolecular interactions are represented. Code for atoms: *Largest*, Ni; *second largest*, Cl; *third largest*, N or O; *smallest*, H; *rest*, C.

and support a D-value of $-0.75\,cm^{-1}$, however, intermolecular H-bonding between terminal chlorides attached to each Ni(II) center and the H-N of the iminodiethanol ligands leads to a small but significant intermolecular exchange (Fig. 6). The key point is that an accurate description of the magnetic and spectroscopic behavior isn't possible without allowing for intermolecular exchange.

SMMs based on cobalt(II) remain controversial and it is reasonable to state that good physical models for the behavior are only just beginning to appear. The earliest reports are somewhat incomplete and it seems best to begin discussion with two reports where characterisation is more complete. These reports (Galloway *et al.*, 2008, Moubaraki *et al.*, 2008) involve study of $[C(NH_2)_3]_8[Co_4(cit)_4]\cdot 8H_2O$. In both cases the magnetic susceptibility measurements show little variation in $\chi_m T$ with temperature, and magnetization against field at low temperature gives a saturation value of *ca.* 6.8 $N\beta$. Moubaraki and co-workers note that this value would be consistent with an effective $S = 1/2$ at each Co(II) center, with a $g_{\text{eff}} = 3.4$; this would imply that the cobalt centers are non-interacting, or at least that the exchange interaction is much weaker than the spin-orbit coupling within individual cobalt(II) centers. Both papers also report a.c. susceptibility measurements and these give convincing evidence that all the $\{Co_4\}$ molecules are undergoing slow relaxation. The papers report different energy barriers and different pre-exponential factors: $U_{\text{eff}}/k = 24$ K, $\tau_o = 3.4 \times 10^{-8}$ s^{-1} (Galloway *et al.*, 2008), $U_{\text{eff}}/k = 13$ K, $\tau_o = 8.4 \times 10^{-7}$ s^{-1} (Moubaraki *et al.*, 2008). The latter energy barrier is based on a wider range of frequencies measured. Direct d.c. relaxation measurements (Galloway *et al.*, 2008) give parameters of $U_{\text{eff}}/k = 21$ K, $\tau_o = 8 \times 10^{-7}$ s^{-1}. Measurements on single crystals down to 40 mK show hysteresis loops that are temperature and sweep rate dependent, but in all cases there is a collapse in magnetization at zero applied field, presumably due to very fast quantum tunneling of magnetization.

The two early reports of cobalt(II) SMMs are related to these later works. The first report (Yang *et al.*, 2002) involves a $\{Co_4\}$ heterocubane, and while the data support a large spin ground state, the published proposal that this has an $S = 6$ ground state with g = 1.97 and a D value of 44 K seems unlikely. The molecule shows hysteresis at low temperature, but the a.c. susceptibility data reported suggest that only a small fraction of the sample is showing slow relaxation.

The other early report involves a $\{Co_6\}$ cage stabilised by citrate (Murrie *et al.*, 2003), which also contains a central heterocubane but here there are two further cobalt(II) centers on the periphery. This material presents a somewhat different problem, due to the fact that as crystallised the compound contains eleven water molecules as solvate. The d.c. susceptibility measurements can be interpreted as due to effective $S = 1/2$ centers ferromagnetically coupled, giving an $S_{\text{eff}} = 3$ ground state with large g-value. At low temperature there is a rise in the out-of-phase susceptibility, but no peaks are observed for the samples as prepared. The behavior changes as the samples desolvate, and the increase in χ" moves to higher temperature, such that a peak is observed and it becomes possible to calculate an energy barrier assuming that a thermally-activated relaxation is the cause for this peak. It is worrying that partial desolvation changes the magnetism in this way, as this could also be related to changes in inter-molecular exchange or the introduction of disorder that might be required for a spin glass.

Intermolecular exchange is a vital consideration in the slow relaxation of an octametallic cobalt(II) complex (Langley *et al.*, 2005). $[HNEt_3][Co_8(chp)_{10}(Hchp)_2(O_3PPh)_2(NO_3)_3]$ was studied as a member of an extensive family of cobalt(II) phosphonates. Variable temperature susceptibility measurements show a decline in the product χ_mT with temperature, which is probably due to spin-orbit coupling of the individual centers. A low temperature maximum in χ_mT could be consistent with ferromagnetic exchange between effective $S = 1/2$ centers. The magnetization does not saturate up to fields of 7 T. a.c. susceptibility studies show a frequency dependent peak in χ", indicating slow relaxation of magnetization, however the height of this peak compared with the values of χ' suggest that only around 10% of the molecules are relaxing slowly. An Arrhenius analysis of the a.c. data for two different samples gives $U_{\mathrm{eff}}/k = 82 \pm 4$ K and $\tau_o = 2 \times 10^{-12}$ or 2×10^{-11} s^{-1}; the pre-exponential factor is far too small for a typical SMM, and the variation between samples is worrying. The value is much more typical of a single chain magnet (SCM) (Coulon *et al.*, 2006) and there is a weak H-bonding path between $\{Co_8\}$ cages within the structure. The explanation given for the behavior is that this H-bonding is influencing the slow relaxation, and this should be considered a type of SCM rather than SMM. Other cobalt(II) phosphonates also show slow relaxation but again it is always only a fraction of the molecules showing the phenomena (Langley *et al.*, 2005).

A recent example where all molecules appear to be undergoing slow relaxation is $[Co_{12}(bm)_{12}(NO_3)(O_2CMe)_6(EtOH)_6](NO_3)_5$, where bm = the anion of 1*H*-benzimidazoyl-2-ylmethanol (Zeng *et al.*, 2007). This beautiful molecule contains three heterocubanes linked together to give a molecule with three-fold crystallographic symmetry (Fig. 7). The variable temperature susceptibility measurements show a slight increase in χ_mT with falling temperature, until 20 K at which point $\chi_m T$ increases rapidly reaching a maximum of 78 emu K mol^{-1} at 2.4 K. The authors

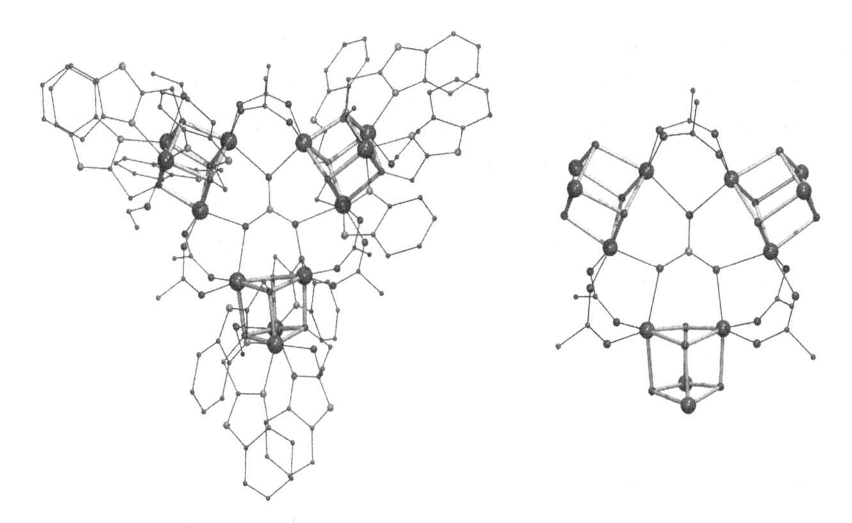

Fig. 7. PovRay representation of $[Co_{12}(bm)_{12}(NO_3)(O_2CMe)_6(EtOH)_6](NO_3)_5$ with hydrogen atoms omitted for clarity (left) and its core (bottom). Code for atoms: *largest grey*, Co; *second largest grey*, O; *second largest light grey*, N; *rest*, C.

interpret this behavior as due to weak ferromagnetic exchange combined with spin-orbit coupling of the individual cobalt(II) centers. The result is an effective $S = 6$ ground state due to coupling of the individual $S_{\text{eff}} = 1/2$ centers. This result can be reconciled with the magnetization at low temperature, assuming a high $g_{\text{eff}} = 4.63$. As in the individual heterocubanes with citrate, there is slow relaxation, which can be fitted to an Arrhenius law to give $U_{\text{eff}}/k = 15\,\text{K}$, $\tau_0 = 1.94 \times 10^{-7}\,\text{s}^{-1}$, but no hysteresis is observed and a collapse of magnetization at zero-field occurs at the lowest temperatures measured.

This paper and the two extremely thorough papers on $[C(NH_2)_3]_8[Co_4(cit)_4] \cdot 8H_2O$ raise a problem, which is probably not simply a question of semantics. The molecules unquestionably show slow relaxation of magnetization, but do not retain magnetization at zero-field. Therefore, should the term "single molecule magnet" be used for such a molecule, where there is no possibility of magnetization without an external field? If a field is required to retain magnetization, then the material is not a magnet.

In addition to these structures based on heterocubanes, there are heptametallic disc-like cobalt cages, some of which show slow relaxation of magnetization. Two early examples, are a heterovalent $\{Co^{II}_6Co^{III}\}$ cage and a homovalent $\{Co_7^{II}\}$ cage (Moragues-Canovás *et al.*, 2006, Zhang *et al.*, 2006) and in both cases there is reasonable evidence of ferromagnetic exchange between the spin centers, which are treated as effective $S = 1/2$ sites. The Co^{III} site is at the center of the disc in the heterovalent example. In the former case low temperature measurements fail to show hysteresis, while in the latter paper a.c. susceptibility measurements show a rise in χ" at 2 K, but no maximum. A more recent study of a $\{Co_7^{II}\}$ cage (Wang *et al.*, 2008) shows very similar results to the previous homovalent, however a $\{Co_4^{II}Co_3^{III}\}$ disc, $[Co_7(HL)_6(NO_3)_3(H_2O)_3](NO_3)_2$, ($H_3L = H_2NC(CH_2OH)_3$, Fig. 8) shows some hysteresis at below 1 K in temperature (Ferguson *et al.*, 2007). Here a Co^{II} site is at the center of the disc, and the valency of the sites alternate around the exterior.

In most of the cases mentioned above a working model for explaining the magnetism comes from assuming spin-orbit coupling leads to an effective $S_{eff} = 1/2$ per cobalt, and then ferromagnetic coupling between these centers. Fits of magnetization then require a high g-value. The problem with such models is that they cannot explain the slow relaxation observed, or

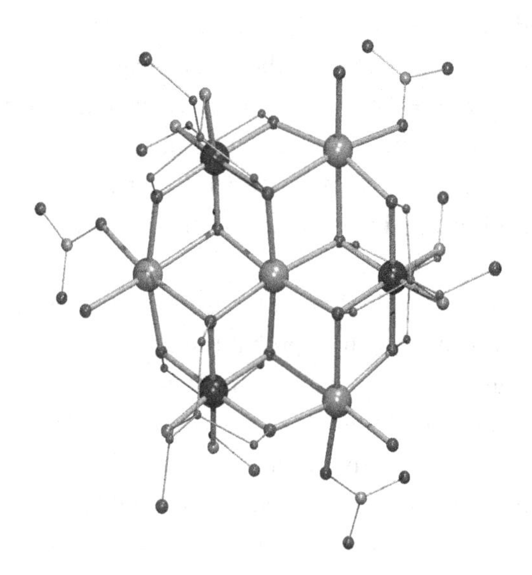

Fig. 8. PovRay representation of $[Co_7(HL)_6(NO_3)_3(H_2O)_3](NO_3)_2$ with hydrogen atoms omitted for clarity. Code for atoms: *largest grey*, Co(III); *largest light grey*, Co(II); *second largest grey*, O; *second largest light grey*, N; *rest*, C.

occasions when slow relaxation is not seen. New theory in the understanding of the magnetism of Co(II) cage complexes is vital.

A very interesting paper has discussed a different heterovalent $\{Co^{II}_3Co^{III}_4\}$ disc (Chibotaru *et al.*, 2008). Here the central cobalt is Co(II), and two further Co(II) sites are found in the ring. As the four Co(III) sites contain low spin Co(III) the magnetism is reduced to considering a liner trimetallic unit of Co(II) centers. The central cobalt site is different to the ring cobalt site, and the approach adopted is to perform *ab initio* calculations to obtain the electronic structure of the two different cobalt sites. By introducing ligand field splitting and then allowing spin-orbit interactions between the states derived from the ligand field splitting, a series of Kramer's doublets are produced. Because of spin-orbit coupling it is not possible to assign every Kramer's doublet to a parent quartet state, i.e. it is not valid to say that doublet X comes from the $^4T_{1g}$ or $^4T_{2g}$. In turn, this means that the energies cannot be understood as zero-field splitting of a given $S = 3/2$ level. The result then allows the magnetic properties of the individual cobalt(II) sites to be calculated and summing these contributions produce a good match for the room temperature $\chi_m T$ and predict a steady decline in $\chi_m T$ with decline in temperature. The observed $\chi_m T$ declines to a minimum at 60 K, and then increases rapidly. Therefore the difference between the observation and the sum of the calculated contributions due to the single ions must be due to exchange. The exchange interaction is evaluated by a method due to Lines (Lines, 1971), and is found to be ferromagnetic. The authors then fit both the variable temperature susceptibility data and magnetization data.

The conclusions of this paper are intriguing: the g-tensor for the ground Kramer's doublet of the complex is massively anisotropic — and the resulting $g_{av} > 6$; the traditional description as coupling effective $S = 1/2$ centers with high g-values is therefore related to this observation. However by detailed analysis much greater insight is possible into the slow relaxation in cobalt(II) compounds. The key problem is that because Co(II) compounds are in the weak-exchange regime, i.e. J is not bigger than spin-orbit coupling, tunneling through energy barrier is a major contribution to relaxation, rather than relaxation by going over the barrier. To prevent tunneling, high symmetry for the cage compound is absolutely vital. The authors use this to explain why their compound is not an SMM — the symmetry is too low — but they also state that symmetry will be the most important ingredient in building SMMs with orbitally degenerate spin centers.

Iron(II) containing SMMs remain rare; the ferrous heterocubanes reported by the Oshio group (Oshio *et al.*, 2004) and a family of enneametallic cages remain the most thoroughly studied examples (Boudalis *et al.*, 2004). Surprisingly the only new non-cyanate example claimed is a diferrous compound (Boudalis *et al.*, 2007) and even here the energy barrier is very small and slow relaxation is only observed by Mössbauer spectroscopy, and possibly by a rise of χ" at 2 K.

6. Slow Relaxation in Complexes Involving 4f-Elements

6.1. *Single atom magnets*

An intriguing class of molecules that show slow relaxation are the *bis*-phthalocyanine (H_2Pc) lanthanide sandwich complexes $(Bu_4N)[Ln(Pc)_2]$, where Ln = Tb, Dy and Ho (Ishikawa *et al.*, 2003). In all cases the Ln^{3+} ion is bound to eight N-donors from the Pc ligands and has a square-antiprismatic coordination geometry. The most surprising observation is for $[Tb(Pc)_2]^-$, where a.c. susceptibility shows out-of-phase peaks at up to 40 K, and Arrhenius analysis of the frequency-dependent data gives a barrier height of 230 cm^{-1}. The Dy^{3+} and Ho^{3+} complexes are also SMMs, but with lower energy barriers (Ishikawa *et al.*, 2005a and 2005b).

For the lanthanoids, the orbital angular momentum is not quenched and therefore we have to consider the total angular momentum of the ions. For the three ions here the ground states are: for Tb^{3+}, $^7F_6$; for Dy^{3+}, $^6H_{15/2}$; for Ho^{3+}, $^5I_8$. The *total* angular momentum is given by the quantum number J, which is 6, 15/2 and 8 respectively for these ground states. The projection of J on to a specific axis is quantised, just as S, and hence we have M_J levels ranging from $+J$ to $-J$ in integer steps. And also, as for S, crystal field effects can remove degeneracy from these M_J levels. In $(Bu_4N)[Tb(Pc)_2]$, NMR and susceptibility experiments show that the $M_J = \pm 6$ sub-states are lowest lying with a gap of 436 cm^{-1} to the $M_J = \pm 5$ sub-states. This is equivalent to an easy-axis magnetic anisotropy, but relying on J rather than S. The other difference is that the energy gaps in $[Tb(Pc)_2]^-$ are much higher than in 3d-SMMs. Smaller gaps to excited states are found for the Dy(III) and Ho(III) complexes.

It is also possible to tune the spacing of the sub-states and hence the energy barriers by changing the ligand field strength (Takamatsu *et al.*, 2007). The complex $[Tb\{Pc(OEt)_8\}_2]^-$ undergoes a two electron oxidation to $[Tb\{Pc(OEt)_8\}_2]^+$ (nb. the ethoxide-derivatised phthalocyanine is used to increase solubility) and this reduces the Ln-N distances and increases the ligand field strength. There is an increase in the energy gap between ground and first excited M_J states and the energy barrier is increased. The thermal barrier calculated from an Arrhenius analysis shifts to $550\,cm^{-1}$. A similar increase in the thermal barrier for spin relaxation is seen for the Dy analogue (Ishikawa *et al.*, 2008).

Isostructural, diamagnetic analogues of $(Bu_4N)[Ln(Pc)_2]$ can also be prepared and doped with the Tb, Ho and Dy complexes and this allows studies of the hysteresis behavior of these SMMs while excluding dipolar interactions between the SMMs (Ishikawa *et al.*, 2005a). The result is that it becomes possible to study how nuclear hyperfine effects influence QTM. Micro-SQUID measurements on single crystals of $(Bu_4N)[Y(Pc)_2]$ doped with 2% Tb- or Ho-doped show large QTM steps near but not at zero-field. The steps are due to anti-crossings between the combined M_J, M_I states. This is the first direct observation of nuclear-spin driven QTM in SMMs.

A further family of lanthanide sandwich compounds: $Na_9[LnW_{10}O_{36}]$ also show slow relaxation, but here it is restricted to Er^{3+} ($^4I_{15/2}$ ground state) and is not seen for Tb^{3+} or Dy^{3+} complexes (AlDamen *et al.*, 2008). This is the reverse of the case for the phthlocyanines. The change is also due to differences in the crystal field; however, in the phthalocyanine system the square-antiprism geometry is axially compressed while in the polyoxotungstate system it is elongated. This suggests that a balance of crystal field with a specific lanthanoid could lead to even higher energy barriers. It seems likely that axial symmetry is required.

6.2. *Polymetallic 4f-complexes*

Given the remarkable results on monometallic phthalocyanine complexes the study of polymetallic lanthanoid complexes seems an area ripe for examination. However the number of reports remains small, and almost all concentrate on polymetallic dysprosium cages. This may be motivated by the fact that Dy(III) has the highest magnetic moment of any single ion, however the most interesting relaxation behavior was found for $[TbPc_2]^-$ cages so it is odd that there are so few studies of polymetallic terbium cages in the literature.

One such study uses a calixarene to bridge between two Tb(III) centers (Kajiwara *et al.*, 2008a). The terbium ions are sufficiently separated that the magnetism can be interpreted as due to non-interacting ions. There is evidence of some slow relaxation, but at such a low temperature that no energy barrier could be derived. The difference from the Pc cases above are likely due to the low site symmetry of the Tb ions in this complex, and possibly to the much weaker crystal field provided by the calix-[4]-arene compared with phthalocyanine.

There are several fascinating reports of magnetic studies of polymetallic dysprosium compounds, chiefly from the Powell group. The earliest, and probably most interesting, describes a $\{Dy_3^{III}\}$ triangle, (Tang *et al.*, 2006). The structure contains two μ_3-hydroxides bridging a regular triangle of metal ions, with the exterior bridged by *o*-vanillin (Fig. 9). The magnetic studies show a decline in the product $\chi_m T$ with a fall in temperature, which is largely due to crystal field effects. However below 30 K the fall is too rapid to be reproduced by consideration of the crystal field alone, and the explanation is that there is anti-ferromagnetic exchange between the Dy^{III} centers. The oddity is that the susceptibility vanishes at low temperature, despite the Dy^{III} ion being an odd-electron system. This unprecedented observation is connected to the triangular arrangement of the spin centers combined with the easy-axis Ising type anisotropy (Luzon *et al.*, 2008).

A dinuclear Dy^{III} complex shows a more normal slow relaxation (Leng *et al.*, 2008). As usual, the product $\chi_m T$ shows a decline with temperature due to crystal field effects; the minimum value is found at 23 K, and below this temperature there is a very rapid increase in $\chi_m T$ consistent with a ferromagnetic exchange between the centers. Magnetization measurements below 10 K show a rapid increase in the value of M at low fields, but no saturation. There is also no hysteresis observed for M vs. H measurements, however there is a frequency dependent peak in the a.c. susceptibility.

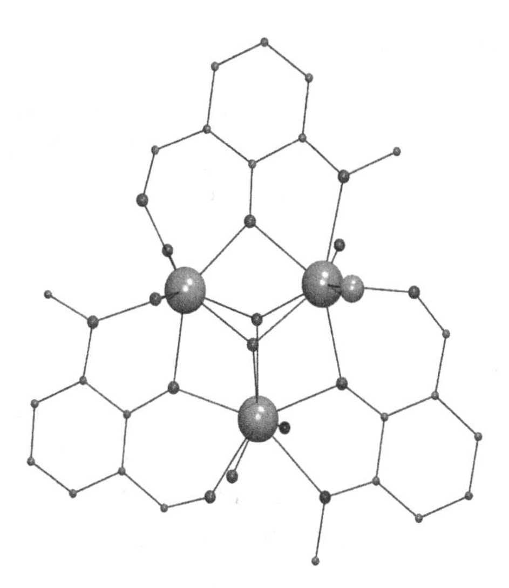

Fig. 9. PovRay representation of $[Dy_3(OH)_2(van)_3Cl(H_2O)_5]^{3+}$ with hydrogen atoms omitted for clarity. Code for atoms: *largest grey*, Dy; *second largest grey*, Cl; *third largest grey*, O; *rest*, C.

This compound shows an attractive crossover from thermal relaxation to quantum tunneling; above 8 K the a.c. data suggest an activation barrier of between 56 and 71 K for the two different solvates studied. Below 2 K the dynamics become temperature independent. As in the cobalt(II) cages described above, there is very fast zero-field relaxation due to tunneling, which explains the lack of hysteresis.

A planar $\{Dy_4\}$ and a square pyramidal $\{Dy_5\}$ cage have also been reported (Zheng *et al.*, 2008, Garner *et al.*, 2008). Structurally these are very beautiful, but magnetically these are less interesting than either the triangle or dinuclear cage discussed above. Both compounds show a decline in $\chi_m T$ with falling temperature which is mainly due to crystal field effects, and possibly due to weak anti-ferromagnetic exchange. In both cases no hysteresis is observed in M vs H, but a rise in χ'' at low temperature is seen, consistent with slow relaxation of magnetization.

These results suggest that most interesting magnetism for dysprosium complexes is likely to be observed either for axially symmetric cages, such as the triangle, or where ferromagnetic exchange is observed. Clearly a great deal more needs to be done.

6.3. *Heterometallic 3d-4f SMMs*

The study of heterometallic 3d-4f complexes goes back to original work by Benelli and Gatteschi and co-workers (Bencini *et al.*, 1985); the early work has been reviewed (Winpenny, 1998 and Sakamoto *et al.*, 2001). Much of this early work was motivated by the observation that exchange between Cu(II) and Gd(III) ions tends to be ferromagnetic, and hence offered a route to high spin molecules.

One of the earliest reports of a 3d-4f SMMs involves a tetranuclear $\{Cu_2 Tb_2\}$ cage, $[Cu(hbhmb)Tb(hfac)_2]_2$ (H_2hbhmb = 1-(2-hydroxybenzamido)-2-(2-hydroxy-3-methoxy-benzylideneamino)-ethane, Fig. 10) (Osa *et al.*, 2004); the Cu(II) and Tb(III) ions alternate within a four-membered ring, and the variable temperature susceptibility data are consistent with a ferromagnetic exchange between the different metals. For this cage a.c. susceptibility measurements show a peak in χ'', and give an energy barrier of 21 K for relaxation of magnetization by a thermal process. The isostructural $\{Ni_2Tb_2\}$ cage shows no out-of-phase susceptibility; the 3d-ion site has a square planar coordination geometry and in this geometry Ni(II) is diamagnetic.

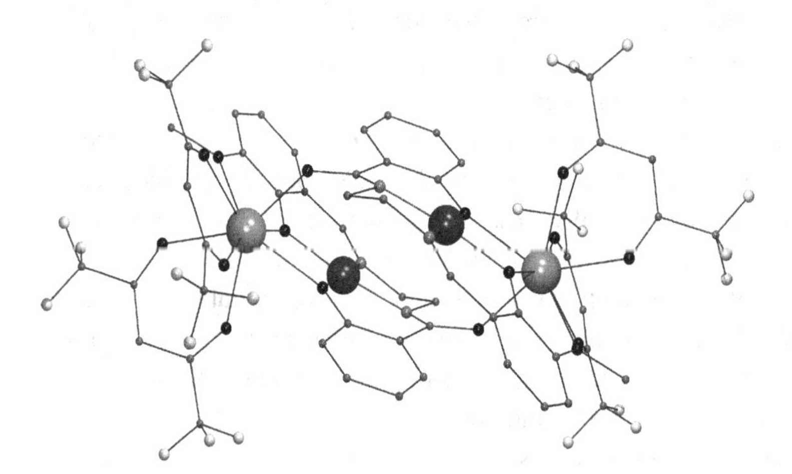

Fig. 10. PovRay representation of $[Cu(hbhmb)Tb(hfac)_2]_2$ with hydrogen atoms omitted for clarity. Code for atoms: *largest grey*, Cu; *largest light grey*, Tb; *second largest black*, O; *second largest grey*, N; *second largest light grey*, F; *rest*, C.

Therefore it appears the SMM behavior of the $\{Cu_2Tb_2\}$ cage must require a 3d-4f interaction and is not simply a property of the terbium ion. The analogous $\{Cu_2Dy_2\}$ cage also shows signs of slow relaxation, but the energy barrier must be much lower given the lack of a peak in the out-of-phase susceptibility. A fuller study of the family of $\{Cu_2Ln_2\}$ complexes, including XMCD studies (Hamamatsu *et al.*, 2007). The conclusions are interesting: one group of lanthanides (Ce, Nd, Sm and Yb) show anti-ferromagnetic exchange; a second group (La, Eu, Pr and Lu) show no magnetic interaction; a third (Gd, Tb, Dy, Ho, Er and Tm) show ferromagnetic exchange. The XMCD measurements show that the moments on the 3d- and 4f-ions are parallel but that the moment on the 4f-ions is lower than that predicted by Hund's rule. Another group has repeated aspects of this work (Costes *et al.*, 2006a).

Costes and colleagues have reported a dinuclear Cu-Tb SMMs (Costes *et al.*, 2006b), using ligands that can be traced back to the Schiff-bases originally used by Benelli and Gatteschi. The energy barrier, as determined by a.c. susceptibility is around 14 K.

A trinuclear Dy-Cu-Dy complex, where the metal centers are bridged by di-2-pyridyl ketoximate, also shows slow relaxation (Mori *et al.*, 2006), but with a higher energy barrier, with $U_{\text{eff}}/\text{k} = 47\,\text{K}$. The authors use an Ising model to show that the exchange between Cu(II) and Dy(III) centers is weakly anti-ferromagnetic with a doubly degenerate ground state with $|M^z| = 19\mu_B$. A tetrametallic $\{Cu_2Dy_2\}$ molecule reported by the same group also shows slow relaxation, but with a much lower energy barrier (Ueki *et al.*, 2006).

Probably the most aesthetically pleasing Cu-Ln SMM is the $\{Dy_3Cu_6\}$ cage, $[Dy_3Cu_6(tfea)_6(OH)_6(H_2O)_{10}]Cl_2$ (H_2tfea = 1,1,1-trifluoro-7-hydroxy-4-methyl-5-azahept-3-en-2-one, Fig. 11) reported by the Luneau group (Aronica *et al.*, 2006). This cage involves a triangular arrangement of four Dy_2Cu_2 cubanes, with each Dy a member of two such cubanes. The variable temperature susceptibility data indicate that there is no significant Cu. . .Cu exchange interaction, and it is not clear whether the Dy-Cu interactions are important. The field dependence of the magnetization of the sample at 2 K shows two steps, with the first step occurring at around 1 T external field, and giving a value of $6\mu_B$. The second step occurs at

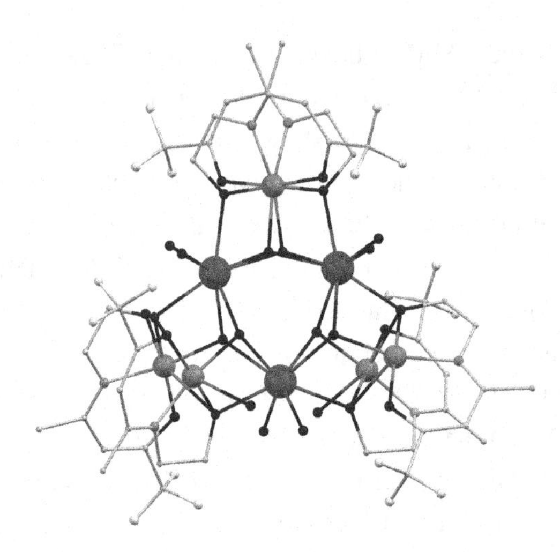

Fig. 11. PovRay representation of $[Dy_3Cu_6(tfea)_6(OH)_6(H_2O)_{10}]Cl_2$ with hydrogen atoms omitted for clarity. Code for atoms: *largest*, Dy; *second largest*, Cu; *third largest black*, O; *third largest grey*, N; *third largest light grey*, F; *rest*, C.

above 3 T, and there is no saturation even at 5 T. The steps suggest that a weak anti-ferromagnetic exchange is being overcome by the external field, leading to a higher spin ground state at higher fields. a.c. susceptibility measurements showed that slow relaxation is present in this molecule, with an energy barrier $U_{eff}/k = 25$ K.

A single crystal magnetic study of a Tb-Cu cage has also been reported (Kajiwara *et al.*, 2008b). This approach is probably more generally needed for 3d-4f complexes, where the anisotropy of the 4f-ions is of vital importance in deciding the physics observed. The key point from the observation is that a very subtle change in the crystal field about the Tb(III) center is sufficient to change the compound from being an easy-axis system, and hence an SMM, to being easy-plane. In designing future 3d-4f SMMs such considerations will have to be taken into account.

One of the earliest Mn-Ln SMMs, is a $\{Mn_8^{III}Ce\}$ cage, with an $S = 16$ ground state (Tasiopolous *et al.*, 2003). However the cerium is present in the +4 oxidation state, and the magnetism is due purely to manganese (III) ions. The synthesis, which involves reaction of a manganese acetate polymer with ceric ammonium nitrate is intriguing as it suggests the Ce(IV) ion acts as a template for the formation of this compound.

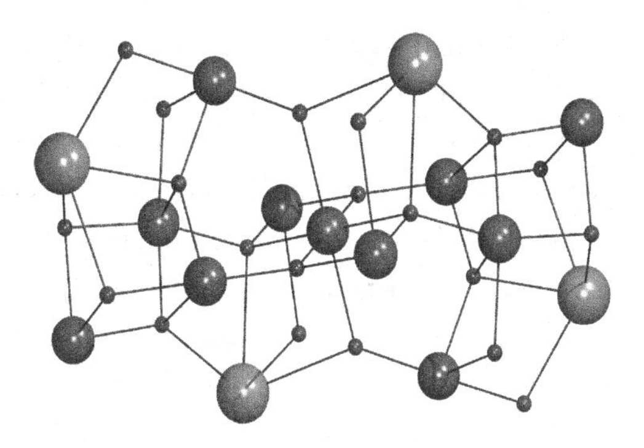

Fig. 12. PovRay representation of the core of $[Mn_{11}Dy_4O_8(OH)_6(OMe)_2(O_2CPh)_{16}(NO_3)_5(H_2O)_3]$. Code for atoms: *largest*, Dy; *second largest*, Mn; *rest*, O.

The cage, $[Mn_{11}Dy_4O_8(OH)_6(OMe)_2(O_2CPh)_{16}(NO_3)_5(H_2O)_3]$, also reported by the Christou group (Mishra *et al.*, 2004), is also an SMM and shows hysteresis in magnetization against field at low temperatures. The synthesis is remarkably simple — an oxo-centered manganese carboxylate triangle is reacted with hydrated dysprosium nitrate. The structure is somewhat irregular (Fig. 12). It is also possible to make the compound with Nd, Gd, Ho and Eu in place of Dy.

A similar reaction, but with the additional of tripodal triol and in MeCN rather than MeOH, leads to tetranuclear $\{Mn_2^{III}Ln_2\}$ cages (Mishra *et al.*, 2005) also behaving as SMM. One of these has an energy barrier of 15 K, and shows very rapid quantum tunneling at zero external field and low temperature. The four metal ions are in a single plane, with the two Mn centers at the center of the structure. Two tripodal ligands are found above and below the plane of metal centers.

Similar reactions using oxo-centered iron triangles reacted with lanthanide nitrates in the presence of triols, also generates some interesting $\{Fe_2Ln_2\}$ compounds (Murugesu *et al.*, 2006). The cages where Ln = Ho and Dy are SMMs, and thus are early examples of Fe-4f SMMs. A more detailed study of dimetallic Fe-Dy compound has been reported (Ferbinteanu *et al.*, 2006). The energy barrier in this compound is low, but a theoretical approach is presented to modelling the exchange interaction between the 3d- and 4f-ions.

The idea of reacting a pre-formed manganese cage with dysprosium nitrate was also exploited to make a $\{Mn_{11}Gd_2\}$ cage (Mereacre *et al.*, 2007). In this case the manganese cage is a mixed-valent hexanuclear cage bound to pivalate ligands and 2-furan carboxylic acid was added as a co-ligand. The structure is described as a "bell-like", with the eleven manganese centers forming the bell, and the two Gd ions acting as the clapper of the bell. Very low T magnetic measurements show that the compound is an SMM, albeit with an energy barrier of 18 K.

One further example of this approach uses the same $\{Mn_6\}$ cage and lanthanide nitrates as starting material, however with addition of *N*-methyl-diethanolamine as the co-ligand (Mereacre *et al.*, 2008). This produces a family of $\{Mn_5Ln_4\}$ clusters. The structure contains two $\{Mn^{III}Mn^{IV}Ln_2\}$ heterocubanes, which share a common Mn^{IV} vertex. Two further Mn^{III} centers are found attached to the structure. The compounds where Ln = Tb or Dy are SMMs with an energy barrier >30 K in each case.

A $\{Mn_6Dy_6\}$ contains a central hexagonal ring of Dy(III) centers, capped by two $\{Mn^{III}_2Mn^{IV}\}$ triangles (Zaleski *et al.*, 2004). Magnetic studies suggest the compound may be an SMM, but this is not as clear cut as in the $\{Mn_{11}Dy_4\}$ cage. The magnetization of the compound does not saturate to high fields and it is unclear what the ground state of the molecule might be; indeed it seems unlikely that there is an isolated ground state. There is a rise in χ" at low temperature and high frequencies, but no maximum is observed in the experiments reported.

A similar set of ligands has been used to make $\{Mn^{III}_6Ln_4\}$ cages; the cage contains a metallocrown of four Mn and two Ln centers, capped externally by two Mn centers, and containing two Ln centers above and below the central loop (Zaleski *et al.*, 2007). The cage with Ln = Dy is an SMM with an estimated energy barrier of 11 K.

A very different approach to the synthesis of 3d-4f compounds has been reported by Chandrasekhar and co-workers (Chandrasekhar *et al.*, 2007). Here phosphorous-based multisite coordination ligands are used to build a linear trinuclear $\{Co_2Gd\}$ cage. The magnetism is extremely well-behaved, with ferromagnetic exchange evident between the metal centers, and low temperature a.c. susceptibility studies allowing estimate of the energy barrier as 27 K. The approach has been adapted to other 3d and 4f-combinations, and a family of $\{Ni_2Ln\}$ cages has been reported (Chandrasekar *et al.*, 2008); the cage containing Dy is again an SMM.

Finally, use of a Schiff-base ligand to make $\{Mn_4^{III}Ln_2\}$ cages. The structure contains a plane of six metal centers (Shiga *et al.*, 2008). Magnetic studies suggest the compound with Ln = Tb is an SMM. Very high frequency EPR studies are also reported. This is the first example of high frequency EPR being used for these 3d-4f SMMs.

7. Metallocyanate Based SMMs

Despite the ubiquity of cyanide as a bridging ligand in molecular magnetism, the first metallocyanate SMM was only reported by the Long group in 2002 (Sokol *et al.*, 2002). The highest barrier in a metallocyanate complex is also reported by the Long group (Freedman *et al.*, 2008) and this area has been reviewed (Beltran and Long, 2005). The use of cyanides allows inclusion of 5d-metals, e.g. rhenium (Freedman *et al.*, 2008; Schelter *et al.*, 2007), and this in turn brings the possibility of redox-switchable SMMs. There has also been an interesting proposal (Glaser *et al.*, 2006) to design SMMs using the symmetry that can be induced by metallocyanates. However, while many new examples are being reported progress towards higher blocking temperatures has stalled.

8. Conclusions

Reviewing the 2005–2008 literature on SMMs shows that the majority of progress is now being made on the hetero-spin systems, especially those involving 4f-elements. The SMMs based on Mn(III) still predominant in the literature, and beautiful new examples are being reported, but it looks as if the physics of these systems is now largely understood and that the energy barrier in Mn(III) SMMs will never rise far beyond the original reports for Mn_{12} cages.

The work on 4f-elements, particularly the phthalocyanine-based complexes reported by Ishikawa (Ishikawa *et al.*, 2003), look as if higher energy barriers to magnetization reversal could be achieved, however the problem there becomes controlling the quantum tunneling processes.

New theoretical methods and use of single crystal experiments to study anisotropic systems also look worthy of a great deal of future work. It is good to see how much work is now appearing on cobalt(II), for example, even where our understanding of the physics remains incomplete. Maybe in systems where traditional means of modelling magnetic data are difficult, we may find new physics and eventually new technological solutions.

References

Accorsi, S., Barra, A.-L., Caneschi, A., Chastanet, G., Cornia, A., Fabretti, A.C., Gatteschi, D., Mortalo, C., Olivieri, E., Parenti, F., Rosa, P., Sessoli, R., Sorace, L. Wernsdorfer, W. and Zobbi, L. (2006). Tuning Anisotropy Barriers in a Family of Tetrairon(III) Single-Molecule Magnets with an S = 5 Ground State. *J. Am. Chem. Soc.* **128**, 4742–4755.

Affronte, M. (2009). Molecular nanomagnets for information technologies. *J. Mater. Chem.* **19**, 1731–1737.

Ako, A.M., Hewitt, I.J., Mereacre, V., Clerac, R., Wernsdorfer, W., Anson, C.E. and Powell, A.K. (2006). A ferromagnetically coupled Mn_{19} aggregate with a record $S = 83/2$ ground spin state. *Angew. Chem.-Int. Edit.* **45**, 4926–4929.

AlDamen, M.A., Clemente-Juan, J.M., Coronado, E., Marti-Gastaldo, C. and Gaita-Arino. A. (2008). Mononuclear Lanthanide Single-Molecule Magnets Based on Polyoxometalates. *J. Am. Chem. Soc.* **130**, 8874–8875.

Aliaga-Alcalde, N., Edwards, R.S., Hill, S.O., Wernsdorfer, W., Folting, K. and Christou, G. (2004). Single-molecule magnets: Preparation and properties of low symmetry $[Mn_4O_3(O_2CPh\text{-}R)_4(dbm)_3]$ complexes with S = 9/2. *J. Am. Chem. Soc.* **126**, 12503–12516.

Andres, H., Basler, R., Gudel, H.U., Aromí, G., Christou, G., Buttner, H. and Ruffle, B. (2000). Inelastic Neutron-Scattering and Magnetic-Susceptibilities of the Single-Molecule Magnets $[Mn_4O_3X(OAc)_3(dbm)_3]$ (X = Br, Cl, Oac, and F) — Variation of the Anisotropy Along the Series. *J. Am. Chem. Soc.* **122**, 12469–12477.

Aromí, G., Bhaduri, S., Artus, P., Folting, K. and Christou, G. (2002). Bridging Nitrate Groups in $[Mn_4O_3(NO_3)(O_2CMe)_3(R_2dbm)_3]$ (R = H, Et) and $[Mn_4O_2(NO_3)(O_2CEt)_6(bpy)_2](ClO_4)$ — Acidolysis Routes to Tetranuclear Manganese Carboxylate Complexes. *Inorg. Chem.* **41**, 805–817.

Aromí, G. and Brechin, E.K. (2006). Synthesis of 3*d* metallic single-molecule magnets. *Struct. Bond.* **122**, 1–67.

Aromí, G., Wemple, M.W., Aubin, S.J., Folting, K., Hendrickson, D.N. and Christou, G. (1998). Modeling the photosynthetic water oxidation complex: Activation of water by controlled deprotonation and incorporation into a tetranuclear manganese complex. *J. Am. Chem. Soc.* **120**, 5850–5851.

Aronica, C., Piley, G., Chastanet, G., Wernsdorfer, W., Jacquot, J.-F. and Luneau, D. (2006). A Nonanuclear Dysprosium(III)–Copper(II) Complex Exhibiting Single-Molecule Magnet Behavior with Very Slow Zero-Field Relaxation. *Angew. Chem. Int. Ed.* **45**, 4659–4662.

Aubin, S.M.J., Wemple, M.W., Adams, D.M., Tsai, H.L., Christou, G. and Hendrickson, D.N. (1996). Distorted $Mn^{IV}Mn_3^{III}$ cubane complexes as single-molecule magnets. *J. Am. Chem. Soc.* **118**, 7746–7754.

Aubin, S.M.J.D., N.R.; Wemple, M.W., Maple, M.B., Christou, G. and Hendrickson, D.N. (1998). Half-Integer-Spin Small Molecule Magnet Exhibiting Resonant Magnetization Tunneling. *J. Am. Chem. Soc.* **120**, 839–840.

Bagai, R. and Christou, G. (2009). The Drosophila of single-molecule magnetism: $[Mn_{12}O_{12}(O_2CR)_{16}(H_2O)_4]$. *Chem. Soc. Rev.* **38**, 1011–1026.

Bahr, S., Milios, C.J., Jones, L.F., Brechin, E.K., Mosser, V. and Wernsdorfer, W. (2009). Quantum tunneling of magnetization in the single-molecule magnet Mn-6. *New J. Chem.* 1231–1236.

Barra, A.-L., Bianchi, F., Caneschi, A., Cornia, A., Gatteschi, D., Gorini, L., Gregoli, L., Maffini, M., Parenti, F., Sessoli, R., Sorace, L. and Talarico, A.M. (2007). New Single-Molecule Magnets by Site-Specific Substitution: Incorporation of "Alligator Clips" into Fe4 Complexes. *Eur. J. Inorg. Chem.* 4145–4152.

Beltran, L.M. and Long, J.R. (2005). Directed Assembly of Metal-Cyanide Cluster Magnets. *Acc. Chem. Res.* **38**, 325–334.

Bencini, A., Benelli, C., Caneschi, A., Carlin, R.L., Dei, A. and Gatteschi, D. (1985). Crystal and molecular-structure of and magnetic coupling in 2 complexes containing gadolinium(III) and copper(II) ions *J. Am. Chem. Soc.* **107**, 8128–8136.

Boudalis, A.K., Donnadieu, B., Nastopoulus, V., Clemente-Juan, J.M., Mari, A., Sanakis, Y., Tuchagues, J.-P. and Perlepes, S.P. (2004). A Nonanuclear Iron(II) Single-Molecule Magnet. *Angew. Chem. Int. Ed.* **43**, 2266–2270.

Boudalis, A.K., Sanakis, Y., Clemente-Juan, J.M., Mari, A. and Tuchagues, J.-P. (2007). Diferrous Single-Molecule Magnet. *Eur. J. Inorg. Chem.* 2409–2415.

Boudalis, A.K., Pissas, M., Raptopoulou, C.P., Psycharis, V., Abarca, B. and Ballesteros, R. (2008). Slow Magnetic Relaxation of a Ferromagnetic Ni_5^{II} Cluster with an $S = 5$ Ground State. *Inorg. Chem.* **47**, 10674–10681.

Brechin, E.K., Yoo, J., Nakano, M., Huffman, J.C., Hendrickson, D.N. and Christou, G. (1999). A new class of single-molecule magnets: mixed-valent $[Mn_4(O_2CMe)_2(Hpdm)_6](ClO_4)_2$ with an $S = 8$ ground state. *Chem. Commun.* 783–784.

Cadiou, C., Murrie, M., Paulsen, C., Villar, V., Wernsdorfer, W. and Winpenny, R.E.P. (2001). "The First Nickel-Based Single Molecule Magnet: Resonant Quantum Tunneling in an S = 12 Molecule", *Chem. Commun.* 2666–2667.

Chandrasekhar, V., Pandian, B.M., Azhakar, R., Vittal, J.J. and Clerac, R. (2007). Linear Trinuclear Mixed-Metal Co^{II}-Gd^{III}-Co^{II} Single-Molecule Magnet: $[L_2Co_2Gd][NO_3].2CHCl_3$ ($LH_3 = (S)P[N(Me)NdCH\text{-}C_6H_3\text{-}2\text{-}OH\text{-}3\text{-}OMe]_3$). *Inorg. Chem.* **46**, 5140–5142.

Chandrasekhar, V., Pandian, B.M., Boomishankar, R., Steiner, A., Vittal, J.J., Houri, A. and Clerac, R. (2008). Trinuclear Heterobimetallic Ni_2Ln complexes $[L_2Ni_2Ln][ClO_4]$ (Ln = La, Ce, Pr, Nd, Sm, Eu, Gd, Tb, Dy, Ho, and Er; $LH_3 = (S)P[N(Me)NdCH\text{-}C_6H_3\text{-}2\text{-}OH\text{-}3\text{-}OMe]_3$): From Simple Paramagnetic Complexes to Single-Molecule Magnet Behavior. *Inorg. Chem.* **47**, 4918–4929.

Chibotaru, L.F., Ungur, L., Aronica, C., Elmoll, H., Pilet, G. and Luneau, D. (2008). Structure, magnetism, and theoretical study of a mixed-valence $Co_3^{II}Co_4^{III}$ heptanuclear wheel: Lack of SMM behavior despite negative magnetic anisotropy. *J. Am. Chem. Soc.* **130**, 12445–12455.

Cornia, A., Costantino, A.F., Zobbi, L., Caneschi, A., Gatteschi, D., Mannini, M. and Sessoli, R. (2006). Preparation of novel materials using SMMs. *Struct. Bond.* **122** 133–161.

Costes, J.-P., Auchel, M., Dahan, F., Peyrou, V., Shova, S. and Wernsdorfer, W. (2006a). Synthesis, Structures, and Magnetic Properties of Tetranuclear Cu^{II}-Ln^{III} Complexes. *Inorg. Chem.* **45**, 1924–1934.

Costes, J.-P., Dahan, F. and Wernsdorfer, W. (2006b). Heterodinuclear Cu-Tb Single-Molecule Magnet. *Inorg. Chem.* **45**, 5–7.

Coulon, C., Miyasaka, H. and Clerac, R. (2006). Single-chain magnets: Theoretical approach and experimental systems. *Struct. Bond.* **122**, 163–206.

del Barco, E., Kent, A.D., Yang, E.C. and Hendrickson, D.N. (2004). Quantum superposition of high spin states in the single molecule magnets Ni_4, *Phys. Rev. Lett.* **93**, Art. 157202.

Edwards, R.S., Maccagnano, S., Yang, E.C., Hill, S., Wernsdorfer, W., Hendrickson, D.N. and Christou, G. (2003). High-frequency electron paramagnetic resonance investigations of tetranuclear nickel-based single-molecule magnets. *Inorg. Chem.* **93**, 7807–7809.

Escuer, A. and Aromí, G. (2006). Azide as a bridging ligand and magnetic coupler in transition metal clusters. *Eur. J. Inorg. Chem.* 4721–4736.

Evangelisti, M., Luis, F., Mettes, F.L., Aliaga, N., Aromi, G., Christou, G. and de Jongh, L.J. (2002). Through quantum tunneling to dipolar order: the effect of varying magnetic anisotropy in three structurally related Mn-4 molecular clusters. *Polyhedron* 2169–2173.

Feng, P.L., Beedle, C.C., Koo, C., Lawrence, J., Hill, S. and Hendrickson, D.N. (2008). Origin of magnetization tunneling in single-molecule magnets as determined by single-crystal high-frequency EPR. *Inorg. Chim. Acta* **361**, 3465–3480.

Feng, P.L., Koo, C., Henderson, J.J., Nakano, M., Hill, S., del Barco, E. and Hendrickson, D.N. (2008). Single-molecule-magnet behavior and spin changes affected by crystal packing effects. *Inorg. Chem.* **47**, 8610–8612.

Ferbinteanu, M., Kajiwara, T., Choi, K.-Y., Nojiri, H., Nakamoto, A., Kojima, N., Cimpoesu, F., Fujimura, Y., Takaishi, S. and Yamashita, M. (2006). A Binuclear Fe(III)Dy(III) Single Molecule Magnet. Quantum Effects and Models. *J. Am. Chem. Soc.* **128**, 9008–9009.

Ferguson, A., Parkin, A., Sanchez-Benitez, J., Kamenev, K., Wernsdorfer, W. and Murrie, M. (2007). A mixed-valence Co_7 single-molecule magnet with C_3 symmetry. *Chem. Commun.* 3473–3475

Ferguson, A., Lawrence, J., Parkin, A., Sanchez-Benitez, J., Kamenev, K., Brechin, E.K., Wernsdorfer, W., Hill, S. and Murrie, M. (2008). Synthesis and characterisation of a Ni_4 single-molecule magnet with S_4 symmetry. *Dalton Trans.* 6409–6414.

Freedman, D.E., Jenkins, D.M., Iavarone, A.T. and Long, J.R. (2008). A Redox-Switchable Single-Molecule Magnet Incorporating $[Re(CN)_7]^{3-}$. *J. Amer. Chem. Soc.* **130**, 2884–2885.

Galloway, K.W., Whyte, A.M., Wernsdorfer, W., Sanchez-Benitez, J., Kamenev, K.V., Parkin, A., Peacock, R.D. and Murrie, M. (2008). Cobalt (II) citrate cubane single molecule magnet. *Inorg. Chem.* **47**, 7438–7442.

Gamer, M.T., Lan, Y., Roesky, P.W., Powell, A.K. and Clerac, R. (2008). Pentanuclear Dysprosium Hydroxy Cluster Showing Single-Molecule-Magnet Behavior. *Inorg. Chem.* **47**, 6581–6583.

Gatteschi, D., Sessoli, R. and Villain, R. (2006a). *Molecular Nanomagnets*, Oxford University Press, Oxford.

Glaser, T., Heidemeier, M., Weyhermüller, T., Hoffmann, R.D., Rupp, H., Müller, P. (2006). Property-oriented rational design of single-molecule magnets: A C_3-symmetric Mn_6Cr complex based on three molecular building blocks with a spin ground state of $S_t = 21/2$. *Angew. Chem. Int. Ed.* **45**, 6033–6037.

Hamamatsu, T., Yabe, K., Towatari, M., Osa, S., Matsumato, N., Re, N., Pochaba, A., Mrozinski, J., Gallani, J.-L., Barla, A., Imperia, P., Paulsen, C. and Kappler, J.-P. (2007). Magnetic Interactions in CuII-LnIII Cyclic Tetranuclear Complexes: Is It Possible to Explain the Occurrence of SMM Behavior in CuII-TbIII and CuII-DyIII Complexes? *Inorg. Chem.* **46**, 4458–4468.

Hill, S., Edwards, R.S., Aliaga-Alcalde, N. and Christou, G. (2003). Quantum coherence in an exchange-coupled dimer of single-molecule magnets. *Science* **302**, 1015–1018.

Inglis, R., Jones, L.F., Karotsis, G., Collins, A., Parsons, S., Perlepes, S.P., Wernsdorfer, W. and Brechin, E.K. (2008). Enhancing SMM properties via axial distortion of $Mn(III)_3$ clusters. *Chem. Commun.* 5924–5926.

Inglis, R., Jones, L.F., Mason, K., Collins, A., Moggach, S.A., Parsons, S., Perlepes, S.P., Wernsdorfer, W. and Brechin, E.K. (2008). Ground Spin State Changes and 3D Networks of Exchange Coupled $[Mn(III)_3]$ Single-Molecule Magnets. *Chem. Eur. J.* **14**, 9117–9121.

Ishikawa, N., Sugita, M., Ishikawa, T., Koshira, S. and Kaizu, Y. (2003). Lanthanide Double-Decker Complexes Functioning as Magnets at the Single-Molecular Level. *J. Am. Chem. Soc.* **125**, 8694–8695.

Ishikawa, N., Sugita, M. and Wernsdorfer, W. (2005a). Nuclear Spin Driven Quantum Tunneling of Magnetization in a New Lanthanide Single-Molecule Magnet: Bis(Phthalocyaninato)holmium Anion. *J. Am. Chem. Soc.* **127**, 3650–3651.

Ishikawa, N., Sugita, M. and Wernsdorfer, W. (2005b). Quantum Tunneling of Magnetization in Lanthanide Single-Molecule Magnets: Bis(phthalocyaninato)terbium and Bis(phthalocyaninato)dysprosium Anions. *Angew. Chem. Int. Ed.* **44**, 2931–2935.

Ishikawa, N., Mizuno, Y., Takamatsu, S., Ishikawa, T. and Koshihara, S. (2008). Effects of Chemically Induced Contraction of a Coordination Polyhedron the Dynamical Magnetism of Bis(phthalocyaninato)dysprosium, a Single-4f-Ionic Single-Molecule Magnet with a Kramers Ground State. *Inorg. Chem.* **47**, 10217–10219.

Jamet, M., Wernsdorfer, W., Thirion, C., Mailly, D., Dupuis, V., Melinon, P. and Perez, A. (2001). Magnetic anisotropy of a single cobalt nanocluster. *Phys. Rev. Lett.* **86**, 4676–4679.

Kajiwara, T., Hasegawa, M., Ishii, A., Katagiri, K., Baatar, M., Takaishi, S., Iki, N. and Yamashita, M. (2008a). Highly Luminescent Superparamagnetic Diterbium(III) Complex Based on the Bifunctionality of *p-tert*-Butylsulfonylcalix[4]arene. *Eur. J. Inorg. Chem.* 5565–5568.

Kajiwara, T., Nakano, M. Takaishi, S. and Yamashita, M. (2008b). Coordination-Tuned Single-Molecule-Magnet Behavior of Tb^{III}-Cu^{II} Dinuclear Systems. *Inorg. Chem.* **47**, 8604–8606.

Langley, S.J., Helliwell, M., Sessoli, R., Rosa, P., Wernsdorfer, W. and Winpenny, R.E.P. (2005). Slow relaxation of magnetization in an octanuclear cobalt(II) phosphonate cage complex. *Chem. Commun.* 5029–5031.

Lecren, L., Li, Y.G., Wernsdorfer, W., Roubeau, O., Miyasaka, H. and Clerac, R. (2005). $[Mn_4(hmp)_6(CH_3CN)_2(H_2O)_4]^{4+}$: A new single-molecule magnet with the highest blocking temperature in the Mn_4/hmp family of compounds. *Inorg. Chem. Commun.* **8**, 626–630.

Lecren, L., Roubeau, O., Coulon, C., Li, Y.G., Le Goff, X.F., Wernsdorfer, W., Miyasaka, H. and Clerac, R. (2005). Slow relaxation in a one-dimensional rational assembly of antiferromagnetically coupled [Mn_4] single-molecule magnets. *J. Am. Chem. Soc.* **127**, 17353–17363.

Lecren, L., Roubeau, O., Li, Y.G., Le Goff, X.F., Miyasaka, H., Richard, F., Wernsdorfer, W., Coulon, C. and Clerac, R. (2008). One-dimensional coordination polymers of antiferromagnetically-coupled [Mn_4] single-molecule magnets. *Dalton Trans.* 755–766.

Lecren, L., Wernsdorfer, W., Li, Y.G., Roubeau, O., Miyasaka, H. and Clerac, R. (2005). Quantum tunneling and quantum phase interference in a [$Mn_2^{II}Mn_2^{III}$] single-molecule magnet. *J. Am. Chem. Soc.* **127**, 11311–11317.

Leng, P.-H., Burchall, T.J., Clérac, R. and Murugesu, M. (2008). Dinuclear Dysprosium(III) Single-Molecule Magnets with a Large Anisotropic Barrier *Angew. Chem. Int. Ed.* **47**, 8848–8851.

Lines, M.E. (1971). Orbital angular momentum in theory of paramagnetic clusters. *J. Chem. Phys.* **55**, 2977–2984.

Lis, T. (1980). Preparation, Structure, and Magnetic Properties of a Dodecanuclear Mixed-Valence Manganese Carboxylate. *Acta Cryst.* **B36**, 2042–2046.

Luzon, J., Bernot, K., Hewitt, I.J., Anson, C.E., Powell, A.K. and Sessoli, R. (2008). Spin chirality in a molecular dysprosium triangle: The archetype of the noncollinear ising model. *Phys. Rev. Lett.* **100**, art. 247205.

Mannini, M., Pineider, F., Sainctavit, P., Danieli, C., Otero, E., Sciancalepore, C., Talarico, A.M., Arrio, M.-A., Cornia, A., Gatteschi, D. and Sessoli, R. (2009). Magnetic memory of a single-molecule quantum magnet wired to a gold surface. *Nature Materials* **8**, 194–197.

Mereacre, V.M., Ako, A.M., Clerac, R., Wernsdorfer, W., Filoti, G., Bartlome, J., Anson, C.E. and Powell A.K. (2007). A Bell-Shaped $Mn_{11}Gd_2$ Single-Molecule Magnet. *J. Am. Chem. Soc.* **129**, 9248–9249.

Mereacre, V., Ako, A.M., Clerac, R., Wernsdorfer, W., Hewitt, I.J., Anson, C.E. and Powell A.K. (2008). Heterometallic [Mn_5-Ln_4] Single-Molecule Magnets with High Anisotropy Barriers. *Chem. Eur. J.* **14**, 3577–3584.

Milios, C.J., Inglis, R., Vinslava, A., Bagai, R., Wernsdorfer, W., Parsons, S., Perlepes, S.P., Christou, G. and Brechin, E.K. (2007). Toward a magnetostructural correlation for a family of Mn_6 SMMs. *J. Am. Chem. Soc.* **129**, 12505–12511.

Milios, C.J., Raptopoulou, C.P., Terzis, A., Lloret, F., Vicente, R., Perlepes, S.P. and Escuer, A. (2004). Hexanuclear manganese(III) single-molecule magnets. *Angew. Chem. Int. Ed.* **43**, 210–212.

Milios, C.J., Vinslava, A., Wood, P.A., Parsons, S., Wernsdorfer, W., Christou, G., Perlepes, S.P. and Brechin, E.K. (2007). A single-molecule magnet with a "twist". *J. Am. Chem. Soc.* **129**, 8–9.

Mishra, A., Wernsdorfer, W., Abbound, K.A. and Christou, G. (2004). Initial Observation of Magnetization Hysteresis and Quantum Tunneling in Mixed Manganese-Lanthanide Single-Molecule Magnets. *J. Am. Chem. Soc.* **126**, 15648–15649.

Mishra, A., Wernsdorfer, W., Parsons, S., Christou, G. and Brechin, E.K. (2005). The search for 3d–4f single-molecule magnets: synthesis, structure and magnetic properties of a [$Mn_2^{III}Dy_2^{III}$] cluster. *Chem. Commun.* 2086–2088.

Miyasaka, H., Nakata, K., Lecren, L., Coulon, C., Nakazawa, Y., Fujisaki, T., Sugiura, K., Yamashita, M. and Clerac, R. (2006). Two-dimensional networks based on Mn-4 complex linked by dicyanamide anion: From single-molecule magnet to classical magnet behavior. *J. Am. Chem. Soc.* **128**, 3770–3783.

Moragues-Canovas, M., Helliwell, M. Ricard, L., Riviere, E., Wernsdorfer, W., Brechin, E.K. and Mallah, T. (2004). An Ni4 Single-Molecule Magnet: Synthesis, Structure and Low-Temperature Magnetic Behavior. *Eur. J. Inorg. Chem.* 2219–2222.

Moragues-Canovas, M., Talbot-Eeckelaers, C.E., Catala, L., Lloret, F., Wernsdorfer, W., Brechin, E.K. and Mallah, T. (2006). Ferromagnetic cobalt metallocycles. *Inorg. Chem.* **45**, 7038–7040.

Mori, F., Nyui, T., Ishida, T., Nogami, T., Choi, K.Y. and Nojiri, H. (2006). Oximate-bridged trinuclear Dy-Cu-Dy complex behaving as a single-molecule magnet and its mechanistic investigation. *J. Am. Chem. Soc.* **128**, 1440–1441.

Moubaraki, B., Murray, K.S., Hudson, T.A. and Robson, R. (2008). Tetranuclear and octanuclear cobalt (II) citrate cluster single molecule magnets. *Eur. J. Inorg. Chem.* 4525–4529.

Moushi, E.E., Stamatatos, T.C., Wernsdorfer, W., Nastopoulos, V., Christou, G. and Tasiopoulos, A.J. (2009). A Mn_{17} Octahedron with a Giant Ground-State Spin: Occurrence in Discrete Form and as Multidimensional Coordination Polymers. *Inorg. Chem.* **48**, 5049–5051.

Murrie, M., Teat, S.J., Stoeckli-Evans, H. and Güdel H.U. (2003). Synthesis and Characterization of a Cobalt(II) Single-Molecule Magnet. *Angew. Chem. Int. Ed.* **42**, 4653–4656.

Murugesu, M., Mishra, A., Wernsdorfer, W., Abboud, K.A. and Christou, G. (2006). Mixed 3d/4d and 3d/4f metal clusters: Tetranuclear $Fe_2^{III}M^{III}$ ($M^{III} = Ln$; Y) and $Mn_2^{IV}M^{III}$ (M = Yb; Y) complexes, and the first Fe/4f single-molecule magnets. *Polyhedron* **25**, 613–625.

Osa, S., Kido, T., Matsumato, N., Re, N., Pochaba, A. and Mrozinski, J. (2004). A Tetranuclear 3d-4f Single Molecule Magnet: $[Cu^{II}LTb^{III}(hfac)_2]_2$. *J. Am. Chem. Soc.* **126**, 420–421.

Oshio, H., Hoshimo, N., Ito, T. and Nakano, M. (2004). Single-Molecule Magnets of Ferrous Cubes: Structurally Controlled Magnetic Anisotropy. *J. Am. Chem. Soc.* **126**, 8805–8812.

Park, K., Pederson, M.R., Richardson, S.L., Aliaga-Alcalde, N. and Christou, G. (2003). Density-functional theory calculation of the intermolecular exchange interaction in the magnetic Mn-4 dimer. *Phys. Rev. B* **68**, art. 020405.

Ramsey, C.M., Del Barco, E., Hill, S., Shah, S.J., Beedle, C.C. and Hendrickson, D.N. (2008). Quantum interference of tunnel trajectories between states of different spin length in a dimeric molecular nanomagnet. *Nature Phys.* **4**, 277–281.

Rebilly, J.N., Charron, G., Riviere, E., Guillot, R., Barra, A.L., Serrano, M.D., van Slageren, J. and Mallah. T. (2008). Large magnetic anisotropy in pentacoordinate Ni^{II} complexes. *Chem. Eur. J.* **14**, 1169–1177.

Roubeau, O. and Clerac, R. (2008). Rational Assembly of High-Spin Polynuclear Magnetic Complexes into Coordination Networks: the Case of a $[Mn_4]$ Single-Molecule Magnet Building Block. *Eur. J. Inorg. Chem.* 4325–4342.

Ruiz, E., Cirera, J., Cano, J., Alvarez, S., Loose, C. and Kortus, J. (2008). Can large magnetic anisotropy and high spin really coexist? *Chem. Commun.* 52–54.

Sakamoto, M., Manseki, K. and Okawa, H. (2001). d-f Heteronuclear complexes: synthesis, structures and physicochemical aspects. *Coord. Chem. Rev.* **219**, 379–414.

Schleter, E.J., Karadas, F., Avendano, C., Prosvirin, A.V., Wernsdorfer, W. and Dunbar, K.R. (2007). A Family of Mixed-Metal Cyanide Cubes with Alternating Octahedral and Tetrahedral Corners Exhibiting a Variety of Magnetic Behaviors Including Single Molecule Magnetism. *J. Amer. Chem. Soc.* **129**, 8139–8149.

Sessoli, R., Tsai, H.L., Schake, A.R., Wang, S.Y., Vincent, J.B., Folting, K., Gatteschi, D., Christou, G. and Hendrickson, D.N. (1993). High-Spin Molecules — $[Mn_{12}O_{12}(O_2CR)_{16}(H_2O)_4]$. *J. Am. Chem. Soc.* **115**, 1804–1816.

Shah, S.J., Ramsey, C.M., Heroux, K.J., DiPasquale, A.G., Dalal, N.S., Rheingold, A.L., del Barco, E. and Hendrickson, D.N. (2008). Molecular Wheels: New Mn_{12} Complexes as Single-Molecule Magnets. *Inorg. Chem.* **47**, 9569–9582.

Shiga, T., Hoshino, N., Nakano, M., Nojiri, H. and Oshio, H. (2008). Syntheses, structures, and magnetic properties of manganese–lanthanide hexanuclear complexes. *Inorg. Chim. Acta* **361**, 4113–4117.

Sokol, J.J., Hee, A.G. and Long, J.R. (2002). A Cyano-Bridged Single-Molecule Magnet: Slow Magnetic Relaxation in a Trigonal Prismatic MnMo6(CN)18 Cluster. *J. Am. Chem. Soc.* **124**, 7656–7657.

Sreerama, S.G. and Pal, S. (2002). A novel carboxylate-free ferromagnetic trinuclear μ_3-oxo-manganese(III) complex with distorted pentagonal-bipyramidal metal centers. *Inorg. Chem.* **41**, 4843–4845.

Stamatatos, T.C. and Christou, G. (2009). Azide Groups in Higher Oxidation State Manganese Cluster Chemistry: From Structural Aesthetics to Single-Molecule Magnets. *Inorg. Chem.* **48**, 3308–3322.

Stamatatos, T.C., Foguet-Albiol, D., Lee, S.C., Stoumpos, C.C., Raptopoulou, C.P., Terzis, A., Wernsdorfer, W., Hill, S.O., Perlepes, S.P. and Christou, G. (2007). "Switching on" the properties of single-molecule magnetism in triangular Manganese(III) complexes. *J. Am. Chem. Soc.* **129**, 9484–9499.

Stamatatos, T.C., Foguet-Albiol, D., Stoumpos, C.C., Raptopoulou, C.P., Terzis, A., Wernsdorfer, W., Perlepes, S.P. and Christou, G. (2005). Initial example of a triangular single-molecule magnet from ligand-induced structural distortion of a $[Mn_3^{III}O]^{7+}$ complex. *J. Am. Chem. Soc.* **127**, 15380–15381.

Takamatsu, S., Ishikawa, T., Koshihara, S. and Ishikawa, N. (2007). Significant Increase of the Barrier Energy for Magnetization Reversal of a Single-4f-Ionic Single-Molecule Magnet by a Longitudinal Contraction of the Coordination Space. *Inorg. Chem.* **46**, 7250–7252.

Tang, J., Hewitt, I., Madhu, N.T., Chastanet, G., Wernsdorfer, W., Anson, C.E., Benelli, C., Sessoli, R. and Powell, A.K. (2006). Dysprosium Triangles Showing Single-Molecule Magnet Behavior of Thermally Excited Spin States. *Angew. Chem. Int. Ed.* **45**, 1729–1733.

Tasiopoulos, A.J., Wernsdorfer, W., Moulton, B., Zaworotko, M.J. and Christou, G. (2003). Template Synthesis and Single-Molecule Magnetism Properties of a Complex with Spin $S = 16$ and a [Mn8O8]8+ Saddle-Like Core. *J. Am. Chem. Soc.* **125**, 15274–15275.

Tasiopoulos, A.J., Wernsdorfer, W., Abboud, K.A. and Christou, G. (2004). A Reductive-Aggregation route to $[Mn_{12}O_{12}(OMe)_2(O_2CPh)_{16}(H_2O)_2]^{2-}$ Single-Molecule Magnets Related to the $[Mn_{12}]$ Family. *Angew. Chem., Int. Ed.* **43**, 6338–6342.

Thomas, L., Lionti,.F, Ballou, R., Gatteschi, D., Sessoli, R. and Barbara, B. (1996). Macroscopic quantum tunneling of magnetization in a single crystal of nanomagnets. *Nature* **383**, 145–147.

Tuan, N.A., Katayama, S.I. and Chi, D.H. (2009). A systematic study of influence of ligand substitutions on the electronic structure and magnetic properties of Mn-4 single-molecule magnets. *Phys. Chem. Chem. Phys.* **11**, 717–729.

Ueki, S., Okazama, A., Ishida, T., Nogami, T. and Nojiri, H. (2007). Tetranuclear heterometallic cycle Dy_2Cu_2 and the corresponding polymer showing slow relaxation of magnetization reorientation. *Polyhedron* **26**, 1970–1976.

Waldmann, O., Ako, A.M., Güdel, H.U. and Powell, A.K. (2008). Assessment of the anisotropy in the molecule Mn-19 with a high-spin ground state $S = 83/2$ by 35 GHz electron paramagnetic resonance. *Inorg. Chem.* **47**, 3486–3488.

Wang, X.-T., Wang, B.-W., Wang, Z.-M., Zhang, W. and Gao. S. (2008). Azide and oxo-bridged ferromagnetic clusters: Three face-shared tetracubane Ni(II)/Co(II) hexamers and a wheel-shaped SMM-like Co(II) heptamer. *Inorg. Chim. Acta* **361**, 3895–3902.

Wemple, M.W., Adams, D.M., Folting, K., Hendrickson, D.N. and Christou, G. (1995). Incorporation of Fluoride into a Tetranuclear $Mn/O/RCO_2$ Aggregate - Potential Relevance to Inhibition by Fluoride of Photosynthetic Water Oxidation. *J. Am. Chem. Soc.* **117**, 7275–7276.

Wemple, M.W., Adams, D.M., Hagen, K.S., Folting, K., Hendrickson, D.N. and Christou, G. (1995). Site-specific Ligand Variation in Manganese-Oxide Cubane Complexes, and Unusual Magnetic Relaxation Effects in $[Mn_4O_3X(OAc)_3(dbm)_3]$ ($X = N^{3-}$, OCN^-; Hdbm = dibenzoylmethane). *J. Chem. Soc., Chem. Commun.* 1591–1593.

Wernsdorfer, W., Aliaga-Alcalde, N., Hendrickson, D.N. and Christou, G. (2002). Exchange-biased quantum tunneling in a supramolecular dimer of single-molecule magnets. *Nature* **416**, 406–409.

Winpenny, R.E.P. (1998). The structures and magnetic properties of complexes containing 3d- and 4f-metals. *Chem. Soc. Rev.* **27**, 447–452.

Yang, E.-C., Hendrickson, D.N., Wernsdorfer, W., Nakano, M., Zakharov, L.N., Sommer, R.D., Rheingold, A.L., Ledezma-Gairaud, M. and Christou, G. (2002). Cobalt single-molecule magnet. *J. Appl. Phys.* **91**, 7382–7384.

Yoo, J., Brechin, E.K., Yamaguchi, A., Nakano, M., Huffman, J.C., Maniero, A.L., Brunel, L.C., Awaga, K., Ishimoto, H., Christou, G. and Hendrickson, D.N. (2000). Single-molecule magnets: A new class of tetranuclear manganese magnets. *Inorg. Chem.* **39**, 3615–3623.

Yoo, J., Wernsdorfer, W., Yang, E.C., Nakano, M., Rheingold, A.L. and Hendrickson, D.N. (2005). One-dimensional chain of tetranuclear manganese single-molecule magnets. *Inorg. Chem.* **44**, 3377–3379.

Yoo, J., Yamaguchi, A., Nakano, M., Krzystek, J., Streib, W.E., Brunel, L.C., Ishimoto, H., Christou, G. and Hendrickson, D.N. (2001). Mixed-valence tetranuclear manganese single-molecule magnets. *Inorg. Chem.* **40**, 4604–4616.

Zaleski, C.M., Depperman, E.C., Kampf, J.W., Kirk, M.L. and Pecoraro, V.L. (2004). Synthesis, Structure, and Magnetic Properties of a Large Lanthanide–Transition-Metal Single-Molecule Magnet. *Angew. Chem. Int. Ed.* **43**, 3912–3914.

Zaleski, C.M., Kampf, J.W., Mallah, T., Kirk, M.L. and Pecoraro, V.L. (2007). Assessing the Slow Magnetic Relaxation Behavior of $Ln_4^{III}Mn_6^{III}$ Metallacrowns. *Inorg. Chem.* **46**, 1954–1956.

Zeng, M.-H., Yao, M.-Y., Liang, H., Zhang, W.-Y. and Chen, X.-M. (2007). A Single-Molecule-Magnetic, Cubane-Based, Triangular Co_{12} Supercluster. *Angew. Chem. Int. Ed.* **46**, 1832–1835.

Zhang, Y.-Z., Wernsdorfer, W., Pan, F., Wang, Z.-M. and Gao, S. (2006). An azido-bridged disc-like heptanuclear cobalt(II) cluster: towards a single-molecule magnet. *Chem. Commun.* 3302–3304.

Zheng, Y.-Z., Lan, Y., Anson, C.E. and Powell, A.K. (2008). Anion-Perturbed Magnetic Slow Relaxation in Planar {Dy4} Clusters. *Inorg. Chem.* **47**, 10813–10815.

Chapter 3

THE NANOSCOPIC V_{15} CLUSTER: A UNIQUE MAGNETIC POLYOXOMETALATE

BORIS TSUKERBLAT and ALEX TARANTUL

Department of Chemistry,
Ben-Gurion University of the Negev,
84105 Beer-Sheva, Israel

1. The Unique Magnetic Polyoxometalate V_{15}

In a review (Pope and Müller, 1991) about polyoxometalates (POMs) one can read: "*Polyoxometalates form a class of inorganic compounds that is unmatched in terms of molecular and electronic structural versatility, reactivity, and relevance to analytical chemistry, catalysis, biology, medicine, geochemistry, materials science, and topology.*" The importance of POMs for molecular magnetism and materials science can be recognized from the book by Gatteschi *et al.* entitled "*Molecular Nanomagnets*" (Gatteschi *et al.*, 2006) (for general related books and reviews of the POM subject see also Refs. Müller *et al.*, 1998; Clemente-Juan and Coronado, 1999). In this context, it was argued in the title of a subsequent review that POMs represent a "*Source of Unusual Spin Topologies*" (Gatteschi *et al.*, 2001). Polyoxometalate-based molecular spin systems can be divided into two categories. Either the early transition metal centers of the POMs are in oxidation states characterized by unpaired electrons, which is the case for reduced polyoxovanadate(IV)-type species, or the POMs form a matrix for embedded magnetic centers, mainly of the 3d row (see special chapters in Refs. Müller *et al.*, 1998; Gatteschi *et al.*, 2001, 1994). The polyoxovanadates are especially interesting as some pure metal-oxide surfaces.

The unique cluster anion* present in $K_6[V^{IV}_{15}As_6O_{42}(H_2O)]\cdot 8H_2O$ containing 15 V^{IV} ions ($S_i = 1/2$) and exhibiting layers of different magnetizations was discovered and first investigated two decades ago (Müller and Döring, 1988; Gatteschi *et al.*, 1991; Barra *et al.*, 1992; Gatteschi *et al.*, 1993). The synthesis of this fascinating cluster opened a new trend in molecular magnetism closely related to the promising field of single molecule magnets that is expected to give a revolutionary impact on the design of new memory storage devices of molecular size and quantum computing.

Studies of the adiabatic magnetization and quantum dynamics show that the V_{15} cluster exhibits the hysteresis loop of magnetization (Barbara, 2003; Chiorescu *et al.*, 2000; Chiorescu *et al.*, 2000; Miyashita, 1996; Nojiria *et al.*, 2004; Miyashita, 1995) of molecular origin and can be referred to as a mesoscopic system on the border line between the classical and quantum world. The studies of the static magnetic susceptibility (Barra *et al.*, 1992; Gatteschi *et al.*, 1993), energy pattern (Platonov *et al.*, 2002; Kostyuchenko and Popov, 2008; Raghu *et al.*, 2003; Miyashita *et al.*, 2003; De Raedt *et al.*, 2004; De Raedt *et al.*, 2003; De Raedt *et al.*, 2004; Konstantinidis and Coffey, 2002; Machida and Miyashita, 2005), *ab initio* electronic structure calculations (Kortus *et al.*, 2001; Kortus *et al.*, 2001; Boukhvalov *et al.*, 2003; Boukhvalov *et al.*, 2004; Barbour *et al.*, 2006) and inelastic neutron scattering (INS) (Chaboussant *et al.*, 2002; Chaboussant *et al.*, 2004) showed that the low lying part of the energy spectrum is well isolated from the remaining spin levels and can be understood as a result of interaction between three moieties consisting of five strongly coupled spins giving rise to spin $S_i = 1/2$ of each moiety. The model of spin triangle for the low lying

---

*The related solid compound $K_6[V_{15}As_6O_{42}(H_2O)]\cdot 8(H_2O)$ shows bands according to electronic transitions at 365, 540(sh) and 800 nm (Döring, 1990). The band positions are practically identical to those in aqueous solution, whereas the characteristic VO^{2+} d-d bands at ca. 540 and 800 nm can only be observed at higher concentrations and correspond to the expected ones (see Ref. Lever, 1984 Table 6.4, page 390). The characteristic Raman bands (1000 (vs), 976 (w), 949 (m), 861 (m), 792 (m) and 416 (m) cm^{-1}; excitation line 514.5 nm) allow the identification of V_{15} in solution (Bertaina *et al.*, 2008).

excitations suggested in Refs. Barra *et al.*, 1992; Chiorescu *et al.*, 2000 includes isotropic Heisenberg-Dirac-Van Vleck (HDVV) exchange interaction and antisymmetric (AS) exchange proposed by Dzyaloshinsky, 1957 and Moriya, 1960 as an origin of spin canting in magnetic materials.

The understanding of the role of the AS exchange in spin frustrated systems dates back to the seventies (see review article (Tsukerblat *et al.*, 1987) and references therein). AS exchange was shown to result in a zero-field splitting (ZFS) of the frustrated ground state of the triangular half-integer spin systems, in magnetic anisotropy, essential peculiarities of the EPR spectra and a wide range of phenomena related to hyperfine interactions (Tsukerblat *et al.*, 1987; Belinskii *et al.*, 1972; Belinskii *et al.*, 1973; Tsukerblat *et al.*, 1974; Tsukerblat and Belinskii, 1983; Tsukerblat *et al.*, 1975; Belinskii *et al.*, 1974; Tsukerblat *et al.*, 1983; Tsukerblat *et al.*, 1984; Tsukerblat *et al.*, 1984; Fainzilberg *et al.*, 1985; Tsukerblat *et al.*, 1985; Fainzilberg *et al.*, 1980; Fainzilberg *et al.*, 1981; Fainzilberg *et al.*, 1981; Fainzilberg *et al.*, 1982). Recently, large AS exchange was evoked (Yoon *et al.*, 2004) to interpret the unusual properties of tricopper clusters. Then AS exchange in the spin-frustrated trinuclear Cu(II) cluster was intensively studied in the Refs. Belinsky, 2008; Belinsky, 2008. Some conclusions made in the earlier papers have been recently extended, in particular, those regarding the ZFS and Kramers theorem (see Ref. Tsukerblat *et al.*, 2006).

In this chapter we give a short overview of the magnetic interactions in the V_{15} system with emphasis on the manifestations of the AS exchange and structural instabilities in spin-frustrated systems. The model of triangle is analyzed and applied to the study of the energy levels crossover in magnetic fields of different directions as well as in the phenomena like low-temperature EPR, high field magnetization, structural instabilities arising from the spin-vibronic interaction and the Jahn-Teller effect (JTE).

2. Structure and Superexchange Pathways

The molecular cluster V_{15} has a distinct layered quasi-spherical structure (Müller and Döring, 1988; Gatteschi *et al.*, 1991). Fifteen V^{IV} ions ($s_i = 1/2$) are placed in a large central triangle sandwiched by two distorted hexagons possessing overall $\mathbf{D}_3$ symmetry (Figs. 1–3).

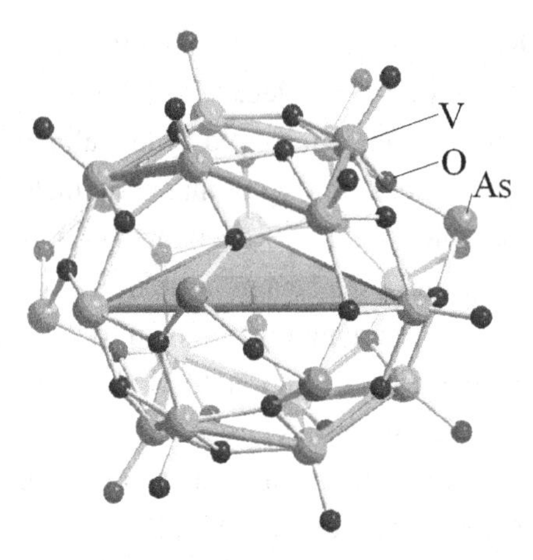

Fig. 1. The cluster anion $[V^{IV}_{15}As_6O_{42}(H_2O)]^{6-}$: ball-and-stick representation without the central water molecule emphasizing the V_3 triangle (Müller and Döring, 1988).

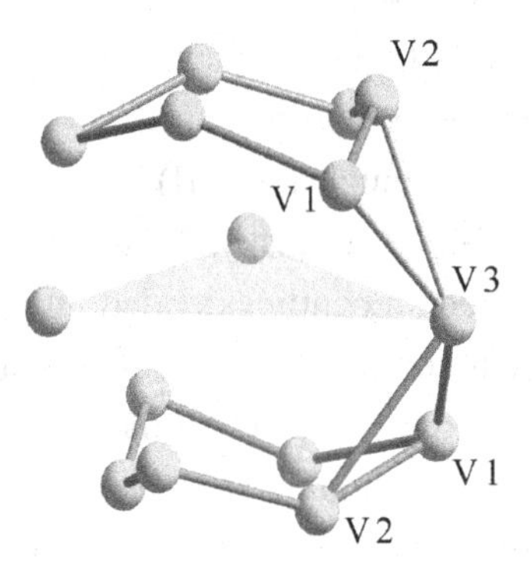

Fig. 2. Scheme of the V_{15} metal network. Three different kinds of vanadium sites, i.e. V1, V2 and V3 exist in the cluster.

Five different pathways for the antiferromagnetic isotropic superexchange can be distinguished as schematically shown in Fig. 3 where the corresponding exchange parameters are also indicated. Since vanadium ions occupy low-symmetry positions, the orbital degeneracy is removed and therefore the leading term in the exchange Hamiltonian can be represented by the isotropic HDVV model:

$$H_0 = -2 \sum_{\langle i,k \rangle} J_{ik} \mathbf{S}_i \mathbf{S}_k, \tag{1}$$

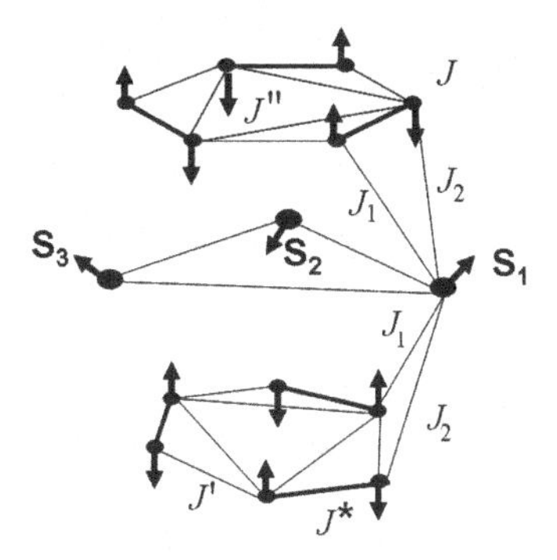

Fig. 3. Schematic structure of dominant exchange pathways and spin arrangement in V_{15}.

where the summation is extended over all pairs of the magnetic ions and J_{ik} are the exchange parameters. First calculation of the energy levels of the V_{15} system within the isotropic exchange (HDVV) model was given in Ref. Barra *et al.*, 1992 with the aid of the irreducible tensor operators (ITO) technique and the set of the exchange parameters J^*, J_1, J', J_2, J'' have been estimated by analyzing the temperature dependence of the magnetic susceptibility. Later on new sets of these parameters were deduced from the adiabatic magnetization measurements in ultrahigh fields (Platonov *et al.*, 2002; Kostyuchenko and Popov, 2008) and derived from the *ab initio* calculations (Kortus *et al.*, 2001; Boukhvalov *et al.*, 2003; Boukhvalov *et al.*, 2004). Using the MAGPACK software package (Borras-Almenar *et al.*, 2001) based on the ITO technique, the energy levels have been calculated in Ref. Tsukerblat *et al.*, 2006 with three sets (I,II,III) of these five parameters so far proposed (Table 1) assuming (for those sets only) also that $J_1 \approx J'$ and $J_2 \approx J''$ as suggested in Ref. Barra *et al.*, 1992. A general pattern of the energy pattern is shown in Fig. 4 where the energy levels are given for the set III of the parameters. One can see that two low lying levels corresponding to the full spins $S = 1/2$ and $S = 3/2$ are very close in all three cases, the corresponding gaps prove to be $3.2\,\text{cm}^{-1}$ (set I), $3.54\,\text{cm}^{-1}$ (set II) and $1.5\,\text{cm}^{-1}$ (set III). The next excited level is found to be a spin doublet at $621.07\,\text{cm}^{-1}$, $528.72\,\text{cm}^{-1}$ and $345.04\,\text{cm}^{-1}$ correspondingly. The results show that in all cases the two low lying levels are well separated from the excited ones that are much higher and can be

Table 1. Isotropic exchange parameters (in cm^{-1}) of V_{15}.

Sets of parameters	Reference	J^*	J_1	J'	J_2	J''
I	(Barra *et al.*, 1992)	−556	−27	−27	−125	−125
II	(Gatteschi *et al.*, 1993)	−556	−104	−104	−208	−208
III	(Platonov *et al.*, 2002)	−170	−28	−28	−56	−56
IV	(Kostyuchenko and Popov, 2008)	−240	−42	−104	−82	−195
V	(Kortus *et al.*, 2001)	−2340	−111	183	−188	−128
VI	(Boukhvalov *et al.*, 2004)	−632	−152	−31	−188	−95

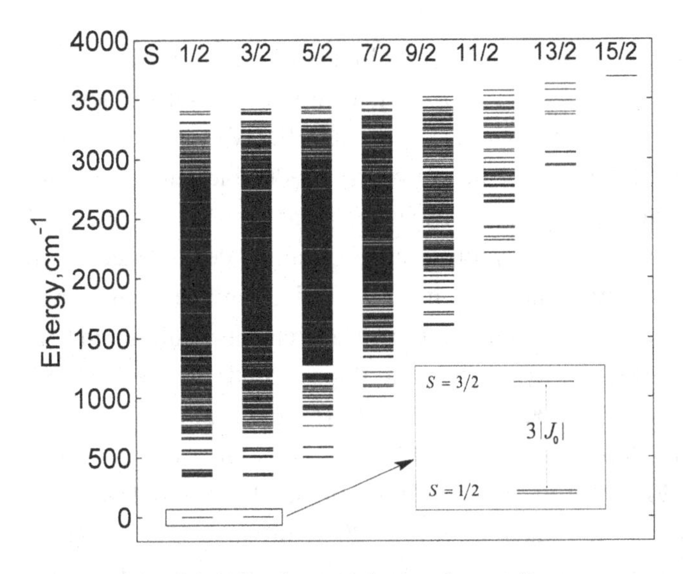

Fig. 4. Energy levels of the V_{15} cluster calculated according to parameters from Ref. Platonov *et al.*, 2002. The levels are grouped according to the total spin S. Inset: the two lowest levels $S = 1/2$ and $S = 3/2$.

viewed as an energy pattern of an isolated spin-1/2 triangular unit (Fig. 4, inset) as was proposed in Ref. Barra *et al.*, 1992 and substantiated within the perturbation theory.

The full Hilbert space for 15 spins involves $2^{15} = 32\,768$ states and although the exchange problem is tractable even if anisotropic contributions are taken into account (Barra *et al.*, 1992; Gatteschi *et al.*, 1993; Platonov *et al.*, 2002; Kostyuchenko and Popov, 2008; De Raedt *et al.*, 2004; De Raedt *et al.*, 2003; De Raedt *et al.*, 2004; Konstantinidis and Coffey, 2002; De Raedt *et al.*, 2003) including the *ab initio* calculations (Kortus *et al.*, 2001; Kortus *et al.*, 2001; Boukhvalov *et al.*, 2003; Boukhvalov *et al.*, 2004), a simplified model of a spin triangle (Barra *et al.*, 1992; Gatteschi *et al.*, 1993) gives accurate results for the low lying set of levels that manifest themselves in the low temperature behavior of the system. In fact, the parameter J^* is the leading one and the parameters J_1, J_2, J', J'' are significantly smaller and seem to be of the same order so that each spin V3 of the central triangle is coupled to a pair of strongly antiferromagnetically interacting spins V1, V2 from the lower and upper hexagons as shown in Figs. 2 and 3.

The cluster can be viewed as consisting of three pentanuclear subunits, each subunit (Fig. 2) consists of the inner triangle ion V3 and two pairs V1–V2 from the upper and lower hexagons. At low temperatures the total magnetic moment of the ions V1 and V2 is quenched due to the strong antiferromagnetic coupling between them. Only ion V3 is active and its spin determines the spin of the whole subunit. So the subunit can be considered as an effective quasiparticle of spin $s = 1/2$ placed in the corner of a central triangle and the entire system can be viewed as an effective trinuclear ring of spins $s_i = 1/2$. It should be noted that although hexagons do not contribute to the total magnetic moment of the cluster, the actual exchange pathways between the subunits do pass through the hexagons. Recent experimental data on NMR (Sec. 10) provide also a direct evidence of the fact that at low temperatures the hexagons are inactive and the properties of V_{15} are entirely defined by the spins of the inner triangle. Of course the model of spin triangle is well applicable at relatively low temperatures when the excited levels (at about 600 cm^{-1}) are not populated.

3. Exchange Interactions within the Triangle Model

3.1. *Isotropic exchange within the triangle model*

The isotropic superexchange within the model of triangle can be described by the conventional HDVV Hamiltonian for a symmetric triangle:

$$H_0 = -2J_0(\boldsymbol{S}_1\boldsymbol{S}_2 + \boldsymbol{S}_2\boldsymbol{S}_3 + \mathbf{S}_3\boldsymbol{S}_1), \qquad (2)$$

where $S_i = 1/2$ and J_0 is the parameter of the antiferromagnetic exchange ($J_0 < 0$); for the sake of convenience we use a positive value $J = -J_0$ and the basis will be labeled as $|S_1S_2(S_{12})S_3SM\rangle \equiv |(S_{12})SM\rangle$. The energy pattern involves two energy levels

$$\varepsilon(S) = JS(S+1) \tag{3}$$

with $S = 1/2$ and $S = 3/2$ that are separated by an energy gap of $3J$ (the constant term in Eq. (3) is omitted). It is essential that they do depend upon the full spin S of the system and are independent of S_{12} where $\mathbf{S}_{12} = \mathbf{S}_1 + \mathbf{S}_2$ represents the intermediate spin in the three-spin coupling scheme (for an $s = 1/2$ triangle the quantum number $S_{12} = 0$ and 1). This leads to the two-fold "accidental" degeneracy of two $S = 1/2$ doublets (four-fold degeneracy including two spin projection for each $S = 1/2$) in a seeming contradiction to the Kramers theorem. The analysis of the HDVV Hamiltonian (see review article (Tsukerblat *et al.*, 1987) and references therein) revealed that the "degeneracy doubling" in the ground spin-frustrated state $(S_{12})S = (0)1/2$, $(1)1/2$ is related to the exact orbital degeneracy so that the ground term is the orbital doublet ^2E in the trigonal ($\mathbf{D}_3$) symmetry while the excited spin level $S = 3/2$ corresponds to the orbital singlet $^4A_2$. This can be symbolically indicated as:

$$2D^{(1/2)} \Rightarrow {}^2E, \quad 2D^{(3/2)} \Rightarrow {}^2A_2$$

where $D^{(S)}$ are the irreducible representations of the rotation group R_3 numerating spin states of the systems. In this view it should be noted that the term "accidental" degeneracy means that the spin model itself does not provide full information about the nature of the degeneracy.

It was concluded (Tsukerblat *et al.*, 1987) that the AS exchange acts within the $(S_{12})S = (0)1/2$, $(1)1/2$ manifold like first order spin-orbital interaction within the ^2E term and gives rise to two doublets in agreement with the Kramers theorem (Tsukerblat, 2006).

3.2. *'Accidental' degeneracy and spin-frustration*

Turning back to the isotropic superexchange, Eq. (2), in a triangular unit one can see that the V_{15} cluster belongs to the so-called spin frustrated systems. Spin frustration implies that in a classical spin representation a set of the competing spin-spin interactions cannot be 'saturated'. An equilateral spin triangle with antiferromagnetic exchange coupling forcing the spins to align antiparallel represents a widely used example. This kind of spin array is typical for a variety of polynuclear coordination compounds (see discussion in Refs. Schnack, 2006; Schnack, 2007; Schnalle and Schnack, 2009) with interesting and unusual properties in low dimensional lattices. In particular, many theoretical studies on this issue are focused on $s = 1/2$ Kagomé antiferromagnets based on triangular lattices (Matan *et al.*, 2006; Anderson, 1987; Shores *et al.*, 2005; Bartlett *et al.*, 2004; Bartlett and Nocera, 2005; Grohol *et al.*, 2005; Inami *et al.*, 2000; Elhajal *et al.*, 2002; Bulaevskii *et al.*, 2008).

Due to the trigonal geometry of the vanadium triangle the antiparallel spin alignment is possible within any selected pair (Fig. 5), whereas the third spin (encircled) cannot be paired to minimize the energy of the exchange with spins 1 and 2 simultaneously, and in this sense proves to be frustrated. In this classical consideration the spin frustration appears as a terminological problem implying only that the classical 'up-down' picture is inappropriate for the description of this situation. The real contents of the concept of spin frustration can be revealed through the analysis of the specific physical phenomena related to this spin arrangement.

In this view let us focus on the exact quantum-mechanical analysis of the HDVV Hamiltonian. As was mentioned in Sec. 3.3.1 the 'degeneracy doubling' with respect to the intermediate spin within the spin coupling scheme in the ground manifold $(S_{12})S = (0)1/2$, $(1)1/2$ is associated with the exact orbital degeneracy in the multi-electron triangular system so that the ground term is represented by the orbital doublet ^2E of the trigonal point group. The study of the more complicated systems containing triangular faces (for example, triangular clusters with half-integer spins $s \geq 1/2$) shows that spin frustration is inherently related to the orbital degeneracy of

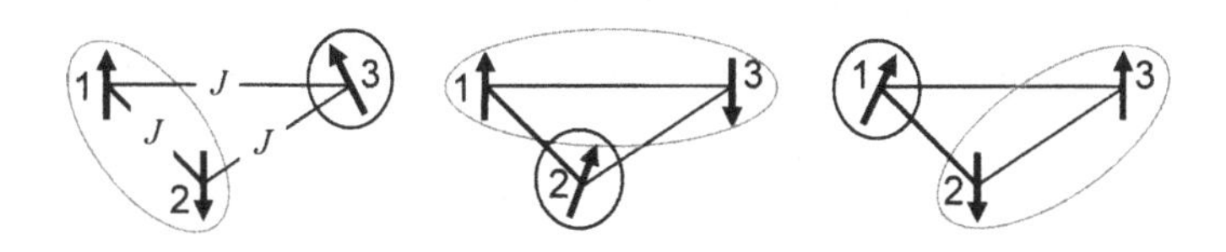

Fig. 5. Scheme of a classical spin frustration in a triangular unit.

the spin multiplets. On the other hand, the orbital degeneracy produces a strong magnetic anisotropy due to the contribution of the orbital magnetic component. At the same time orbitally degenerate systems undergo the Jahn-Teller (JT) instability that tends to remove the degeneracy related to a high symmetric configuration of the ions. All named factors are competitive and their physical consequences constitute the real contents of the concept of spin frustration in degenerate systems that will be discussed in the next Sections.

It should be mentioned that spin frustration has pronounced physical consequences not only in degenerate systems. For example, a spin-1 triangle for which the antiferromagnetic ground state is an orbital singlet $^1A_1$ represents also a spin frustrated system. Detailed discussion of spin frustration in different systems and its physical manifestations is given in Refs. Schnack, 2006; Schnack, 2010.

3.3. *Pseudo-angular momentum representation*

By applying the symmetry operations of the $\mathbf{D}_3$ point group to the basis functions $|(0)1/2 \pm 1/2\rangle$ and $|(1)1/2 \pm 1/2\rangle$ (that generate interexchange of spins) one can find that they correspond to x, y basis of the irreducible representation E. Therefore their circular superpositions (Tsukerblat *et al.*, 2006) correspond to the projections $M_L = +1$ and $M_L = -1$ of the fictitious angular momentum $L = 1$ (pseudo-angular momentum representation). We will use a short notation $U_{M_L}(SM_S)$ for these functions $|S = 1, M_S = \pm 12\rangle$.

$$
\begin{aligned}
u_{\pm 1}(1/2, \pm 1/2) &= \mp 1/\sqrt{2}(|(0)1/2, \pm 1/2\rangle \pm i|(1)1/2, \pm 1/2\rangle), \\
u_{\pm 1}(1/2, \mp 1/2) &= \mp 1/\sqrt{2}(|(0)1/2, \mp 1/2\rangle \pm i|(1)1/2, \mp 1/2\rangle).
\end{aligned}
\tag{4}
$$

Using this conception one can introduce the functions $U_S(M_J)$ belonging to a definite full spin S and projections $M_J = M_L + M_S$ of the full pseudo-angular momentum, so that $U_{1/2}(\pm 3/2) = u_{\pm 1}(1/2, \pm 1/2)$ and $U_{1/2}(\pm 1/2) = u_{\pm 1}(\mp 1/2)$. The singlet corresponds thus to $M_L = 0$, and the components are labeled as $u_0(3/2, M_S) \equiv U_{3/2}(M_J)$ with $M_S = \pm 1/2$ and $M_S = \pm 3/2$, so that $M_J = \pm 1/2$ and $\pm 3/2$.

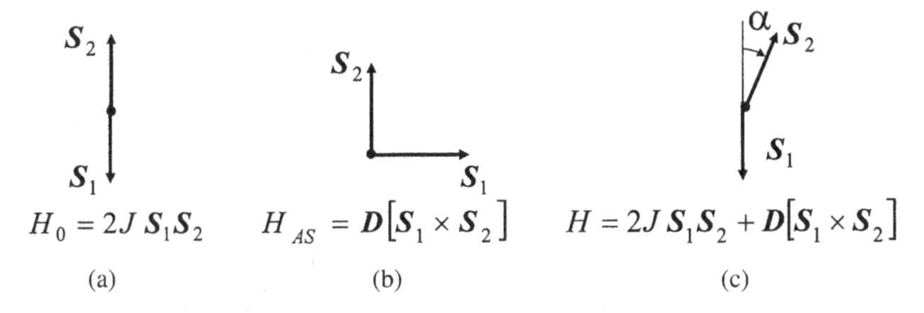

Fig. 6. Classical spin alignment in a spin dimer (a) caused by antiferromagnetic isotropic exchange; (b) AS exchange; (c) combined isotropic and AS exchange interactions.

3.4. *Antisymmetric exchange, zero-field splitting*

AS exchange has been introduced by Dzyaloshinsky, 1957 and Moriya, 1960 to explain the origin of spin canting in magnetic materials. In fact, the isotropic exchange in the classical representation aligns spins of a dimer parallel or antiparallel depending of the sign of J (Fig. 6(a)). The AS exchange (expressed through the vector product $[S_1 \times S_2]$) tends to align spins in the perpendicular directions (Fig. 6(b)). A competition of the isotropic and AS exchange interactions usually results in a relatively small canting angle of the two spins in a dimeric unit. In particular, for $S_1 = S_2 = 1/2$ the canting angle is defined as $\alpha = D/(2|J|)$ as shown in Fig. 6(c) for an antiferromagnetic dimer. From the quantum-mechanical point of view the AS exchange leads to a small mixing of different spin multiplets giving rise to a zero-field splitting (proportional to D^2/J) and magnetic anisotropy in the spin triplet.

The Hamiltonian of AS exchange preserving trigonal symmetry is given in Ref. Tsukerblat *et al.*, 2006:

$$\begin{aligned} H_{AS} &= D_n([S_1 \times S_2]_Z + [S_2 \times S_3]_Z + [S_3 \times S_1]_Z) \\ &\quad + D_l\left([S_1 \times S_2]_X - \frac{1}{2}[S_2 \times S_3]_X + \frac{\sqrt{3}}{2}[S_2 \times S_3]_Y \right. \\ &\quad \left. - \frac{1}{2}[S_3 \times S_1]_X - \frac{\sqrt{3}}{2}[S_3 \times S_1]_Y\right) \end{aligned}$$

$$+ D_t \left([S_1 \times S_2]_Y - \frac{\sqrt{3}}{2}[S_2 \times S_3]_X - \frac{1}{2}[S_2 \times S_3]_Y \right.$$

$$\left. + \frac{\sqrt{3}}{2}[S_3 \times S_1]_X - \frac{1}{2}[S_3 \times S_1]_Y \right) \qquad (5)$$

Here the spin operators are related to the molecular frame, the parameter D_n is associated with the normal (Z-axis) component of AS exchange, and D_l and D_t are those for the in-plane parts (see details in Ref. Tsukerblat *et al.*, 2006). The matrix of H_{AS} can be explicitly calculated on the basis of $|(S_{12})SM\rangle$ using the ITO technique.

It is important that the 'normal part' of the AS exchange operates only within the basis of two 'accidentally' degenerate doublets $(S_{12})S = (0)1/2$, $(1)1/2$, meanwhile the two 'in-plane' contributions (terms of the Hamiltonian associated with the parameters D_l and D_t) lead only to a mixing of the ground spin doublets $(0)1/2$, $(1)1/2$ with the excited spin quadruplet $(1)3/2$ separated from the two low lying spin doublets by the gap $3J$.

AS exchange leads to a splitting of the two $S = 1/2$ levels into two Kramers doublets corresponding to $M_J = \pm 1/2$ and $M_J = \pm 3/2$. Usually the isotropic exchange is a leading interaction, so it is useful to represent the zero-field energies as series in $D_\perp^2/J$. The zero-field splitting of the two spin doublets within this approximation is found as:

$$\Delta \equiv \varepsilon(M_J = \pm 3/2) - \varepsilon(M_J = \pm 1/2) \cong \sqrt{3}D_n - D_\perp^2/8J \qquad (6)$$

One can see that the zero-field splitting appears as a first order effect with respect to the normal component of AS exchange and contains also a second order correction (always negative) arising from the mixing of $(S_{12})1/2$ and $(1)3/2$ multiplets through the in-plane components of AS exchange. It can be said that in-plane components of the AS exchange are reduced by the isotropic exchange so that under the realistic conditions the parameter $D_\perp$ is effectively small. At the same time this part of the AS exchange leads to the avoided crossing of the magnetic sublevels of $S = 1/2$ and $S = 3/2$ multiplets in a high field; at the crossing points the in-plane components of AS exchange act as a first order perturbation (Tsukerblat *et al.*, 2006).

The excited $S = 3/2$ level shows also a zero-field splitting but this splitting $\Delta_1 = D_{\perp}^2/8J$ is not affected by the parameter D_n and represents solely a second order effect with respect to the in-plane part of the AS exchange. For this reason the zero-field splitting of the excited quadruplet is expected to be smaller (if D_n and $D_{\perp}$ are comparable) than the splitting of the two $S = 1/2$ doublets. The sign of Δ determines the ground state. In fact, in the cases of $\Delta > 0$ and $\Delta < 0$ the ground states are the doublets with $|M_J| = 1/2$ and $|M_J| = 3/2$, respectively. The Zeeman sublevels are enumerated by the quantum number M_J as shown in Fig. 10 in the case of $\Delta > 0$, the fine structure of $S = 3/2$ is shown in the inset. According to the general symmetry rule the levels with the same M_J show an avoided crossing, while those with different M_J exhibit the exact crossing.

3.5. *Ab initio calculations*

Along with the semiempirical models based on the spin-Hamiltonian approach, the *ab initio* calculations of the parameters are also required for the in-depth understanding of the origin of the magnetic properties of V_{15}.

The papers (Kortus *et al.*, 2001; Kortus *et al.*, 2001) report the *ab initio* DFT calculations of V_{15} which confirm the fact that the system's ground state possesses the total spin of $S = 1/2$ while the first excited level is a quadruplet $S = 3/2$. These first principle calculations also confirm that the total spin of the cluster in its ground state is entirely determined by the inner triangle. Surprisingly, Ref. Kortus *et al.*, 2001 states that one of three isotropic couplings in the hexagons is ferromagnetic as opposed to all other authors who assume only antiferromagnetic exchange interactions (see Table 1, Sec. 2). But even in spite of this unexpected result the calculations in Ref. Kortus *et al.*, 2001 still show the total spin of $S = 1/2$ in the ground state. The electronic structure of V_{15} has been also studied using local spin density approximations in Refs. Boukhvalov *et al.*, 2003; Boukhvalov *et al.*, 2004; Barbour *et al.*, 2006. The electronic structure calculations motivated the study of the optical spectroscopy study on the V_{15} system (Saito and Miyashita, 2001). The reflectance and optical conductivity of solid V_{15} over a wide energy range have been reported in the

Ref. (Choi *et al.*, 2003). In particular, the energies of the d-d transitions (1.2 eV) and O-V p-d transitions (3.7, 4.3 and 5.6 eV) have been determined. This study provides an important information about the on-site Coulomb repulsion U that is a key parameter in the Anderson's theory of the kinetic exchange.

4. Zeeman Levels, Magnetic Anisotropy

The Hamiltonian of the system in a constant magnetic field is given by

$$H = H_0 + H_{AS} + g\beta \mathbf{HS}, \tag{7}$$

where H_0 and H_{AS} are the isotropic and AS interactions given by Eqs. (2) and (5) respectively and the last is the Zeeman term where β is the Bohr magneton and g is the gyromagnetic factor. For the sake of simplicity, the anisotropy of g will be neglected.

Due to the actual axial symmetry of the system reflected in the pseudo-angular momentum classification of the states, the matrix of the full Hamiltonian can be blocked into four second order matrices each corresponding to a definite projection M_J of the total pseudo-angular momentum. The eigen-functions of the system are found in terms of the superpositions of states with the same M_J originating from $S = 1/2$ and $S = 3/2$ multiplets that correspond to the *jj*-coupling scheme in axial symmetry. In absence of the field the full pattern consists of four Kramers doublets, two of which possess $M_J = \pm 1/2$ and two doublets with $M_J = \pm 3/2$ (shown in bold, Fig. 7) with two zero-field inter-doublet splittings Δ and Δ_1.

When the magnetic field is oriented along C_3 axis (preserving thus the axial symmetry) the energy levels are enumerated by the definite values of M_J (Fig. 7). This allows to find the analytical solution for the energy levels providing an arbitrary interrelation between parameters (Tsukerblat *et al.*, 2006):

$$\begin{aligned}
\varepsilon_{1,2}(H) &= -\frac{1}{4}\sqrt{\left(\sqrt{3}D_n \pm 2g\beta H + 6J\right)^2 + 3D_\perp^2} - \frac{\sqrt{3}}{4}D_n \\
\varepsilon_{3,4}(H) &= -\frac{1}{4}\sqrt{\left(\sqrt{3}D_n \pm 2g\beta H - 6J\right)^2 + 9D_\perp^2} + \frac{\sqrt{3}}{4}D_n \mp g\beta H
\end{aligned}$$

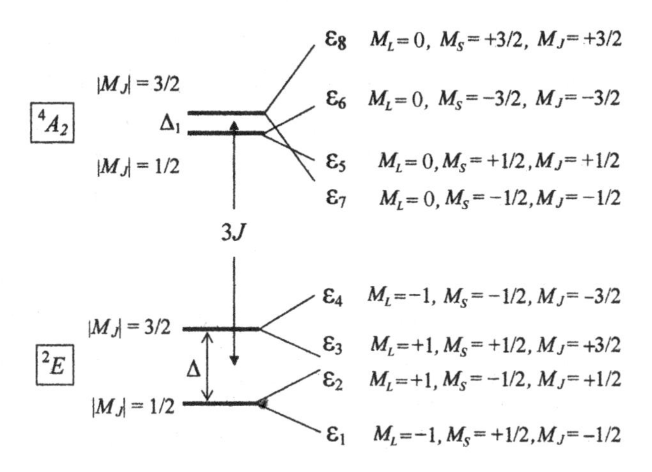

Fig. 7. Labeling of Kramers doublets and Zeeman sublevels in a parallel magnetic field, $D_n > 0$. Inter-doublet zero-field splittings Δ and Δ_1 are given.

$$\varepsilon_{5,6}(H) = \frac{1}{4}\sqrt{\left(\sqrt{3}D_n \mp 2g\beta H + 6J\right)^2 + 3D_{\perp}^2} - \frac{\sqrt{3}}{4}D_n$$

$$\varepsilon_{7,8}(H) = \frac{1}{4}\sqrt{\left(\sqrt{3}D_n \pm 2g\beta H - 6J\right)^2 + 9D_{\perp}^2} + \frac{\sqrt{3}}{4}D_n \mp g\beta H \quad (8)$$

The field dependence of these levels is shown in Fig. 10. One can see that they do depend upon two effective parameters of AS exchange, D_n and $D_{\perp}(D_{\perp}^2 = D_t^2 + D_l^2)$, rather than upon three initial parameters D_n, D_l and D_t of the Hamiltonian.

The Zeeman pattern shows an obvious anisotropy, that is most pronounced for the $S = 1/2$ doublets in the low-field range. Figure 8 shows

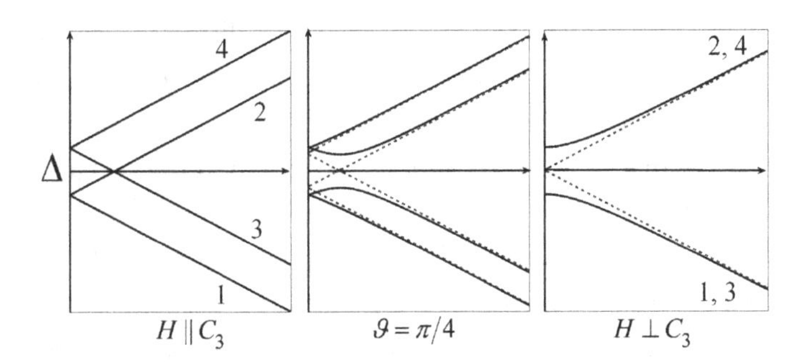

Fig. 8. Low-lying energy levels of an antiferromagnetic trigonal system (arbitrary units) for different orientations of the magnetic field.

how the energy pattern changes as a function of the orientation of the magnetic field with respect to C_3 axis. As it follows from Fig. 8 the magnetic anisotropy proves to be axial with the C_3 being the easy axis of magnetization while the plane of the triangle is the hard magnetic plane. The $S = 3/2$ quadruplet shows also anisotropic Zeeman splitting that will be discussed in more details in Sec. 5.

In Sec. 6 a special role of the normal and in-plane parts of AS in the Zeeman pattern will be emphasized. The Zeeman pattern calculated within the full basis set of 15 vanadium spins is given in Refs. De Raedt *et al.*, 2004; De Raedt *et al.*, 2003; De Raedt *et al.*, 2004; Konstantinidis and Coffey, 2002; Machida and Miyashita, 2005. One should mention Ref. Raghu *et al.*, 2003 that provides an analysis of the low-lying states in V_{15} with the use of spin-density methods.

5. Electron Paramagnetic Resonance

5.1. *EPR spectrum of V_{15}: Role of antisymmetric exchange and selection rules*

In order to reveal a special rule of the AS exchange in EPR (Tsukerblat *et al.*, 1987) let us start with a simple case of the HDVV model for a symmetric triangle. In this case the ground state $2D^{(1/2)}$ is "accidentally" degenerate. Within the HDVV model the two transitions $M_S = \pm 1/2 \leftrightarrow M_S = \mp 1/2$ induced by an alternating field can be observed giving rise to a single line in EPR as shown in Fig. 9(a). In the framework of the isotropic model one can introduce also a static scalene distortion through an additional term $2\eta \mathbf{S}_1\mathbf{S}_2$ in the HDVV Hamiltonian. The ground state in a distorted system is split into two Kramers doublets (separated by the gap $2|\eta|$) labeled by the certain values of the intermediate spins S_{12} ($S_{12} = 0$ and 1 for a spin 1/2 system). In this case the inter-doublet transitions are strictly forbidden and only the intra-doublet transitions with conservation of S_{12} are allowed so that the only EPR line can be observed (Fig. 9(b)). Since both transitions correspond to the same resonance field the temperature populations of the levels do not affect the observed spectrum. In this sense the deformation of the system in the framework of the isotropic exchange model does not affect the EPR pattern.

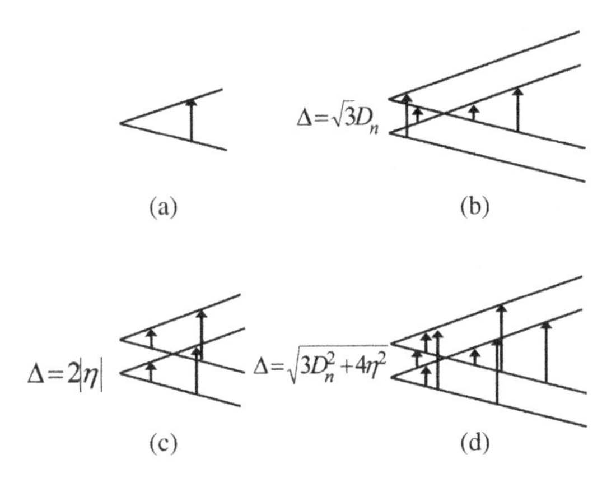

Fig. 9. (a) Single EPR transition in the spin doublet. (b) The normal component of the AS exchange allows the inter-doublet transitions only. (c) Scalene structural deformation allows only the intra-doublet transitions. (d) In the case of AS exchange and distortion, both kinds of transitions are allowed.

When the ZFS (gap $\sqrt{3}|D_n|$) of the "accidentally" degenerate ground state of the strictly trigonal system is caused by the AS exchange (Sec. 3.4) the wave-functions of the Kramers doublets are represented by the circular superposition of the states with different S_{12} (Eq. (4)). This immediately leads to the conclusion that the inter-doublet EPR transitions are allowed giving rise to two lines while the intra-doublet transitions are forbidden (Fig. 9(c)). Finally, when both static distortions and AS exchange coexist the ZFS in the ground manifold is represented by a combined gap $\Delta = \sqrt{4\eta^2 + 3D_n^2}$ and both inter-doublet and intra-doublet types of transitions are allowed (Fig. 9(d)). The relative intensities of these lines depend on the ratio $|\eta/D_n|$.

Experiments on EPR (Sec. 5.2), INS (Sec. 9.2) and NMR (Sec. 10) give an evidence that the AS exchange and distortion coexist in the V_{15} cluster. Later on we give qualitative and quantitative discussions of the mutual effect of these two perturbations. In this Section we discuss the EPR selection rules in the model where only AS exchange is involved.

Within the pseudo-angular momentum approach one can conclude that in accordance with the general principles of quantum mechanics, the selection rule $M_J \rightarrow M_J \pm 1$ for the linearly polarized microwave field $\boldsymbol{H}_{osc} \perp C_3$ defines the allowed transitions as shown in Fig. 10. Using the analytical

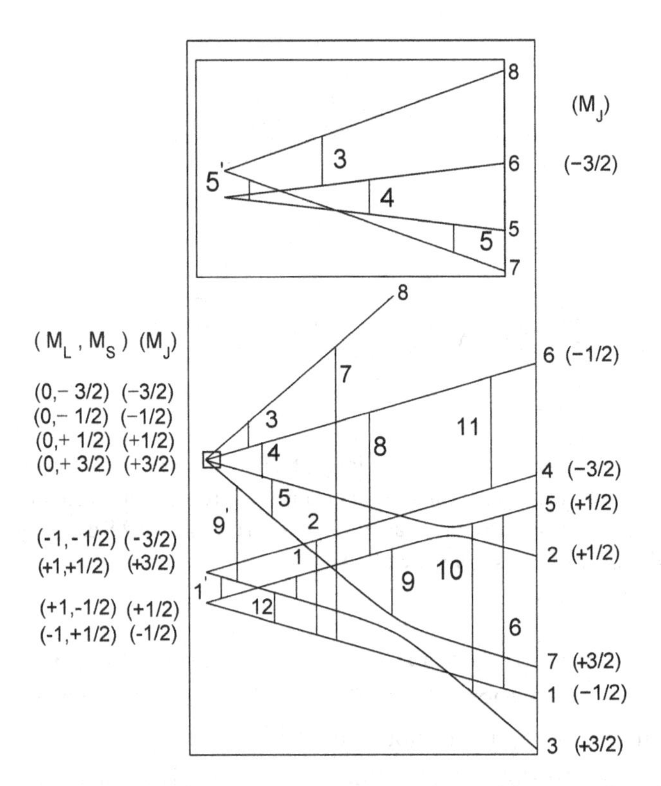

Fig. 10. Energy pattern of V_{15} within the three-spin model ($D_n > 0$) and allowed EPR transitions in the parallel field. Inset: magnified zero-field and Zeeman splitting of $S = 3/2$ level in the parallel field.

solutions for the Zeeman energies, one can evaluate the resonance fields for the EPR transitions. A representative scheme of the transitions and EPR spectra simulated for different microwave frequencies (Tsukerblat *et al.*, 2006), including those used in the high-frequency EPR experiments at ultra-low temperatures (Sakon *et al.*, 2004; Sakon *et al.*, 2005). It is assumed that $g = 1.96$ and $J = 0.847\ \text{cm}^{-1}$ which is consistent with the experimental data (Barbara, 2003). The ZFS in the ground manifold $|\Delta|$ is set to $0.14\ \text{cm}^{-1}$ (see Sec. 9.1). Since the normal and in-plane contributions of AS exchange cannot be distinguished directly from the experimental data on the INS, the ratio $D_\perp / D_n$ is varied (providing a fixed value of $|\Delta|$) in order to reveal the influence of different components of the AS exchange on the EPR pattern. The transformations of the spectrum with the increase of the ratio $D_\perp / D_n$ are also shown in the proposed EPR schemes (Tsukerblat *et al.*, 2006).

Let us consider separately two cases: (i) $D_n \neq 0$, $D_\perp = 0$ and (ii) $D_n \neq 0$, $D_\perp \neq 0$. Since the normal part of the AS exchange does not mix different spin levels one can discuss the case (i) within the Russell-Saunders scheme when the interaction with the alternating field $g\beta\hat{S}_x\mathbf{H}_{osc}$ does not change the orbital quantum number M_L. This implies the following selection rules for the linearly polarized microwave radiation that are strictly valid within the Russell-Saunders approximation: the EPR transitions $M_J \rightarrow M_J \pm 1$ are allowed with the conservation at the same time of the full spin S, projection of the orbital angular momentum M_L and for $M_S \rightarrow M_S \pm 1$:

$$(\Delta S = 0,\ \Delta M_L = 0,\ \Delta M_S = \pm 1,\ \Delta M_J = \pm 1).$$

The allowed intramultiplet transitions are schematically shown in Fig. 10 for three frequency domains: $h\nu < \Delta$, $\Delta < h\nu < 3J$ and $h\nu > 3J$. One can see that the spectrum consists of three lines, one line arises from three strong transitions **3, 4, 5** within $S = 3/2$ multiplet with the resonance field $H_{3,4,5} = h\nu/g\mu_B$ and the two remaining lines correspond to two interdoublet transitions within two $S = 1/2$ levels (Figs. 10 and 11). It should be stressed that the intermultiplet ($S = 1/2 \leftrightarrow S = 3/2$) transitions are strictly forbidden when $D_\perp = 0$ as well as intradoublet transitions in the two $S = 1/2$ levels split by AS exchange (this has been proved in Refs. Tsukerblat *et al.*, 1987; Tsukerblat and Belinskii, 1983). Two situations, namely $h\nu < \Delta$ and $h\nu > \Delta$ within the case (i) should be distinguished. Providing $h\nu < \Delta$ (Fig. 11(a)) the two interdoublet transitions **1** and **1′** have the following resonance fields:

$$H_{1'} = (\sqrt{3}D_n - h\nu)/g\mu_B \quad \text{and} \quad H_1 = (\sqrt{3}D_n + h\nu)/g\mu_B, \qquad (9)$$

One can see that the separation between these lines $H_1 - H_{1'} = 2h\nu/g\beta$ increases with the increase of microwave frequency. The full spectrum is asymmetric with the line at $H_{3,4,5} = h\nu/g\beta$ being closer to the line at $H_{1'}$. The difference in the resonance fields

$$H_1 - H_{3,4,5} = \sqrt{3}D_n/g\beta \qquad (10)$$

is independent of the microwave frequency ν while the difference

$$H_{1'} - H_{3,4,5} = (2h\nu - \sqrt{3}D_n)/g\beta \qquad (11)$$

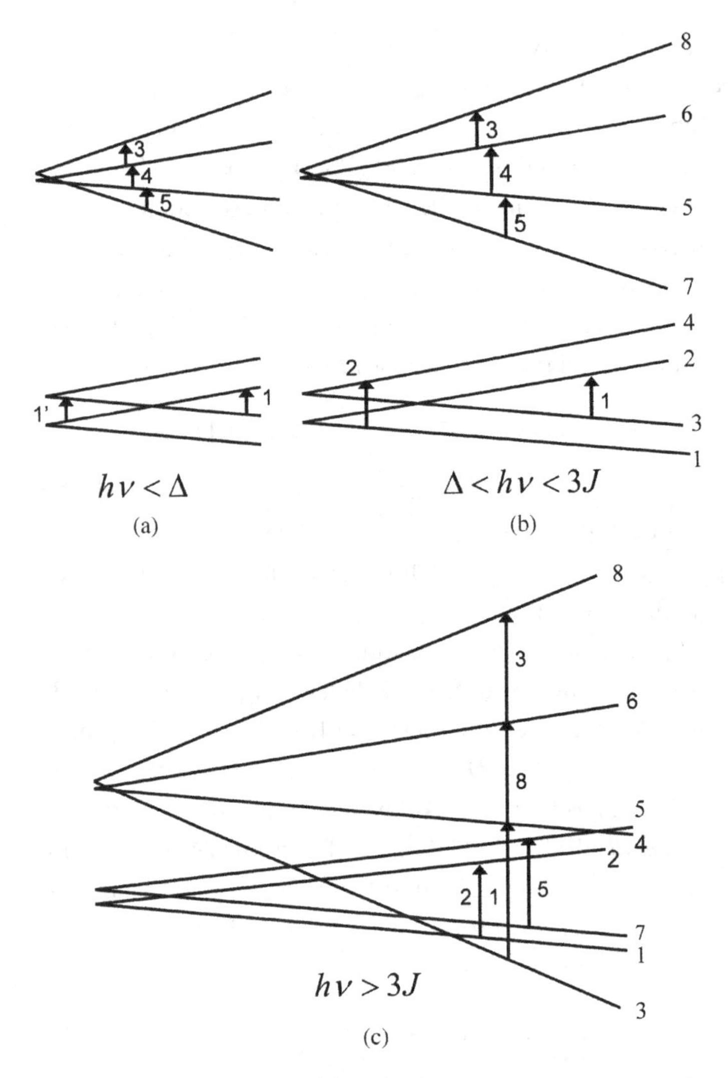

Fig. 11. Scheme of the EPR transitions in the different frequency regions when $D_{\perp} = 0$: (a) $h\nu < \Delta$; (b) $h\nu > \Delta < 3J$; (c) $h\nu > 3J$.

increases with increase of ν. At the relatively high temperatures (when the full intensity follows the low $I \propto T^{-1}$) the ratio of the intensities of the three lines is 1:3:1. In the case of $h\nu > \Delta$ the interdoublet transitions 1 and 2 (Figs. 11(b) and 11(c)) correspond to the resonance fields:

$$H_1 = \left(h\nu + \sqrt{3}D_n\right)/g\beta \quad \text{and} \quad H_2 = \left(h\nu - \sqrt{3}D_n\right)/g\beta. \tag{12}$$

In this case the spectrum consists of the central peak at $H_{3,4,5} = \sqrt{3}D_n/g\beta$ and two equally spaced side-lines at H_1 and H_2 with the ratio of the intensities 1:3:1. It is remarkable that in the case under consideration the full width of the spectrum $H_1 - H_2 = 2\sqrt{3}/g\beta$ is directly related to the AS exchange and is independent of the frequency ν.

In the general case (ii) when both components of AS exchange are nonzero ($D_n \neq 0$ $D_t \neq 0$), different spin levels are mixed and therefore the system can be adequately described by the *jj*-coupling scheme. The two new essential features of the EPR pattern arise from the mixing of $S = 1/2$ and $S = 3/2$ spin levels by the in-plane part of the AS exchange. First, due to axial zero-field splitting $\Delta_1 = D_\perp^2/8J$ of the excited $S = 3/2$ level the transitions **3, 4, 5** at $h\nu > \Delta_1$ have different resonance fields: $H_{3,5} = (h\nu \mp \Delta_1)/g\beta$, $H_4 = g\beta$. Providing $h\nu < \Delta_1$ the line **3** does not exist and the line **5′** corresponds to $H_{5'} = (\Delta_1 - h\nu)/g\beta$ (see Fig. 10, inset). This leads to a peculiar triplet fine structure of the central peak in the patterns of the EPR so far discussed.

The second order effect of mixing through the in-plane AS exchange is relatively small in the wide range of the fields except of the vicinity of the avoided crossing points (Fig. 10) where this component acts as a first order perturbation. The second important consequence of the in-plane AS exchange is that this interaction allows new transitions (obeying the general selection rule $\Delta M_J = \pm 1$) that are forbidden in the Russell-Saunders scheme, namely, the intermultiplet transitions **6–12** and **9′** as shown in Fig. 10. The intensities of these newly allowed transitions depend on the extent of the mixing of $S = 1/2$ and $S = 3/2$ multiplets in the magnetic field and in a wide range of the field they are relatively weak.

5.2. *Discussion of the experimental EPR data*

Detailed EPR measurements at ultra-low temperatures in the range 0.5–4.2 K in a parallel (along C_3 axis) field have been reported in Refs. Sakon *et al.*, 2004; Sakon *et al.*, 2005. Interpretation of the precise measurement of the temperature dependence for the signal intensity and absorption power confirmed that the ground state of V_{15} is a Kramers doublet with agreement with all previous studies. The authors

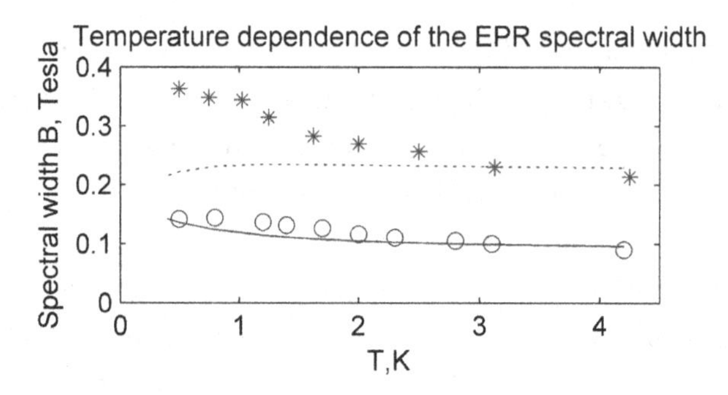

Fig. 12. Theoretical and experimental (Ref. Sakon *et al.*, 2004) temperature dependencies of the EPR spectral width for V_{15}. The theoretical curves are scaled at $T=3K$, $\Delta=0.14\,\mathrm{cm}^{-1}$, parallel field. Theoretical curves: 57.831 GHz (solid), 108 GHz (dots); experimental data: 57.831 GHz (circles), 108 GHz (stars).

of Ref. Sakon *et al.*, 2004 report also the high-frequency (58 and 108 GHz) measurements of the EPR line width at ultra-low temperatures. The transmission spectrum observed at 2.1 *Tesla* represents a relatively broad slightly asymmetric peak that becomes broader when the temperature decreases (Fig. 12).

Since the observed structureless EPR peak does not provide an unambiguous information about the fine structure of the absorption line we will discuss a simple approximation within which only the normal part of the AS exchange is taken into account. Since the fine structure of the absorption line is unresolved it is reasonable to assume that the observed peak can be considered as an envelope of the broadened individual absorption lines arising from the allowed transitions. In the case of $\nu=57.831$ GHz (frequency region $\Delta < h\nu < 3J - \Delta/2$) the superposition involves the central lines **3**, **4**, **5** at $H_{3,4,5}=2.11$ *Tesla* and two sidebands **1** and **2** at $H_1=1.96$ *Tesla* and $H_2=2.26$ *Tesla*. At the frequency $\nu=108$ GHz the full spectrum is assumed to consist of the central peak 1, 3, 8 and sidelines 2 and 5 ($H_{1,3,8}=3.93$ *Tesla*, $H_2=3.78$ *Tesla* and $H_5=4.08$ *Tesla*).

In order to approximately estimate the role of the AS exchange in the broadening of the EPR peak the central second moments of these discrete spectral distributions can be considered:

$$\left\langle (H-\bar{H})^2 \right\rangle = \sum_{i=1}^{3} I_i (H_i-\bar{H})^2 \Big/ \sum_{i=1}^{3} I_i, \tag{13}$$

where I_i are the intensities of the lines at a given temperature and $\bar{H}$ is the first moment (center of gravity) of the spectral distribution:

$$\bar{H} = \sum_{i=1}^{3} I_i H_i \Big/ \sum_{i=1}^{3} I_i . \tag{14}$$

The full width of the observed peak includes also contributions arising from the broadening of the individual lines. In order to take them into account, at least qualitatively, it is reasonable to normalize the full width (obtained with the aid of Eqs. (13) and (4)) vs. temperature at $T = 3$ K. As one can see the evaluated temperature dependence of the spectral width is in a reasonable agreement with the experimental data at $\nu = 58$ GHz (Fig. 12). At the same time at $\nu = 108$ GHz the splitting of the lines due to the AS exchange plays probably a secondary role in the broadening of the observed EPR peak, especially at ultra-low temperatures. This can also be illustrated by plotting the EPR pattern at $T = 0.5$ K for the two employed frequencies (Fig. 13). One can see that if AS exchange plays a major role in formation of the bandwidth at low temperature then the second moment at $\nu = 108$ GHz is expected to be less than that for $\nu = 58$ GHz due to lower intensity of the sidebands. However, the observed width is greater and that requires a more comprehensive analysis. Calculations of the second moments in a more general model with non-zero in-plane AS exchange give similar results. In view of these results one might assume that the broadening of the EPR peak can be attributed to the spin-phonon interaction. The hyperfine interaction

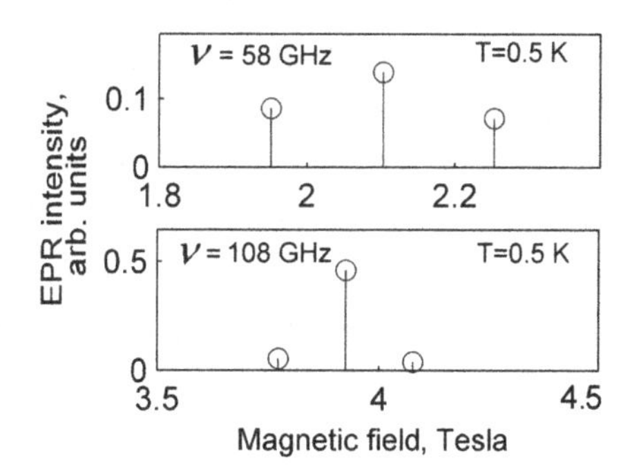

Fig. 13. Calculated EPR lines for $\boldsymbol{H} \| C_3$, $T = 0.5$ K for $\nu = 58$ GHz (top) and $\nu = 108$ GHz (bottom) ($J = 0.847\,\text{cm}^{-1}$, $\Delta = 0.14\,\text{cm}^{-1}$).

should be also mentioned as an essential contribution to the broadening of the EPR line in the vanadium cluster.

The low-temperature EPR experiments on the single V_{15} crystal reported in Ref. Kajiyoshi *et al.*, 2007 provide the direct observation of the anisotropy of the system that manifests itself in dependence of the resonant field (at a given constant frequency) on the direction of the magnetic field. This result can be considered as an evidence for the existence of the AS exchange in V_{15} since just this interaction should be responsible for the anisotropy. The data from Ref. Kajiyoshi *et al.*, 2007 may be used for the estimation of the relative importance of the AS exchange and static distortion in the formation of low lying levels of V_{15}.

Ref. Kajiyoshi *et al.*, 2007 reports also a direct observation of the zero-field energy gap Δ between the two Kramers doublets, but the estimated value of about $0.02\,\mathrm{cm}^{-1}$ (30 mK) is by one order smaller than $0.21-0.28\,\mathrm{cm}^{-1}$ estimated in the INS experiments (Chaboussant *et al.*, 2002; Chaboussant *et al.*, 2004; Chaboussant *et al.*, 2004; Chaboussant *et al.*, 2004).

Ref. Kajiyoshi *et al.*, 2007 describes measurements of the resonant frequency vs. field in the both perpendicular and parallel field orientations. In the second case (Fig. 14) one can see a number of closely located resonance peaks (circles) which are assigned in Ref. Kajiyoshi *et al.*, 2007 to the transitions of the types **1′**, **2** and ***I*** (Fig. 14, inset; see also Figs. 9, 10 and 11). There is also a separate group of peaks (squares in Fig. 14) measured at higher fields, beginning from about 0.1 *Tesla*. These peaks raised questions in the interpretation in Ref. Kajiyoshi *et al.*, 2007. In fact, these peaks can be recognized as the transitions of the type **1** (Fig. 14, inset) if instead of a small gap $0.02\,\mathrm{cm}^{-1}$ one assumes a larger value $\Delta \approx 0.1\,\mathrm{cm}^{-1}$ which is consistent with different estimation for Δ (Sec. 9.1). Indeed, the transition **1** appears at the fields higher than that of the level crossing point H_0 (Fig. 14, inset). If one assumes the above mentioned reasonable value of ZFS, the level crossing point is just at $H_0 = \Delta/g\beta \approx 0.1$ *Tesla* which is in a full agreement with the experiment. According to Ref. Kajiyoshi *et al.*, 2007 the 'distortion-type' transition ***I*** is weaker than the 'AS exchange-type' resonant peaks **1**, **1′** and **2**.

The experimental data on temperature dependence of the EPR lines intensities in V_{15} are given in Refs. Machida *et al.*, 2005; Ajiro *et al.*, 2001.

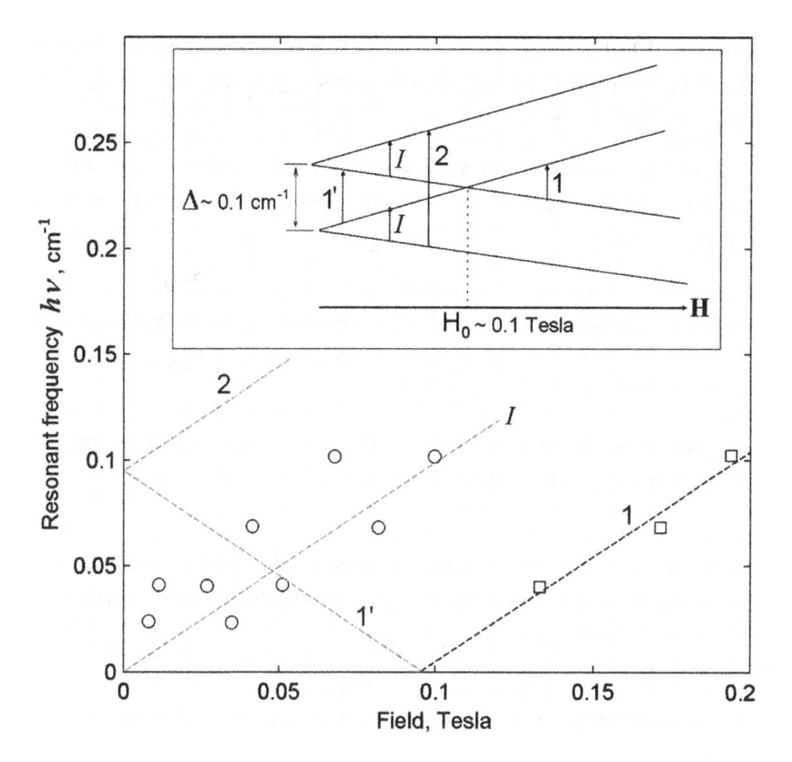

Fig. 14. Experimental data (Kajiyoshi *et al.*, 2007) on resonance peaks in EPR of V_{15} (circles and squares), parallel field. Dashed lines: resonant dependencies $h\nu$ versus field H for different transitions assuming a zero-field splitting gap $\Delta \approx 0.1\,cm^{-1}$ Inset: the low-lying Kramers doublets and EPR transitions.

Table 2. Experimental data on g-factors of V_{15}.

Reference	$g_{\parallel}$	$g_{\perp}$
(Gatteschi *et al.*, 1991; Barra *et al.*, 1992)	1.98	1.95
(Ajiro *et al.*, 2001)	1.98	1.968
(Vongtragool *et al.*, 2003)	1.981	1.953

Analysis of these results confirms the fact that the ground state consists of two Kramers doublets. Refs. Gatteschi *et al.*, 1991; Barra *et al.*, 1992; Ajiro *et al.*, 2001; Vongtragool *et al.*, 2003 provide measurements of the anisotropy of the g-factor (Table 2) at ultra-low temperatures. These data allow to estimate the strength of the AS exchange. One can estimate the normal component of AS exchange which is operative at low fields that are much below the point of crossover $S=1/2 \leftrightarrow S=3/2$ (Secs. 4, 6.1). An approximate estimate for an effective AS exchange parameter is given by (Moriya, 1960): $D \approx J(\Delta g/g_{av})$ where $\Delta g = g_{\|} - g_{\perp}$ and $g_{av} = (1/3)g_{\|} + (2/3)g_{\perp}$. Taking, for example, measurements of g from Ref. Vongtragool *et al.*, 2003 and $J = 0.847\,\mathrm{cm}^{-1}$ (Barbara, 2003), one obtains $D_n \approx 0.01\,\mathrm{cm}^{-1}$ which is smaller but of the same order as the estimation of D_n in Refs. Tarantul *et al.*, 2006; Tarantul *et al.*, 2007.

The anisotropy of g-factor is found to be temperature dependent (Ajiro *et al.*, 2001). This remarkable observation is probably related to the fact that at higher temperatures the spins of the hexagons become unpaired and contribute to the total g-factor of the cluster. An interesting phenomenon occurs at room temperature when the anisotropy of g changes the sign ($g_{\perp} > g_{\|}$).

The transmission spectra measured with the novel efficient technique of frequency swept magnetic resonance spectroscopy (FSMRS) are reported in Ref. Vongtragool *et al.*, 2003. This experiment demonstrates strong asymmetry of the spectral line at high fields (much larger than the crossover field) that can be explained by the anisotropy of the g-factor. This asymmetric shape can also be explained by the asymmetric location and unequal intensities of the sideband resonant peaks 2 and 5, Fig. 11(c). No transitions have been observed (Vongtragool *et al.*, 2003) at zero field in the range 0.15–1 cm^{-1}. This fact raises additional questions about the actual value of ZFS (Sec. 9.1). No $S=1/2 \leftrightarrow S=3/2$ transitions are reported in Refs. Vongtragool *et al.*, 2003; Gatteschi *et al.*, 2006. Such transitions are partially allowed only due to the *in-plane* component of AS exchange and since this part of AS exchange produces weak effect in a wide range of the field except for the vicinity of the crossover point, one should not expect to observe these transitions.

6. Static Magnetization

6.1. *The theoretical model*

In the important particular case when the in-plane part of AS exchange is absent ($D_n \neq 0$, $D_l = D_t = 0$) the special symmetry properties of the matrix of the AS exchange allow to find the exact solution for the Zeeman energy levels:

$$\begin{aligned}\varepsilon_1 &= \varepsilon_3 = -(3/2)J - (1/2)\sqrt{(g\beta H)^2 + 3D_n^2},\\ \varepsilon_2 &= \varepsilon_4 = -(3/2)J + (1/2)\sqrt{(g\beta H)^2 + 3D_n^2},\\ \varepsilon_{5,6} &= (3/2)J \mp (1/2)g\beta H, \quad \varepsilon_{7,8} = (3/2)J \mp (3/2)g\beta H.\end{aligned} \tag{15}$$

In Eq. (15) H is the field in any direction in the plane, let us say $H = H_X$ and, correspondingly, the g-factor is $g \equiv g_\perp$. The levels ε_i with $i = 1, 2, 3, 4$ are related to $S = 1/2$ while $i = 5, 6, 7, 8$ are the numbers of Zeeman sublevels for $S = 3/2$ as shown in Fig. 15(a) in the case of the isotropic model. The energy pattern for the case $D_n \neq 0$, $D_l = D_t = 0$ is shown in Fig. 15(b). Three peculiarities of the energy pattern that are closely related to the magnetic behavior should be noted: (1) the ground state involving two $S = 1/2$ levels shows zero-field splitting into two Kramers doublets separated by the gap $\Delta = \sqrt{3}D_n$; (2) at low fields $g\beta H \leq \Delta$ the Zeeman energies are double degenerate and show quadratic dependence on the

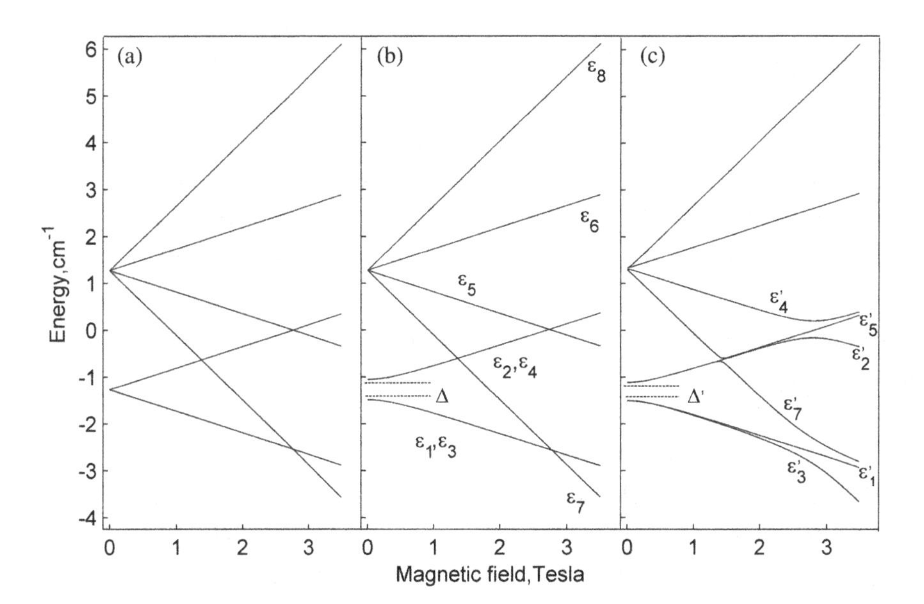

Fig. 15. Energy pattern of the triangular vanadium unit in the magnetic field ($H \perp C_3$), $J = 0.847\,\text{cm}^{-1}$ (a) $D_n = 0$, $D_\perp = 0$; (b) $D_n = 0.3\ J$, $D_\perp = 0$; (c) $D_n = 0.3\ J$, $D_\perp = 0.6\,J$.

field:

$$\begin{aligned} \varepsilon_1 &= \varepsilon_3 = -\sqrt{3}D_n/2 - (g\beta H)^2/4\sqrt{3}D_n, \\ \varepsilon_2 &= \varepsilon_4 = -\sqrt{3}D_n/2 + (g\beta H)^2/4\sqrt{3}D_n. \end{aligned} \tag{16}$$

This behavior is drastically different from that in the isotropic model and from the linear magnetic dependence in the parallel field and can be considered as a breaking of the normal AS exchange by the perpendicular field (Tsukerblat *et al.*, 1987); (3) the magnetic sublevels arising from $S = 3/2$ cross the sublevels belonging to $S = 1/2$ spin levels. At high field the levels $\varepsilon_{1,3}$ and $\varepsilon_{2,4}$ exhibit again linear magnetic dependence. One can see that a strong perpendicular field restores the linear behavior of Zeeman levels but without ZFS, that is the perpendicular field reduces the normal part of AS coupling. This effect of this reduction was understood in the first studies of the AS exchange in spin frustrated triangular systems (Tsukerblat *et al.*, 1987).

When the AS exchange in its general form is involved, the energy pattern shows the new peculiarities (Fig. 15(c)). The low field part of the spectrum is not affected by the in-plane part of AS exchange and is very close to that in Fig. 15(b) due to the fact that the in-plane part of the AS exchange is not operative within the ground manifold and the effect of $S = 1/2 \rightarrow S = 3/2$ mixing is small at low fields due to the large gap $3J \gg |D_\perp|$. At the same time in the vicinity of the crossing points the effect of the normal AS exchange is negligible but the in-plane AS exchange acts as a first order perturbation giving rise to the avoided crossing as shown in Fig. 15(c). Let us note that at low fields that are far from the anticrossing region the energies can be well described by Eqs. (16) even in the general case of AS exchange, although they are deduced for a particular model when $D_n \neq 0$, $D_l = D_t = 0$. In order to obtain three closely spaced low lying levels in the region of the anticrossing field it is reasonable to use the perturbation theory for the in-plane part of AS exchange and the basis formed by three eigen-functions of H_0 whose eigen-values have a crossing point at $g\beta H = 3\,J$, namely, $|(0)1/2, -1/2\rangle$, $|(1)1/2, -1/2\rangle$, $|(1)3/2, -3/2\rangle$ (Fig. 15(a)).

The secular equation can be solved due to some additional symmetry. The following approximate expressions were found (Tarantul *et al.*, 2006; Tarantul *et al.*, 2007) for the energy levels ε_i' in this region of the field:

$$\varepsilon_1' = -\frac{3}{2}J - \frac{1}{2}g\beta H, \tag{17}$$

$$\varepsilon_{3,7}' = -g\beta H \mp \frac{1}{8}\sqrt{(4g\beta H - 12J)^2 + 18D_\perp^2} \tag{18}$$

These expressions provide a rather good accuracy and will be used in the description of the magnetization of V_{15} vs. field and temperature. Let us consider some general features of the field dependence of magnetization related to the AS exchange by plotting the results of sample calculations. The most spectacular is the low temperature limit for which one can find the following expressions for magnetization (per molecule) that work well at low field (Eq. (19)) and at high field in the vicinity of anticrossing of the low lying levels (Eq. (20) (see Ref. Tarantul *et al.*, 2007).

$$\mu(H) = \frac{g^2\beta^2 H}{2\sqrt{3D_n^2 + g^2\beta^2 H^2}}, \tag{19}$$

$$\mu(H) = g\beta + \frac{2g\beta(g\beta H - 3J)}{\sqrt{2(g\beta H - 3J)^2 + 18D_\perp^2}}. \tag{20}$$

Figure 16 (dashed line) shows $\mu(H)$ dependence in the framework of the isotropic model only and when AS exchange is taken into account too. This figure indicates also the regions of the field for which each component of the AS exchange produces the most significant effect on magnetization. The normal part of the AS exchange results in the broadening of the low field step in $\mu(H)$ as shown by the solid line. This broadening is closely related to the quadratic Zeeman effect at low perpendicular field (see Fig. 15(b)). The in-plane part of the AS exchange leads to the broadening of the second step. When anticrossing in the region $H = 3J/g\beta$ appears due to coupling of $S = 1/2$, $M_S = -1/2$ $S = 3/2$, $M_S = -3/2$ levels through in-plane AS exchange, it obviously gives rise to a smooth switch from $S = 1/2$ to $S = 3/2$ as shown in Fig. 16.

6.2. *Discussion of the experimental magnetization data*

The low-temperature adiabatic magnetization vs. field applied in the plane of the V_3 triangle ($\boldsymbol{H} \perp C_3$) exhibits steps whose broadening and shapes

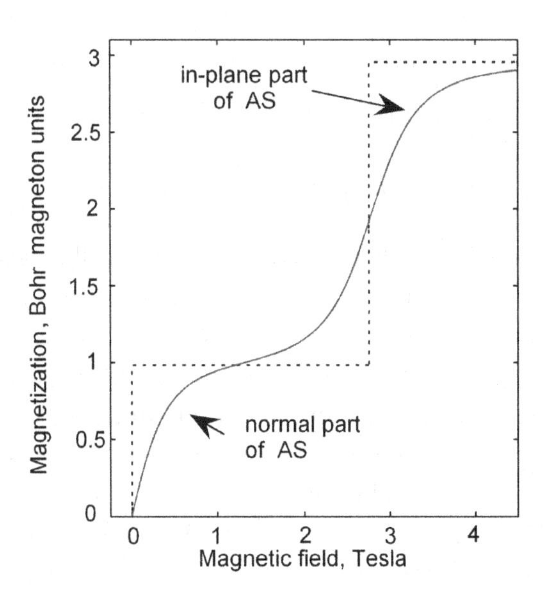

Fig. 16. Magnetization at $T = 0$ in perpendicular field. Dashed line: $J = 0.847\,\text{cm}^{-1}$, $D_n = 0$, $D_\perp = 0$; Solid line: $J = 0.847\,\text{cm}^{-1}$, $D_n = 0.3\,\text{J}$; $D_\perp = 0.6\,\text{J}$.

are temperature dependent (Figs. 16 and 17). Analysis of the experimental data in (Chiorescu *et al.*, 2000) has been performed in the framework of the HDVV model supplemented by a small quadrupolar anisotropy ($J_{XX} = J_{YY} \neq J_{ZZ}$). Agreement between the calculated curves and experimental data proved to be quite good for $T = 0.9$ K and 4.2 K but the low temperature data require a more comprehensive explanation. Modeling of the magnetization curves with consideration of the AS exchange (Tarantul *et al.*, 2006; Tarantul *et al.*, 2007) gives perfect agreement for the whole range of temperatures. Such a modeling allowed to estimate the AS exchange parameters. The best fit gives the following set of parameters: $J = 0.855\,\text{cm}^{-1}$, $g = 1.94$, $D_\perp = 0.238\,\text{cm}^{-1}$, $D_n = 0.054\,\text{cm}^{-1}$. The calculated curves and experimental data are in a full agreement in the whole range of the temperature and field (Fig. 17).

The ZFS obtained from these data, $\Delta = \sqrt{3}D_n = 0.09\,\text{cm}^{-1}$ proves to be in a good agreement with some other estimations (Sec. 9.1). It should be noted that the accuracy of the fit for the low fields (where D_n is more important) is lower than that for the high field magnetization (where the effect of $D_\perp$ dominates). For this reason the inclusion of the parameter η would lead to an excessive flexibility of the model.

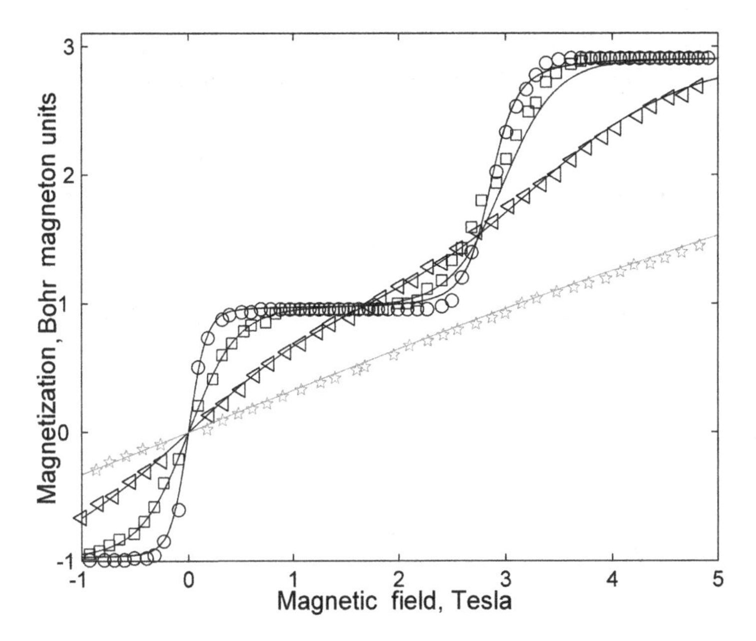

Fig. 17. Experimental data (from Ref. Chiorescu *et al.*, 2000) and theoretical curves of static magnetization with account for the AS exchange ($\boldsymbol{H} \perp C_3$); Experimental data $T = 0.1$ K (circles), $T = 0.3$ K (squares), $T = 0.9$ K (triangles), $T = 4.2$ K (stars). Solid lines: calculated.

High field properties of V_{15} are studied in Refs. Platonov *et al.*, 2002; Kostyuchenko and Popov, 2008; Zvezdin *et al.*, 2001; Mischenko *et al.*, 2003. In the crossover regions corresponding to $S = 3/2 \leftrightarrow S = 5/2$, $S = 5/2 \leftrightarrow S = 7/2$ etc. the model of spin triangle fails and the full spectrum of the system should be taken into account. By comparing the theoretical models with the measurements that were actually performed at ultra-high fields, the authors of Refs. Platonov *et al.*, 2002; Kostyuchenko and Popov, 2008 obtained the estimations for the isotropic exchange parameters (sets (III), (IV), Table 1, Sec. 2).

7. Dynamic Properties, Relaxation, Spin Dynamics

7.1. *Relaxation mechanisms and magnetic hysteresis*

The static magnetic properties reflect the structure of the spin levels while the dynamical properties provide an important information about the relaxation mechanisms in spin system. In a sweeping magnetic field the V_{15} cluster shows a hysteresis loop of magnetization whose form depends on the sweeping rate and the temperature. The experimental data on the hysteresis curves and relaxation times τ_H are given in Refs. Chiorescu *et al.*, 2000; Chiorescu *et al.*, 2000; Barbara *et al.*, 2000 (see Fig. 18). The hysteresis measurements were confined to the low-field area 0–0.7 *Tesla*. The authors proposed a comprehensive theoretical explanation based on the Landau-Zener (Landau and Lifshitz, 1966) two-level model (Fig. 18(c)) with zero-field gap Δ.

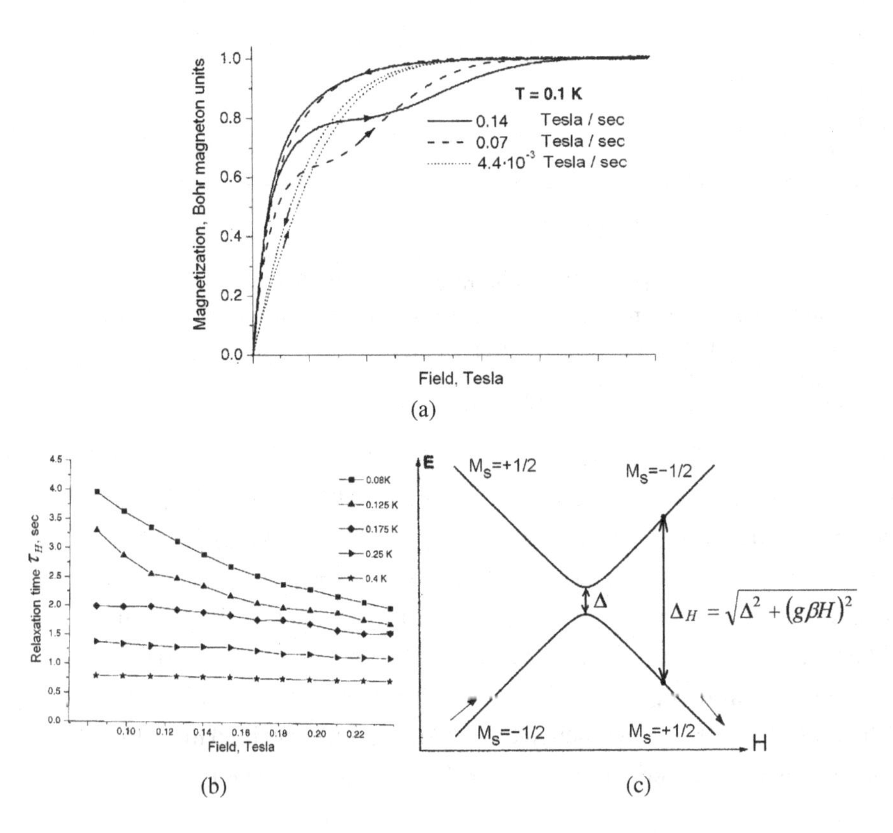

Fig. 18. Dynamical magnetic properties of V_{15} (modified from Ref. Chiorescu *et al.*, 2000). (a) Magnetic hysteresis loops in the vanadium cluster at temperature 0.1 K and different sweeping rates of magnetic field. (b) Relaxation times as a function of field at different temperatures. (c) Two-level energy scheme with anticrossing point and zero-field gap Δ to derive the theoretical model used in Refs. Chiorescu *et al.*, 2000; Chiorescu *et al.*, 2000; Barbara *et al.*, 2000; Chiorescu *et al.*, 2003; Barbara *et al.*, 2002.

The key point is that the thermal contact between the cryostat (whose temperature is *T*) and the V_{15} specimen is not sufficiently good to ensure the fast thermal equilibrium for the system. The specimen is then to be treated as *non-equilibrium* phonon bath. By solving the non-equilibrium rate equations (Abragam and Bleaney, 1970) for the phonon-induced transition $|1/2, 1/2> \leftrightarrow |1/2, -1/2>$ it was shown that at any field H the number of phonons having energy equal to the gap Δ_H (Fig. 18(c)) is different from that in the thermodynamic equilibrium at temperature T. In other words, the *phonon temperature* T_{ph} is different from that of the cryostat. The *spin temperature* T_S for the pair of levels $|M_S = +1/2\rangle$ and $|M_S = -1/2\rangle$ is defined from the ratio

$$\frac{n(|M_S = -1/2\rangle)}{n(|M_S = +1/2\rangle)} = \exp\left(-\frac{\Delta_H}{kT_s}\right), \tag{21}$$

where k is the Boltzmann constant. Providing fast sweeping field, all three temperatures T, T_{ph} and T_S are different. When the field H begins to rise from $H = 0$ in a positive direction, $T_S > T$ and the ground state $|M_S = +1/2\rangle$ is *underpopulated* compared to its equilibrium population. So, when the field is swept to the right, most of the dynamical magnetization curve lies below the static one. However, when the field sweeps in the opposite direction, the ground state $|M_S = +1/2\rangle$ is strongly *overpopulated*, so this part of the hysteresis loop lies above the static magnetization curve. The overpopulation results from a strong deficiency in phonons that could excite the cluster to jump to the upper level; such a lack of phonons is known as a 'phonon bottleneck' (Ref. Abragam and Bleaney, 1970, Sec. 10).

In the case of a relatively low rate of relaxation the system proves to be out of the thermodynamic equilibrium. The hysteresis loop in a sweeping field can be described by the Landau-Zener theory. The higher the sweeping rate, the more difficult for the phonon ensemble to adjust itself to that fast change, so the depth of the loop is increasing. The authors of the Refs. Chiorescu *et al.*, 2000; Chiorescu *et al.*, 2000; Barbara *et al.*, 2000 fitted their numeric simulations to both hysteresis curves and relaxation times measurements. For the ZFS gap Δ the hysteresis data exhibits a value about 50 mK (0.03 cm^{-1}) while the relaxations times provide the higher numbers of 200–300 mK (0.14–0.21 cm^{-1}). These large values are closer to the estimations made in the INS measurements (Chaboussant *et al.*, 2002; Chaboussant *et al.*, 2004; Chaboussant *et al.*, 2004; Chaboussant *et al.*, 2004) (see Table 3 in Sec. 9.1).

In further development (Chiorescu *et al.*, 2003; Barbara *et al.*, 2002) the experiments in the two different regimes are performed: non-adiabatic, with some extent of thermal coupling between the specimen and environment, and with the thermally isolated sample. In the second case, contrary to the non-adiabatic regime, the hysteresis was observed at *slow* sweeping rates. However, at high rates the *adiabatic Landau-Zener transitions* lead to the reversible magnetization due to insufficiency of phonons. These experiments are important as an instrument to control the spin-lattice relaxation times (along with the chemical design and other means to control other sources of decoherence, Sec. 7.3 and references therein). In regard to the zero-field gap value Δ, it is ultimately estimated from the model fits to be 80 mK (0.06 cm^{-1}).

The discrepancy between the measured and calculated relaxation times at low fields was noted in Refs. Chiorescu *et al.*, 2003; Barbara *et al.*, 2002. The real relaxation rates are shown to be a few times larger then the estimated ones. The conclusion is made that the spin-phonon interaction is not the main trigger of relaxation and therefore the authors pay special attention to the interaction between the nuclear spins and the spins of the vanadium ions which quickly fluctuate due to the frustration of the system's ground state.

It should be mentioned that although the theoretical model used in Refs. Chiorescu *et al.*, 2000; Chiorescu *et al.*, 2000; Barbara *et al.*, 2000; Chiorescu *et al.*, 2003; Barbara *et al.*, 2002 provides a very good agreement with the experiment and physical background for the dynamical behavior of V_{15}, the actual low-lying energy spectrum of V_{15} consists of two Kramers doublets and consequently is more complicated then simplified two-level pattern. Therefore there is a need for a more detailed model for the study of the dynamical properties of V_{15}.

The different possible sources of decoherence in transitions between the states $M_S = \pm 1/2$ of ground-state doublets, namely, direct spin-phonon relaxation, two-phonon Raman process, the Orbach process via excited

$S = 3/2$ level, hyperfine field of fluctuating nuclear spins and intercluster dipolar interactions have been considered in Ref. Dobrovitski *et al.*, 2000. The coherence times in all these processes have been numerically estimated and in agreement with conclusions of Refs. Chiorescu *et al.*, 2003; Barbara *et al.*, 2002 it was assumed that the spin-lattice relaxation (of any particular mechanism) is not the fastest one. The role of the leading decoherence process is attributed to the fluctuating field created by the intercluster dipolar interactions. This process got especially comprehensive theoretical treatment. It turns out, however, that even the dipolar interactions provide a rather low decoherence rate. In fact, about five million tunneling events occur before the coherence is destroyed.

Unlike previously mentioned sources, Ref. Nojiria *et al.*, 2004 presents the measurement and discussion of that part of hysteresis in the V_{15} cluster which occurs at higher fields and is driven by Landau-Zener transitions in the $M_S = 1/2 \rightarrow M_S = 3/2$ crossover point. Some additional theoretical analysis of Landau-Zener transitions under a sweeping magnetic field can be found in Refs. Miyashita, 1996; Miyashita, 1995; Miyashita and Nagaosa, 2001.

Alternative models of the dynamical magnetization properties which do not assume the concept of a “phonon bottleneck” are given in Refs. Mischenko *et al.*, 2003; Rudra *et al.*, 2001. The magnetization as a response to a sweeping magnetic field in the thermal environment is investigated in Ref. Saito and Miyashita, 2001 by the use of quantum master equations. The authors came to conclusion that the magnetic plateau appears with almost no connection to the properties of a heat bath in the quasiadiabatic transitions and therefore the emergence of that plateau is quite universal. This phenomena was called ‘magnetic Foehn effect’ (see the detailed discussion in Ref. Saito and Miyashita, 2001).

7.2. *Spin dynamics in the muon scattering experiment*

In the muon scattering experiment reported in Ref. Salman *et al.*, 2008 the positively charged polarized muons were implanted into the powder V_{15} sample before they underwent beta-decay into positrons and neutrinos. The description of the muon spin resonance (μSR) method is given in the book by Gatteschi *et al.*, 2006: "the muons are spin polarized with their spin antiparallel to the beam direction. The positrons are emitted preferentially along the muon spin direction. When the muon is implanted in the sample the presence of a transverse magnetic field causes the precession of the muons and a change of the spin polarization. Short-lived muons have no time to change their spin polarization, therefore the emitted positrons will be captured by the backward detector, while the long-lived muons will invert their polarization giving rise to positrons which are captured by the forward detector". The difference between the backward and forward positron emission intensities is described by the so-called asymmetry function $A(t)$ which is the following:

$$A(t) = A_0 \exp\left(-\sqrt{t/T_1}\right) + A_{Bg} \tag{22}$$

where A_0 is the initial asymmetry and A_{Bg} is the background signal.

The change in the spin polarization of muons is triggered by the interaction (either exchange or dipolar) of their spins with electronic spins in the crystal sample. So the asymmetry decay time T_1 in Eq. (22) is closely related to the exponential correlation functions of fluctuations of the electronic spins

$$\begin{aligned} \langle S_z(t)S_z(0)\rangle &= \frac{1}{3}S(S+1)\exp\left(-\frac{1}{\tau_z}\right) \\ \langle S_+(t)S_-(0)\rangle &= \frac{2}{3}S(S+1)\exp(i\omega_e t)\exp\left(-\frac{1}{\tau_\pm}\right) \end{aligned} \tag{23}$$

where τ_z and $\tau_\pm$ are the correlation times of the parallel and perpendicular (to the applied magnetic field) spin components, respectively, and $\omega_e = g\beta H$ is the electronic resonance frequency.

The measurements of the muons asymmetry decay rate not only provide information about the correlation times but also help to reveal out which kind of fluctuations, parallel or perpendicular, is prevailing. The result is that at low temperatures the fluctuations are predominantly perpendicular and have the correlation time of about 6 nsec. It is interesting that at very low temperatures this estimated correlation time is found to be almost temperature independent and can be traced down to millikelvin temperatures. Most probably the origin of these fluctuations is the coupling of V_{15} molecular spins to the nuclear spin bath (Salman *et al.*, 2008). The above mentioned value of the correlation time thus corresponds to a reasonable hyperfine level broadening of the order of a few tens of millikelvin.

7.3. *Rabi oscillations and implementation of molecular magnets in quantum computing*

An essential concept in the theory of quantum computation (Nielsen *et al.*, 2000) is the qubit (quantum bit) which is a coherent superposition of two quantum states. Last decade was notable for realizations of qubits based on such different physical systems like trapped cold ions (Schmidt-Kaler *et al.*, 2003; Leibfried *et al.*, 2003), quantum dots (Loss and DiVincenzo, 1998), semiconductors (Kane, 1998; Stoneham *et al.*, 2003) and doped fullerenes (Morton, 2006). In this regard a special attention has been paid to molecular magnets (Loss and Leuenberger, 2001; Stepanenko *et al.*, 2008; Troiani *et al.*, 2005; Affronte *et al.*, 2006; Affronte *et al.*, 2007; Carretta *et al.*, 2007; Winpenny, 2008; Ardavan *et al.*, 2007; Timco *et al.*, 2008; Lehmann *et al.*, 2007) that have been proposed as the leading candidates for use as nanoscale qubits. A potentiality of molecular magnets to be implemented in quantum computing is based on the following substantial advantages:

(a) unlike quantum dots the molecular magnets of a specified chemical composition are absolutely identical and therefore they have identical physical characteristics (energy levels, magnetic exchange parameters, etc.);
(b) by a proper chemical synthesis under a specific condition the magnetic clusters can be engineered to have a desired set of physical characteristics that can be controlled by the due choice of the metal ions, ligands, etc. Moreover, the dipolar coupling (a source of decoherence) can be suppressed through the chemical design (Bertaina *et al.*, 2008; Ardavan *et al.*, 2007);
(c) molecular magnets are relatively large (but still nanoscale!) objects, much larger than single ions and thus much easier for individual addressing for the input or read-out of quantum information;
(d) magnetic molecules can be attached to different types of platforms by grafting individual clusters on solid surfaces (see

Refs. Affronte *et al.*, 2006; Affronte *et al.*, 2007; Zobbi *et al.*, 2005; Condorelli *et al.*, 2004; Coronado *et al.*; Corradini *et al.* for realizations and Ref. Park and Pederson, 2005 for computer simulation), grafting monolayers (Cornia *et al.*, 2003; Fleury *et al.*, 2005; Salman *et al.*, 2007) or embedding the isolated molecules in the amorphous media (Bertaina *et al.*, 2008; Bogani *et al.*, 2007).

But the key question formulated by Winpenny and coworkers (Ardavan *et al.*, 2007) is the following: will decoherence times in molecular magnets permit quantum information processing? First prediction of the coherent states in V_{15} (based on the estimations of the relaxation rates) have been made in Ref. Dobrovitski *et al.*, 2000 where it was explicitly stated: '... quantum coherence in a V_{15} molecule is not suppressed and, in principle, can be detected experimentally.' Although this prediction has been made almost a decade ago, until recently such a coherent behavior was not observed experimentally. The evidence of coherent states in the molecular magnet V_{15} was reported in Ref. Bertaina *et al.*, 2008 in which the long living coherent Rabi oscillations (decoherence time of 18 μsec) have been observed and analyzed in the V_{15} nanomagnets, placed in a non-magnetic medium. This finding was considered (Stamp, 2008) as a "milestone on the road" towards processing of quantum information. Intercluster dipolar interactions that were found to be the most important source of decoherence (Dobrovitski *et al.*, 2000) was eliminated due to long distances between the V_{15} clusters. Actually the decoherence is attributed to the electron-nuclei spin-spin interaction (see also Ref. Wernsdorfer *et al.*, 2004 in which the nature of the line broadening is discussed on the base of the resonant photon absorption in the GHz range).

An important step has been made in Ref. Mitrikas *et al.*, 2008 where the observation of the Rabi oscillations in Fe_3 system (ground state spin $S = 1/2$) has been reported (phase memory time was found to be 2.2 μsec). The authors of Ref. Mitrikas *et al.*, 2008 have analyzed different types of the relaxation processes and came to a conclusion about dominant role of the Orbach type relaxation. Quantum coherence in a high spin (ferromagnetic) molecular nanomagnet $[Fe_4^{III}(acac)_6(Br\text{-}mp)_2]$, Fe_4 cluster, was proved to exist in Ref. Schlegel *et al.*, 2008. It was shown (Schlegel *et al.*, 2008) the coherence time can be dramatically increased by modification of the matrix in which the molecular nanomagnets are embedded.

Ref. Bertaina *et al.*, 2008 provides proposals for the additional quenching of the decoherence mechanisms. Detailed discussion and, in particular, more details regarding chemical design are given in Ref. Winpenny, 2008 that highlights the results of Ref. Bertaina *et al.*, 2008. A general conclusion of Ref. Bertaina *et al.*, 2008 is that one could reach, in principle, such a decoherence time in a molecular magnet that permits quantum information processing. It should be noted that the full discussion of the sources of decoherence in molecular magnets requires a more detailed study of spin-phonon interactions. In this view a microscopic consideration of spin-phonon coupling parameters starting with the first principles seems to be an appropriate route (see Ref. Tarantul and Tsukerblat, 2010).

8. Spin-vibronic Interaction

It was already mentioned in Sec. 3.2 that the orbital degeneracy of the spin frustrated $S = 1/2$ state makes it vulnerable to different kinds of small perturbations. Until now most attention was paid to the AS exchange, although impact of possible structural deformation has been mentioned in our discussion of EPR.

In this section let us discuss the spin-vibronic coupling that results from the modulation of exchange interactions by the molecule vibrations. This gives rise to the JTE that is a general feature of orbitally degenerate systems (see books (Bersuker and Polinger, 1989; Englman, 1972; Bersuker, 2006) that give a comprehensive presentation of the JTE).

8.1. *Hamiltonian of spin-vibronic coupling*

Let us start the discussion of the vibrational problem with a definition of the vibrational modes and the normal (or symmetry adapted) coordinates. Normal vibrations of the equilateral triangular unit involve a fully symmetric mode A_1 with the corresponding normal coordinate $Q_{A_1} \equiv Q_1$ and the double-degenerate E-type mode with normal coordinates $Q_{E_X} = Q_X$, $Q_{E_Y} = Q_Y$.

These normal coordinates are related to the Cartesian ones by the following transformations:

$$
\begin{aligned}
Q_1 &= \frac{1}{\sqrt{3}}\left[-\frac{1}{2}(\sqrt{3}X_1 + Y_1) + \frac{1}{2}(\sqrt{3}X_2 - Y_2)\right],\\
Q_x &= \frac{1}{\sqrt{3}}\left[-\frac{1}{2}(\sqrt{3}X_1 - Y_1) - \frac{1}{2}(\sqrt{3}X_2 + Y_2)\right],\\
Q_y &= \frac{1}{\sqrt{3}}\left[\frac{1}{2}(X_1 + \sqrt{3}Y_1) + \frac{1}{2}(X_2 - \sqrt{3}Y_2) - X_3\right].
\end{aligned} \tag{24}
$$

The atomic displacements are shown in Fig. 19. The vibronic interaction arises from the modulation of the isotropic and AS exchange interactions by the molecular displacements. In fact, the exchange parameters are the functions of the interatomic distances so the linear (with respect to Qs) terms of the vibronic Hamiltonian can be represented as:

$$
H_{ev} = 2\sum_{ij} \boldsymbol{S}_i\boldsymbol{S}_j \sum_{\alpha\,=\,1,x,y} \left(\frac{\partial J_{ij}(R_{ij})}{\partial R_{ij}}\right)_{\Delta R_{ij}\,=\,0} \cdot \frac{\partial R_{ij}}{\partial Q_\alpha} Q_\alpha, \tag{25}
$$

$$
H'_{ev} = \sum_{ij} [\boldsymbol{S}_i \times \boldsymbol{S}_j] \sum_{\alpha\,=\,1,x,y} \left(\frac{\partial \boldsymbol{D}_{ij}(R_{ij})}{\partial R_{ij}}\right)_{\Delta R_{ij}\,=\,0} \cdot \frac{\partial R_{ij}}{\partial Q_\alpha} Q_\alpha \tag{26}
$$

Here the summation is extended over all pairwise spin-spin interactions ($ij = 12, 23, 31$). Equations (25) and (26) are the contributions to the overall vibronic coupling related to the isotropic and AS exchange interactions, respectively. After all required transformations one can arrive at the following vibronic Hamiltonian H_{ev} (see details in

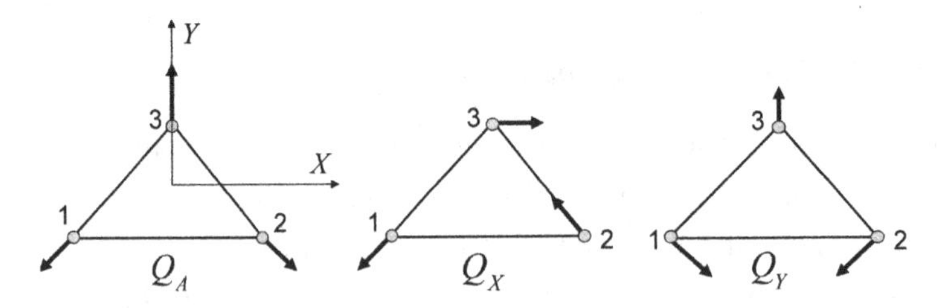

Fig. 19. Atomic displacements in the full symmetric (A_1) and double degenerate (E) modes of the triangular unit.

Ref. Tsukerblat *et al.*, 2007):

$$H_{ev} = \lambda(\hat{V}_1 Q_1 + \hat{V}_x Q_x + \hat{V}_y Q_y) \tag{27}$$

where $\lambda \equiv \sqrt{6}(\partial J_{ij}(R_{ij})/\partial R_{ij})_{\Delta R_{ij}=0}$ is the vibronic parameter associated with the isotropic exchange and the operators $\hat{V}_\alpha$ are the following (Bates and Jasper, 1971) (see Refs. Tsukerblat *et al.*, 1987; Tsukerblat and Belinskii, 1983 and references cited therein):

$$\begin{aligned}
\hat{V}_1 &= \sqrt{\frac{2}{3}}(\boldsymbol{S}_1\boldsymbol{S}_2 + \boldsymbol{S}_2\boldsymbol{S}_3 + \boldsymbol{S}_3\boldsymbol{S}_1), \\
\hat{V}_x &= \frac{1}{\sqrt{6}}(\boldsymbol{S}_2\boldsymbol{S}_3 + \boldsymbol{S}_3\boldsymbol{S}_1 - 2\boldsymbol{S}_1\boldsymbol{S}), \\
\hat{V}_y &= \frac{1}{\sqrt{2}}(\boldsymbol{S}_2\boldsymbol{S}_3 - \boldsymbol{S}_3\boldsymbol{S}_1).
\end{aligned} \tag{28}$$

By applying a similar procedure one can obtain the vibronic contribution associated with the AS exchange. The final expression is the following (Tsukerblat *et al.*, 2007):

$$H'_{ev} = \hat{W}_1 Q_1 + \hat{W}_x Q_x + \hat{W}_y Q_y. \tag{29}$$

The operators $\hat{W}_\alpha$ are expressed in terms of the vector products of spin operators:

$$\begin{aligned}
\hat{W}_1 &= \boldsymbol{\lambda}_{12}[\boldsymbol{S}_1 \times \boldsymbol{S}_2] + \boldsymbol{\lambda}_{23}[\boldsymbol{S}_2 \times \boldsymbol{S}_3] + \boldsymbol{\lambda}_{31}[\boldsymbol{S}_3 \times \boldsymbol{S}_1], \\
\hat{W}_x &= \frac{1}{2}(\boldsymbol{\lambda}_{12}[\boldsymbol{S}_1 \times \boldsymbol{S}_2] + \boldsymbol{\lambda}_{23}[\boldsymbol{S}_2 \times \boldsymbol{S}_3] - 2\boldsymbol{\lambda}_{31}[\boldsymbol{S}_3 \times \boldsymbol{S}_1]), \\
W_y &= \frac{\sqrt{3}}{2}(\boldsymbol{\lambda}_{23}[\boldsymbol{S}_2 \times \boldsymbol{S}_3] - \boldsymbol{\lambda}_{31}[\boldsymbol{S}_3 \times \boldsymbol{S}_1]).
\end{aligned} \tag{30}$$

In Eqs. (30) the values $\boldsymbol{\lambda}_{ij}$ are the vector coupling parameters defined as $\boldsymbol{\lambda}_{ij} \equiv (\partial \boldsymbol{D}_{ij}(R_{ij})/\partial R_{ij})_{\Delta R_{ij}=0}$. Under the condition of trigonal symmetry there are three parameters, namely, the normal part $\lambda_n = \lambda_{ijn}$ and two perpendicular contributions $\lambda_t = \lambda_{ijt}$ and $\lambda_l = \lambda_{ijl}$ where the symbols l and t have the same meaning as in the definition of the AS exchange.

The evaluation of the vibronic matrices can be performed with the aid of the ITO approach (Tsukerblat, 2006; Varshalovich *et al.*, 1988). With this aim each pairwise interaction can be expressed in terms of the zeroth order and first order tensorial products of ITO as:

$$
\begin{aligned}
(\boldsymbol{S}_i\boldsymbol{S}_j) &= -\sqrt{3}\left\{\boldsymbol{S}_i^{(1)} \times \boldsymbol{S}_j^{(1)}\right\}^{(0)},\\
\lambda_{ij}[\boldsymbol{S}_i \times \boldsymbol{S}_j] &= i\sqrt{2}\lambda_- e^{-i\phi}\left\{\boldsymbol{S}_i^{(1)} \times \boldsymbol{S}_j^{(1)}\right\}_1^{(1)} - i\sqrt{2}\lambda_+ e^{i\phi}\left\{\boldsymbol{S}_i^{(1)} \times \boldsymbol{S}_j^{(1)}\right\}_{-1}^{(1)}\\
&\quad - i\sqrt{2}\lambda_n\left\{\boldsymbol{S}_i^{(1)} \times \boldsymbol{S}_j^{(1)}\right\}_0^{(1)}
\end{aligned}
\tag{31}
$$

where $\left\{\boldsymbol{S}_i^{(1)} \times \boldsymbol{S}_j^{(1)}\right\}_m^{(k)}$ is the symbol of the tensor product (Varshalovich *et al.*, 1988), κ is the rank of tensor operator, m enumerates the component of spin ITOs that are related to the sites i and j and $\phi = 0, 2\pi/3, 4\pi/3$ for the sides 12, 23 and 31 of the triangle correspondingly, $\lambda_\pm = \mp(1/1\sqrt{2})(\lambda_l \pm i\lambda_t)$.

8.2. *Adiabatic surfaces*

In order to simplify the further consideration and to get a clear insight on the influence of the JT interaction on the magnetic properties one can assume that the gap $3J$ exceeds considerably the vibronic coupling and AS exchange and therefore the basis set can be reduced to the four low lying spin 1/2 states and the full symmetric mode Q_1 can be excluded. In this view one should note that the role of the A_1 mode is not a simple shift of the Q_1 coordinate. In fact, A_1 vibration is active in the pseudo JTE when a relatively small vibronic contribution of AS exchange is taken into account (a more detailed description will be given elsewhere). In the approximation so far assumed the matrix of the full Hamiltonian $H_{AS} + H_{ev} + H'_{ev} + H_{Zeeman}$ is given in Ref. Tsukerblat *et al.*, 2007. The four eigen-values of this matrix are found as:

$$
\begin{aligned}
&\varepsilon_{1,4}(\rho,\xi)\\
&\quad = \mp\frac{1}{2\sqrt{2}}\hbar\omega\sqrt{2\xi^2 + 2\delta^2 + 3\upsilon^2\rho^2 - 2\sqrt{2}\xi\sqrt{3\upsilon^2\rho^2 + 2\delta^2\cos^2\theta}}\\
&\varepsilon_{2,3}(\rho,\xi)\\
&\quad = \pm\frac{1}{2\sqrt{2}}\hbar\omega\sqrt{2\xi^2 + 2\delta^2 + 3\upsilon^2\rho^2 + 2\sqrt{2}\xi\sqrt{3\upsilon^2\rho^2 + 2\delta^2\cos^2\theta}}
\end{aligned}
\tag{32}
$$

The following dimensionless parameters are introduced: vibronic coupling parameter $\upsilon = (\lambda/\hbar\omega)(\hbar/M\omega)^{1/2}$, zero-field splitting of the ground state $\delta = \sqrt{3}D_n/\hbar\omega \equiv \Delta/\hbar\omega$, applied field $\xi = g\beta H/\hbar\omega$ and coordinates $q_\alpha = (M\omega/\hbar)^{1/2}Q_\alpha$, $H_z = H\cos\theta$. Finally, ρ is the radial component in the plane q_x, q_y defined, as usual, $q_x = \rho\cos\varphi$, $q_y = \rho\sin\varphi$. The adiabatic surfaces are axially symmetric (at an arbitrary direction of the applied field), and the C_3 axis complies with the symmetry of the AS exchange.

In the case of $\delta = 0$ and $\xi = 0$ one faces a two mode pseudo JT problem and obtains simple expressions for a pair of the double degenerate surfaces that are quite similar to that in the pseudo JT $^2E \otimes e$ problem with spin-orbital interaction:

$$U_\pm(\rho)/\hbar\omega = \rho^2/2 \pm (1/2)\sqrt{\delta^2 + 3\upsilon^2\rho^2/2} \tag{33}$$

One can see that in the limit of the isotropic exchange model the surface represents the so-called "Mexican hat" (Fig. 21, top) with the conical intersection at $\rho = 0$ that corresponds to the basic JT $E \otimes e$ problem (Bersuker and Polinger, 1989; Englman, 1972; Bersuker, 2006):

$$U_\pm(\rho)/\hbar\omega = \rho^2/2 \pm \left(\sqrt{3}/2\sqrt{2}\right)|\upsilon|\rho \tag{34}$$

This limiting case corresponding to the well known spin-phonon coupling Hamiltonian (Bates and Jasper, 1971) (see for details (Tsukerblat *et al.*, 1987; Tsukerblat and Belinskii, 1983) and references therein) has recently been considered again in Ref. Popov *et al.*, 2004. Figure 20 illustrates how the nuclear configurations are changed in the course of the rotation along the bottom of the lower sheet of the "Mexican hat"

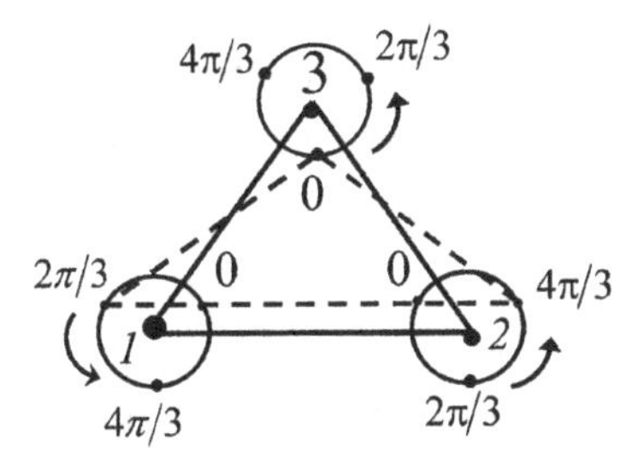

Fig. 20. Distortions of a triangular molecule in course of the rotation along the bottom of the trough of the lower sheet of the adiabatic surface. Instant distorted configurations are depicted by the dashed lines.

(Bersuker and Polinger, 1989; Englman, 1972; Bersuker, 2006). Each instant configuration represents an isosceles triangle while the atoms perform concerted movement along the circles with the phase shifts of $2\pi/3$.

In general, the shape of the surfaces depends on the interrelation between the AS exchange and vibronic coupling that prove to be competitive. In the case of a weak vibronic coupling and/or strong AS exchange $\upsilon^2 < 4|\delta|/3$ the lower surface possesses the only minimum at $q_x = q_y = 0 (\rho = 0)$ so that the symmetric (trigonal) configuration of the system proves to be stable. In the opposite case of a strong vibronic interaction and/or weak AS exchange, $\upsilon^2 > 4|\delta|/3$, the symmetric configuration of the cluster is unstable and the minima are disposed at the ring of the trough of the radius ρ_0:

$$\rho_0 = (1/2)\sqrt{3\upsilon^2/2 - 8\delta^2/3\upsilon^2} \qquad (35)$$

The radius ρ_0 decreases with increasing AS exchange and vanishes at $|\delta| = 3\upsilon^2/4$. An example of this type of the pseudo JT surfaces is shown in Fig. 21, bottom. The depth of the minima ring in the second type (respectively to the top in the low surface) depends on the interrelation between the JT constant and AS exchange and is found to be

$$\varepsilon_0 = \left(3\upsilon^2 - 4\delta^2\right)^2/48\upsilon^2 \qquad (36)$$

while the gap between the surfaces in the minima points $3\upsilon^2/4$ is independent of the AS exchange. Figure 22 shows the case of a strong JT coupling resulting in the instability and the cross-section of the adiabatic potentials.

The results so far described allow to realize how the JTE is interrelated to spin frustration in a triangular unit (like a central triangle in V_{15}) with the orbitally degenerate ground state. Figure 23 illustrates three instant JT conformations and the corresponding networks of the exchange interactions that are assumed to be antiferromagnetic. For example, in the first configuration the side 12 is elongated while the sides 13 and 23 are compressed. Consequently, the exchange interaction between sites 1 and 2 are weakened, while in sides 13 and 23 the exchange is enhanced as shown in Fig. 23. This leads to a definite (rather than frustrated) spin alignment in which spin 1 and 2 are parallel while the pairs 13 and 23 remain antiferromagnetic. This

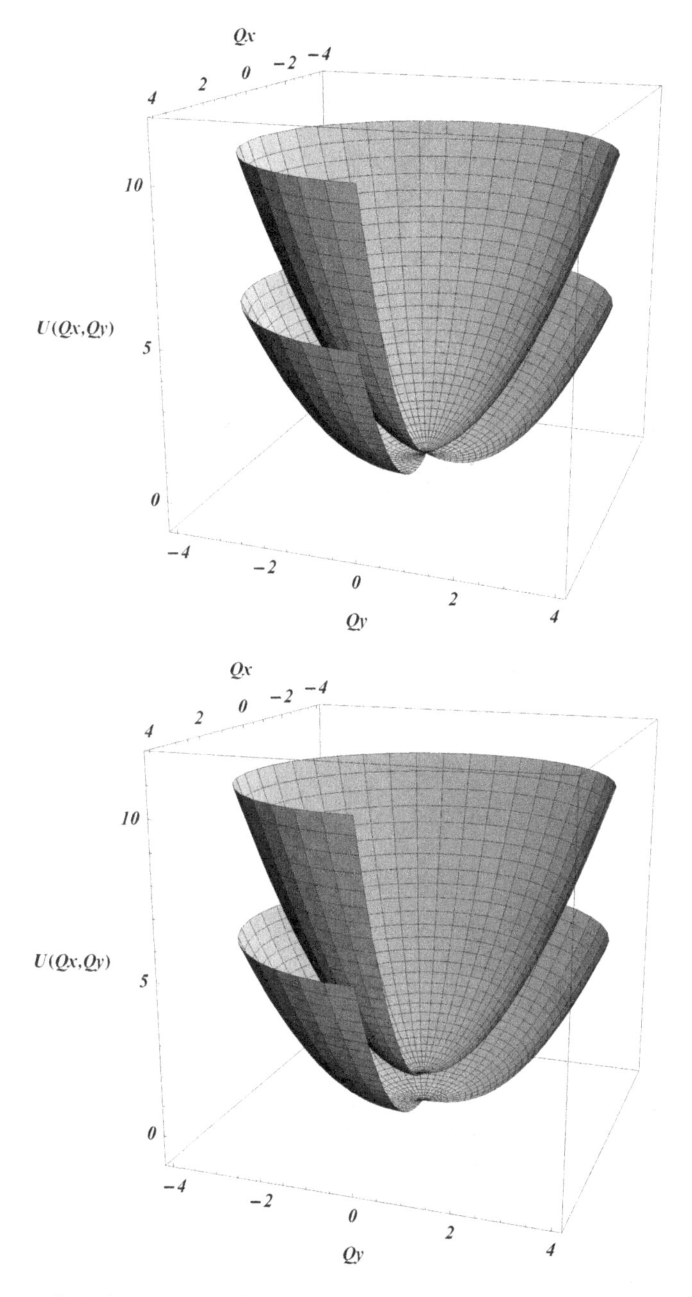

Fig. 21. Adiabatic potentials for the ground state of a triangular exchange system in the space of double degenerate vibrations: strong AS exchange and/or weak vibronic interaction $\delta = 0$, $\upsilon = 2.0$ (top); $\delta = 1.0$, $\upsilon = 1.0$ (bottom).

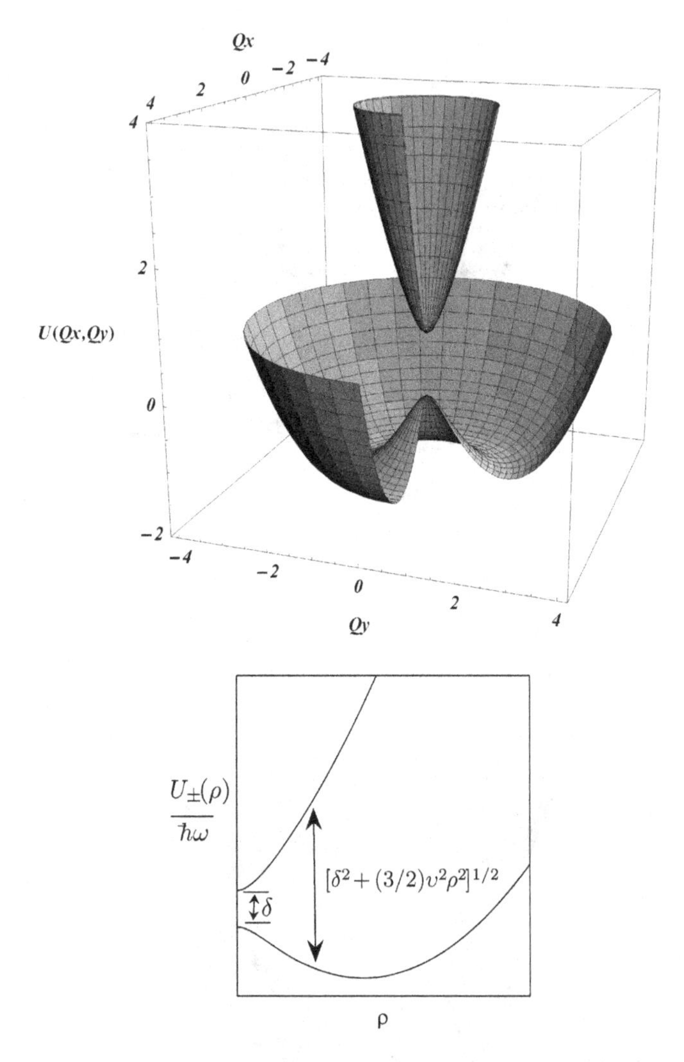

Fig. 22. Adiabatic potentials in the case of weak AS exchange and/or strong vibronic interaction, $\delta = 1.0$, $\upsilon = 3.0$ (top). Section of the adiabatic potentials in the case of JT instability, illustration for the zero-field splitting of the ground state in the vibronically distorted configurations (bottom).

proves to be the most favorable configuration since the energy gain in the pairs 13 and 23 is evidently larger than the energy loss in the pair 12. As one can see this conclusion is valid for all distorted configurations that are different in phases and intermediate the quantum numbers S_{ij}. One can finally

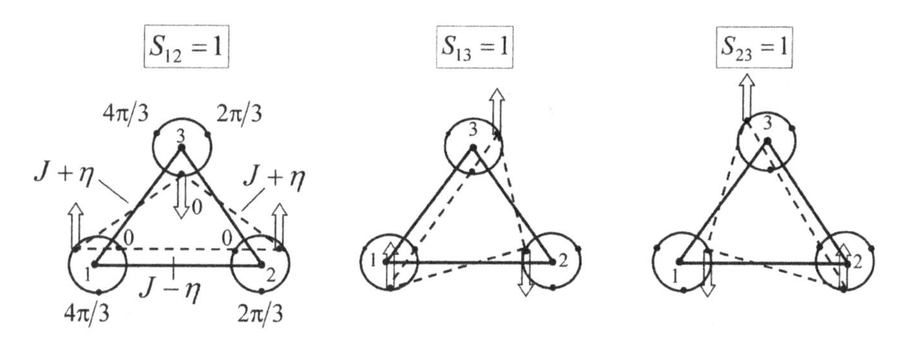

Fig. 23. Illustration for the suppression of spin frustration in a triangular spin cluster by the JT distortions. Instant distorted configurations are depicted by the dashed lines.

conclude that within each spin alignment of the JT conformations the spin frustration is eliminated due to dynamical JT distortions. This seems to be a common result that is applicable to a wide class of spin frustrated systems.

8.3. *Influence of the Jahn-Teller effect on the magnetization*

To clarify the physical consequences of JTE let us assume that the motion of the system is confined to the bottom of the trough. Strictly speaking this is valid providing strong JT coupling but in all cases it gives clear qualitative results and transparent key expressions. Providing $\rho = \rho_0' \equiv \sqrt{3/8}|\upsilon|$ (radius of the minima ring), the value $\sqrt{3\upsilon^2\rho^2/2}$ is simply the JT splitting $E_{JT} = 3\upsilon^2/4$, i.e. the gap between the surfaces in the minima points of the lower surface. The Zeeman sublevels in a weak field (defined by the angle θ) with an accuracy to the second order terms with respect to the field ξ can be found as:

$$
\begin{aligned}
\frac{\varepsilon_{1,3}(\xi)}{\hbar\omega} &= -\frac{1}{2}\delta(\rho_0') \pm \frac{1}{2}\kappa_1(\theta)\xi - \kappa_2(\theta)\xi^2, \\
\frac{\varepsilon_{2,4}(\xi)}{\hbar\omega} &= +\frac{1}{2}\delta(\rho_0') \pm \tfrac{1}{2}\kappa_1(\theta)\xi + \kappa_2(\theta)\xi^2.
\end{aligned}
\qquad (37)
$$

The eigen-values are denoted as $\varepsilon_i(\xi) \equiv \varepsilon_i(\rho_0', \xi)$ and the van Vleck coefficients $\kappa_1(\theta)$ (first coefficient) and $\kappa_2(\theta)$ (second coefficient) (Kahn, 1993) can be directly related to the JT splitting and AS exchange by the following expressions:

$$\begin{aligned}\kappa_1(\theta) &= \sqrt{\frac{E_{JT}^2+\delta^2\cos^2\theta}{E_{JT}^2+\delta^2}},\\ \kappa_2(\theta) &= \frac{\delta^2\sin^2\theta}{4(E_{JT}^2+\delta^2)^{3/2}}\end{aligned}\tag{38}$$

In order to reveal in more details the influence of the JT coupling on the anisotropic properties caused by the AS exchange let us consider the effects of JT coupling in the two principal directions of the magnetic field. In the case of parallel field ($\boldsymbol{H}\|C_3$) one finds that $\kappa_1(0)=1$ and $\kappa_2(0)=0$ so that one obtains a linear Zeeman splitting in a pair of spin doublets in the parallel field. The zero-field splitting (gap between spin doublets) is found to be $\sqrt{E_{JT}^2+\delta^2}$ that is a combined gap comprising contributions of the JT energy ΔE_{JT} and the AS exchange δ. In the case of the perpendicular field one obtains:

$$\begin{aligned}\frac{\varepsilon_{1,3}(\xi)}{\hbar\omega} &= -\frac{1}{2}\sqrt{E_{JT}^2+\delta^2} \pm \frac{E_{JT}}{\sqrt{E_{JT}^2+\delta^2}}\frac{1}{2}\xi - \frac{\delta^2}{4\left(E_{JT}^2+\delta^2\right)^{3/2}}\xi^2\\ \frac{\varepsilon_{2,4}(\xi)}{\hbar\omega} &= +\frac{1}{2}\sqrt{E_{JT}^2+\delta^2} \pm \frac{E_{JT}}{\sqrt{E_{JT}^2+\delta^2}}\frac{1}{2}\xi + \frac{\delta^2}{4\left(E_{JT}^2+\delta^2\right)^{3/2}}\xi^2\end{aligned}\tag{39}$$

Equations (3.39) show that the Zeeman pattern contains both linear and quadratic contributions. The role of the JT coupling can be understood by comparing the Zeeman picture so far obtained with that at $\upsilon=0$. It is important that in the absence of the JT coupling the linear Zeeman terms disappear and the Zeeman energies contain only quadratic terms (with respect to the field). Thus, Fig. 24(a) illustrates two degenerate pairs of the Zeeman levels in a perpendicular field in the symmetric nuclear configuration. In a weak field range they are given by:

$$\begin{aligned}\frac{\varepsilon_1(\xi)}{\hbar\omega} &= \frac{\varepsilon_3(\xi)}{\hbar\omega} = -\frac{|\delta|}{2} - \frac{\xi^2}{4|\delta|}\\ \frac{\varepsilon_2(\xi)}{\hbar\omega} &= \frac{\varepsilon_4(\xi)}{\hbar\omega} = +\frac{|\delta|}{2} + \frac{\xi^2}{4|\delta|}\end{aligned}\tag{40}$$

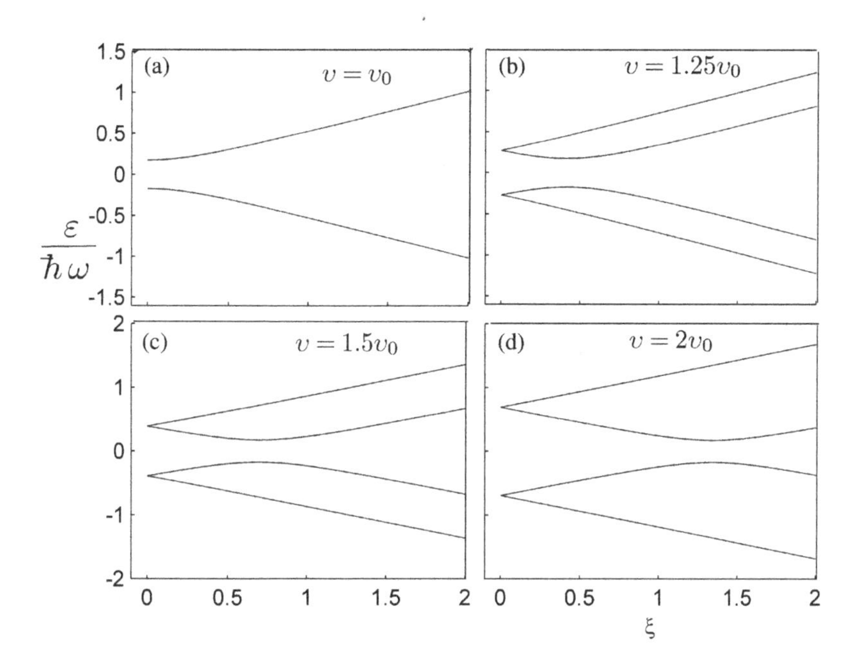

Fig. 24. Influence of the JT interaction on the Zeeman energy pattern in a perpendicular field. (a) $\upsilon^2 \leq \upsilon_0^2 \equiv 4|\delta|/3$, Jahn-Teller interaction is suppressed by AS exchange (b)–(d) influence of the υ/υ_0 on the energy levels.

This can be referred to as the effect of reduction of the magnetization in low magnetic field that is perpendicular to the axis of AS exchange (Tsukerblat *et al.*, 1987; Tsukerblat and Belinskii, 1983). The reduction of the Zeeman energy by the AS exchange gives rise to a small van Vleck type contribution to the magnetic susceptibility at low field $g\beta H \ll D_n$. An essential effect is that the JT interaction leads to the occurrence of the linear terms for the Zeeman energies at low field. This is shown in Fig. 24 that illustrates transformation of the Zeeman levels under the influence of the vibronic coupling obtained with the aid of the general Eqs. (32).

Figure 25 illustrates the influence of JT on the field dependence of the magnetization of a triangular unit that is closely related to the influence of the vibronic coupling on the Zeeman pattern (Fig. 24). The magnetization vs. perpendicular field at $T = 0$ is presented as a function of the vibronic coupling parameter υ that is assumed to satisfy the condition of instability $\upsilon^2 > \upsilon_0^2 \equiv 4|\delta|/3$. One can see that providing $\upsilon = \upsilon_0$ (and of course $\upsilon < \upsilon_0$ that corresponds to a symmetric stable configuration) the magnetization

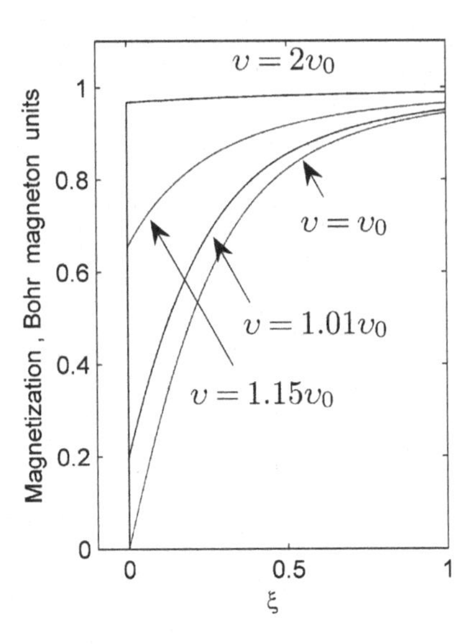

Fig. 25. Influence of the JT interaction on the dependence of magnetization vs. perpendicular field.

slowly increases with the increase of the field (due to reduction of the Zeeman interaction in the low field) and then reaches saturation when the magnetic field is strong enough to break the AS exchange. Increase of the JT coupling leads to the fast increase of the magnetic moments in the region of low field and formation of the step in magnetization caused by the reduction of the magnetic anisotropy (appearance of the linear terms in the Zeeman levels). The height of the step depends on the interrelation between AS exchange and vibronic coupling and can be expressed as:

$$M(H=0) = \frac{g\beta}{2} \frac{E_{JT}}{\sqrt{E_{JT}^2 + \delta^2}} \tag{41}$$

The height of the step increases with the increase of the vibronic coupling. Finally, when the JT coupling is strong enough ($\upsilon = 2\upsilon_0$) one can observe a staircase like behavior of the magnetization with the sharp step in which $M(H)$ jumps from zero to $M(H=0) = g\beta/2$ at zero field (and $T=0$) that is expected for a magnetically isotropic system. The influence of the distortions caused by the JT instability is very pronounced so that the step starts to appear even when the JT coupling slightly exceeds its critical value υ_0 ($\upsilon = 1.01\upsilon_0$ in Fig. 25). Although the semiclassical description in this range of parameters loses its accuracy, the qualitative results are able to draw an adequate physical picture. More accurate quantitative results in this range of the vibronic coupling can be obtained by solving the dynamic pseudo JT problem.

In the case of a strong magnetic field a pseudo-degeneracy of the Zeeman levels appears in the vicinity of the crossover. The influence of JTE on the magnetic properties (Tsukerblat *et al.*, 2007) will be considered later on.

8.4. *Estimation of the vibronic parameters for V_{15}*

No direct experimental data on the vibronic parameters of V_{15} cluster are available, so their approximate estimation may be deduced by analyzing the information obtained for similar systems. According to Ref. White and Geballe, 1979, the dependence of J on the interatomic distance R for MnO and MnF_2 is well approximated by the law

$$J = AR^{-3\gamma_m} \approx AR^{-10} \tag{42}$$

where $\gamma_m \approx 10/3$ is the Grüneisen constant and A is specific for the given compound. Applying this equation to the V_{15} cluster, one can estimate $A = J_0 R_0$ (index 0 denotes the values corresponding to the case when the nuclei are not displaced) and thus, the isotropic spin-vibronic parameter:

$$\lambda = \sqrt{6}\left(\frac{\partial J_{ij}}{\partial R_{ij}}\right)_{\Delta R_{ij}=0} = -10\sqrt{6}AR_0^{-11} \approx -3\frac{\text{cm}^{-1}}{\text{Å}} \tag{43}$$

A similar idea of estimating vibronic parameter for V_{15} is used in Ref. Popov *et al.*, 2004.

Another possible estimation of λ dates back to Ref. (Bates and Jasper, 1971) (see also references therein) in which the ratio $(\partial J/\partial R)_{\Delta R=0}/J$ was found to lie in the range 10–30 Å^{-1} for the Ir ions in the ammonium chloroplatinate and ruby with different doping concentrations. Using this approach, one obtains a value for υ that is of the order of 10^{-2}. With this most optimistic estimation for υ and the value of D_n (Sec. 6.2) one can see that the inequality $\upsilon^2 \ll 4\sqrt{3}D_n/(\hbar\omega)$ holds. Therefore one can conclude that the JT coupling in V_{15} is relatively small and therefore is strongly suppressed by the AS exchange. Presently the information about the vibrational frequencies of V_{15}, in particular, about the JT active ones is very limited. In this respect the recent study of the dielectric properties of V_{15} crystals and the Raman spectra can be useful (Zipse *et al.*, 2005).

The results so far discussed are obtained in the framework of the adiabatic approximation when the full energy of the system is associated with the adiabatic potentials. In Ref. Popkov *et al.*, 2005 the authors reproduce the well-known solution of the dynamic $E \otimes e$ JT problem (see, for example, books (Bersuker and Polinger, 1989; Englman, 1972; Bersuker, 2006) and references therein) and evaluate the probabilities of the electric dipole transitions between the vibronic (mixed electron-vibrational) levels under the action of the electromagnetic field. This approach can be helpful for study of the low lying levels of V_{15}.

9. Role of Structural Deformations

9.1. *Zero-field splitting in a scalene triangular system*

The spin frustrated $S = 1/2$ ground-state is subjected to even small perturbations that are able to remove the orbital degeneracy. Until now only the AS exchange has been considered as a source of ZFS. At the same time a structural deformation of the cluster that gives rise to a deviation from the trigonal symmetry should also be taken into account. Such distortion can originate either from the water molecule located inside the cluster or water molecules that are present in the specimen (Chaboussant *et al.*, 2002; Chaboussant *et al.*, 2004). In a simple case of an isosceles triangle the scalene structural distortion appears as the additional term $2\eta \boldsymbol{S}_1 \boldsymbol{S}_2$ in the HDVV Hamiltonian. The mutual effect of the AS and such a distortion causes a ZFS gap $\Delta = \sqrt{4\eta^2 + 3D_n^2}$ between two Kramers doublets (Tsukerblat *et al.*, 1987). This gap was reported in some experimental works based on different techniques. The results are summarized in Table 3. One can see that different estimations give significantly different results. Although both perturbations, AS exchange and structural deformations give rise to a ZFS, their physical role is essentially different. In fact, the AS exchange is an anisotropic interaction, while the deformation does not break the isotropy of the HDVV model. In this view the interrelation between these two perturbations is an important issue in the description of the physical properties of V_{15}.

Table 3. Different estimations of the ZFS Δ in the V_{15} system.

Reference	Method	Δ mK	Δ cm^{-1}	Section
(Kajiyoshi *et al.*, 2007)	EPR	30	0.02	5.2
(Chiorescu *et al.*, 2003; Barbara *et al.*, 2002)	Hysteresis modeling	80	0.06	7.1
(Tarantul *et al.*, 2007)	Static magnetization	130	0.09	6.2
(Chiorescu *et al.*, 2003; Barbara *et al.*, 2002)	Magnetic relaxation times	200–300	0.14–0.21	7.1
(Chaboussant *et al.*, 2002; Chaboussant *et al.*, 2004; Chaboussant *et al.*, 2004; Chaboussant *et al.*, 2004)	Inelastic neutron scattering	300–400	0.21–0.28	9.2
(Vongtragool *et al.*, 2003)	EPR	*no* transition was found at zero field in the range 0.15–1 cm^{-1} (220–1,440 mK)		5.2

9.2. *Discussion of inelastic neutron scattering experiments*

INS spectra of the powder samples of V_{15} are studied in detail in Refs. Chaboussant *et al.*, 2002; Chaboussant *et al.*, 2004; Chaboussant *et al.*, 2004; Chaboussant *et al.*, 2004. The data include both the scattering spectra (energy transfer) and Q-dependence for some transitions (see Fig. 26 based on Refs. Chaboussant *et al.*, 2002; Chaboussant *et al.*, 2004; Chaboussant *et al.*, 2004; Chaboussant *et al.*, 2004). Both structural deformation and AS exchange are mentioned in Refs. Chaboussant *et al.*, 2002; Chaboussant *et al.*, 2004 as a possible origin of ZFS, while according to Refs. Chaboussant *et al.*, 2004; Chaboussant *et al.*, 2004 only a scalene distortion is present in V_{15}. A comprehensive analysis of the EPR and INS experiments on the frustrated trinuclear clusters including V_{15} is given in Ref. Belinsky, 2009.

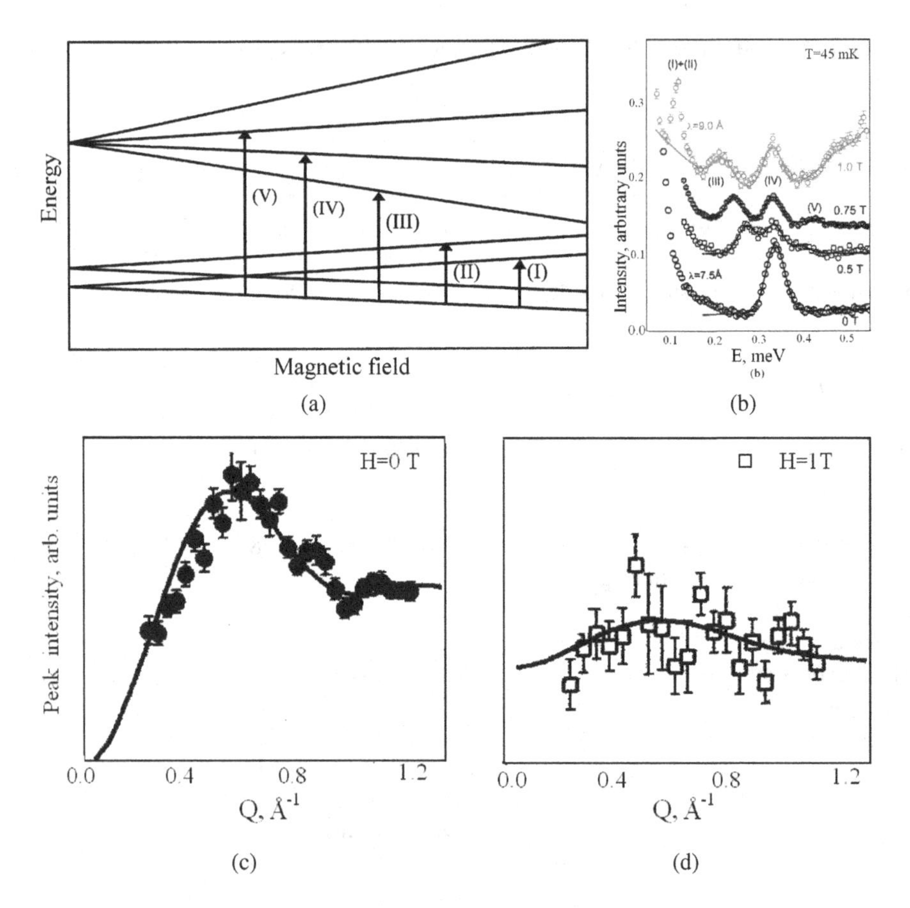

Fig. 26. INS data for V_{15} (modified from Refs. Chaboussant *et al.*, 2004; Chaboussant *et al.*, 2004). (a) interdoublet transitions (I)–(II) and doublet-quadruplet transitions (III), (IV), (V); (b) INS spectra measured at different fields; (c) total Q-dependence of transitions (III + IV +V) measured at zero field: experiment (spheres) and fit (curve); (d) experimentally found Q-dependence of the single transition (I) at 1 *Tesla* and $\lambda = 90$ nm (open squares); theoretical curve calculated in Ref. Chaboussant *et al.*, 2004 (continuous line).

Following Refs. Chaboussant *et al.*, 2004; Chaboussant *et al.*, 2004 one should assume that ZFS of 0.2–0.3 cm^{-1} cannot entirely originate from the AS exchange. In fact, such a relatively large normal part of AS exchange would make energy levels significantly dependent on orientation of the field and, therefore, it should widen the INS spectrum peaks in powder. However no such broadening was actually observed in the experiment.

The maximum I of the intensity in INS spectrum related to the transition in a spin system from the state $|a\rangle$ to the state $|b\rangle$ (Fig. 26(b)) depends on the vector $\boldsymbol{Q}=\boldsymbol{k}_0 - \boldsymbol{k}\,\boldsymbol{k}_0, \boldsymbol{k}$ are the incident and scattered wave-vectors, respectively) mostly through the structure factor $S(\boldsymbol{Q})$. The intensity and $S(\boldsymbol{Q})$ are related as $I(\boldsymbol{Q}) \sim S(\boldsymbol{Q})$ where

$$S(\boldsymbol{Q}) = \sum_{i,j} \exp(\mathrm{i}\boldsymbol{Q}\boldsymbol{R}_{ij}) \sum_{\alpha,\beta} \langle a|S_{i\alpha}|b\rangle \langle b|S_{j\beta}|a\rangle \tag{44}$$

In Eq. (44) S_i is the spin operator of the ion i, $\boldsymbol{R}_{ij}=\boldsymbol{R}_i - \boldsymbol{R}_j, \boldsymbol{R}_i$ are the position vectors of the ions. Exact formulas for the powder-averaged structure factors for the trinuclear clusters with the scalene distortion and for the important case of $R_{12}=R_{23}=R_{31}=R$ are given in Ref. Belinsky, 2009. The results for the transitions from (I) to (V) (Fig. 26(a)) are the following:

$$\begin{aligned} S_{I,av}(Q) &= 1/2, \\ S_{II,av}(Q) &= (1/3)(1-\mu_0), \quad S_{III,av}(Q) = (1/2)(1-\mu_0), \\ S_{IV,av}(Q) &= (1/3)(1-\mu_0), \quad S_{V,av}(Q) = (1/6)(1-\mu_0), \\ \mu_0 &= \sin(QR)/(QR) \end{aligned} \tag{45}$$

Thus, as shown in Ref. Chaboussant *et al.*, 2004, the total Q-dependence of the transitions (III+IV+V) at zero field can be found as $S_{III,IV,V,av}(Q) = 1 - \sin(QR)/(QR)$. This expression describes the experimental data (Fig. 44(c)) with a very good accuracy and this can be considered as an argument in favor of an important role of the scalene distortion in the V_{15} cluster. This agreement shows also that there is no significant mixing between the $S=1/2$ and $S=3/2$ states at low fields. In fact, this mixing should be attributed to the in-plane component of AS (see Sec. 6) which is, in principle, not significant in that range of fields where the experiments were performed.

According to Ref. Belinsky, 2009 the transition I in a distorted system (in absence of the AS exchange) is Q-independent (Eq. (45)). On the contrary, the experimental data exhibit a weak Q-dependence for this transition (Fig. 13(c)). Based on this observation, one can conclude that actually the ZFS most probably contains contributions arising from both distortion and AS exchange.

To summarize the results so far discussed, one should note that the INS measurements in conjunction with other experimental techniques (EPR in Ref. Kajiyoshi *et al.*, 2007, Sec. 5.2, ^{51}V-NMR (Furukawa *et al.*, 2007; Furukawa *et al.*, 2007), Sec. 10) provide evidences in favor of the existence of both distortion and the normal part of AS in the V_{15} complex. However, presently the question about the exact estimation of the ratio of the parameters of these two interactions remains open. At the same time the in-plane part of the AS exchange was convincingly estimated on the basis of the analysis of static magnetization (Refs. Tarantul *et al.*, 2006; Tarantul *et al.*, 2007, Sec. 6.2).

9.3. *Energy pattern of a scalene triangular system*

The full Hamiltonian of a distorted trinuclear system, Eq. (46), includes isotropic exchange (first term, H_0), selected static structural distortion, let us say, along side 12 (Fig. 27), AS exchange H_{AS} and the isotropic Zeeman term:

$$H = 2J\sum_{i,k} \mathbf{S}_i\mathbf{S}_k + 2\eta\mathbf{S}_1\mathbf{S}_2 + \sum_{i,k} \mathbf{D}_{ik}[\mathbf{S}_i \times \mathbf{S}_k] + g\beta\mathbf{H}\sum_i \mathbf{S}_i \qquad (46)$$

where symbols i, k run over all pairs $i, k = (1, 2), (2,3)$ and $(3,1)$. The distortions are assumed to be relatively small ($|\eta| < |J|$) and their influence on the deviation of the AS exchange from the trigonal (actually, axial) symmetry will be neglected, so it is assumed that $D_{ik\alpha} = D_\alpha$, $\alpha = l, t, n$ for any side.

Figure 28(a) shows the influence of the AS exchange on the Zeeman pattern of the distorted system. The static distortion removes "accidental" degeneracy of the ground level (separating its states accordingly to the intermediate spin $S_{12} = 0, 1$) and results in the zero-field splitting $\Delta = 2|\eta|$. Due to the magnetic isotropy of the system in the absence of AS exchange

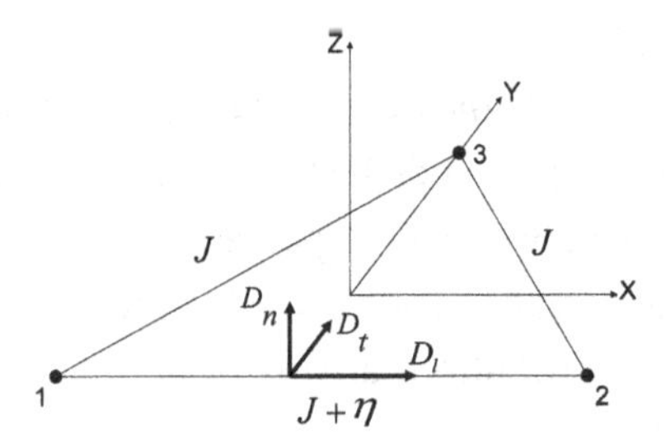

Fig. 27. Model of magnetic exchange interactions acting in a triangular cluster. $J_{31} = J_{23} = J$, $J_{12} = J + \eta$, molecular coordinate system xyz.

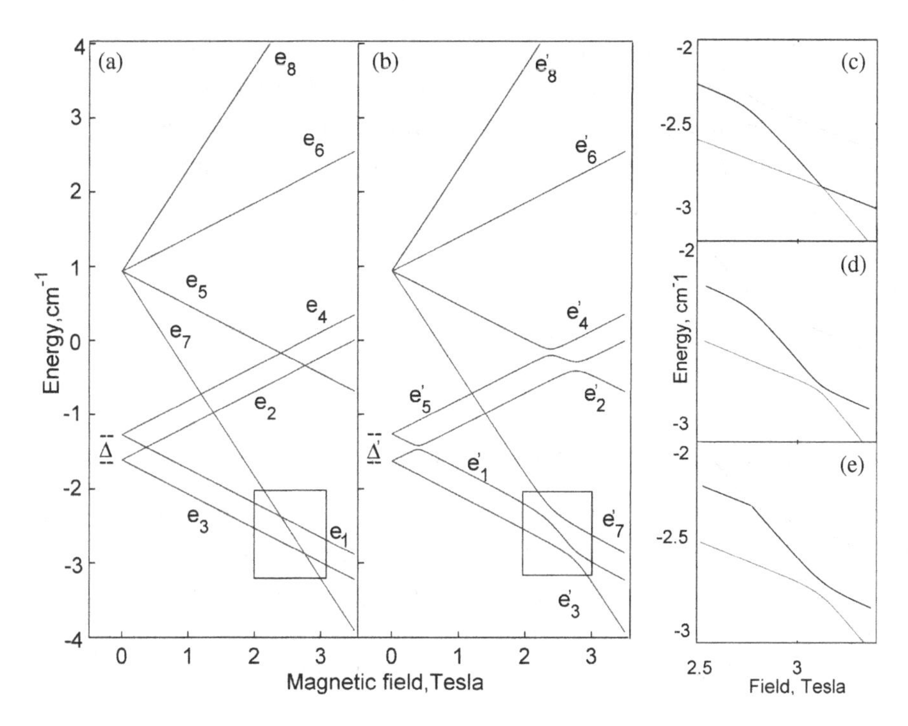

Fig. 28. Energy levels of a scalene triangular system in the perpendicular field (in plane of the triangle) and influence of the parameters D_t, D_l on the anticrossing region (box). In all figures $J = 0.85\,\text{cm}^{-1}$ and $\eta = 0.2J$; (a) $D_n = D_l = D_t = 0$; (b) $D_n = 0.1J$, $D_t = 0.08J$, $D_l = 0.13J$; (c) $D_n = 0.1J$, $D_t = 0$, $D_l = 0.1J$; (d) $D_n = 0.1J$, $D_t = 0.05J$, $D_l = 0.1J$; (d) $D_n = 0.1J$, $D_t = 0.1J$, $D_l = 0$.

one can observe the exact crossing of the magnetic sublevels as shown in Fig. 28(a). Figure 28(b) shows the effect of AS exchange on the Zeeman pattern of the distorted system. The joint action of static deformation and AS exchange results in the ZFS that can be approximately presented as (Tarantul *et al.*, 2008; Tarantul *et al.*, 2008)

$$\Delta = \sqrt{4\eta^2 + 3D_n^2} - \left(D_l^2 + D_t^2\right)/8J \tag{47}$$

and produces avoided crossing and a peculiar non-linear behavior at low field $\boldsymbol{H}$ whose direction is in plane of the triangle.

In the region of avoiding crossing the behavior of the levels (which influences the dynamics of magnetization) crucially depends on the interrelation between D_t, D_l and η (Fig. 28, insets). One can see that the combined action of the distortion and AS exchange results in a new pattern in the anticrossing region, in particular, new rules for the crossing, that are different from those in a symmetric model (Tsukerblat *et al.*, 2006). It is interesting that due to the symmetry lowering the two in-plane components of AS exchange, D_t and D_l, act independently and cannot be combined into the single parameter $D_{\perp} = \sqrt{D_t^2 + D_l^2}$ which has been introduced (Tsukerblat *et al.*, 2006) for a trigonal system.

Similar results for the energy pattern in V_{15} assuming both AS and structural distortion are given in Ref. De Raedt *et al.*, 2003.

9.4. *Magnetic properties of the scalene systems*

Influence of the structural deformation on the shape of the static susceptibility vs. field in the anticrossing region is illustrated in Fig. 29. Within the isotropic model the low-temperature susceptibility has a narrow-pulse shape (Fig. 29(a)) that corresponds to the exact crossing of the $S = 3/2$, $M_S = -3/2$ and $S = 1/2$, $M_S = -1/2$ levels giving rise to the sharp step of magnetization. Figs. 29(a–c) show the effect of the AS exchange for different ratios $2\eta/\Delta$ provided that the zero-field gap Δ is fixed. The peak of χ vs. field becomes smoother with the increase of the

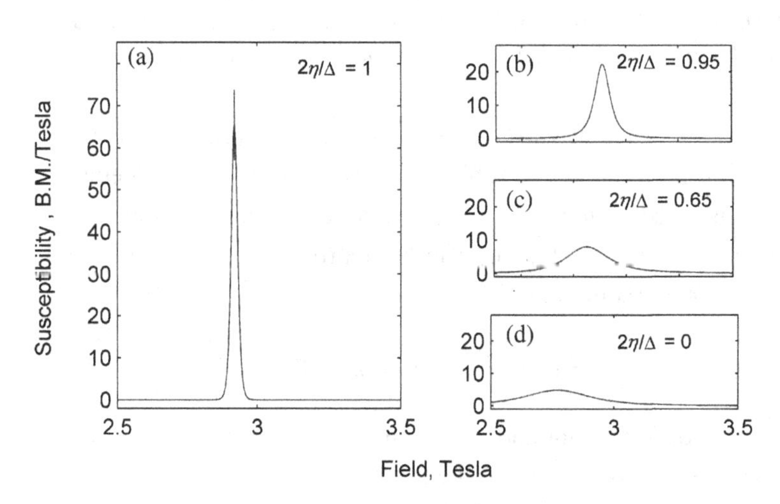

Fig. 29. Magnetic susceptibility in the perpendicular field at $T = 0.01$ K for different ratios $2\eta/\Delta$ with $\Delta = 0.14\,\text{cm}^{-1}$, $J = 0.85\,\text{cm}^{-1}$ and constant ratio $D_n{:}D_t{:}D_l = 1{:}2.12{:}2.12$.

AS exchange. This shows that the structural deformation reduces the AS exchange and consequently the magnetic anisotropy. On the other hand, a strong AS exchange reduces the effect of structural deformation so that these two interactions prove to be competitive.

This observation reveals also the role of the spin-vibronic JT coupling. In the case of strong JT coupling resulting in the instability of the symmetric (trigonal) configuration, the magnetic anisotropy caused by the AS exchange is expected to be reduced and in this limit one can expect a sharp step of magnetization. On the contrary, when AS exchange is strong enough, the symmetric system remains stable and shows the magnetic anisotropy related to AS.

9.5. *Field induced Jahn-Teller instability*

An interesting area of the magnetoelastic instability (instability induced by an external magnetic field) which results in the field induced cooperative phenomena in the molecule based magnets has recently been developed (Waldmann *et al.*, 2006; Waldmann, 2007). This area is expected to be promising in the control of the properties of the molecular magnetic materials by means of external fields.

The V_{15} system has an unusual pattern of the Zeeman levels in the anticrossing area that is drastically different from conventional two-level scheme. In this region the Zeeman sublevels can be affected by the pseudo JTE that appears due to the presence of quasidegenerate levels mixed by the AS exchange and non-symmetric part of the isotropic exchange. The main contribution to the vibronic interaction arises from the modulation of the isotropic exchange by the JT displacements of the E- type (double degenerate mode Q_X, Q_Y, Fig. 19) of spin sites, the corresponding operator is given by Eq. (27). The matrix of the Hamiltonian in the restricted basis set of three eigenfunctions

$$|(1)3/2, M_{\perp} = -3/2\rangle, \quad |(0)1/2, M_{\perp} = -1/2\rangle, \quad |(1)1/2, M_{\perp} = -1/2\rangle$$

is given in Refs. Tarantul *et al.*, 2008; Tarantul *et al.*, 2008. For the sake of clarity the JT distortion will be considered separately from the static one, i.e. in the limit of $\eta = 0$. Adiabatic surfaces of the system for different values of the applied field are given in Fig. 30. In addition to the dimensionless parameters defined in Sec. 8.2, the two additional parameters should be defined: $j_0 = J/\hbar\omega$ and $\gamma_\alpha = (3/4)\sqrt{2}(D_\alpha/\hbar\omega)$, $\alpha = n, l, t$.

In the crossing point ($\xi = 3j_0$) the symmetric (trigonal) system is stable providing weak JT coupling (Fig. 30(a)) and/or relatively strong

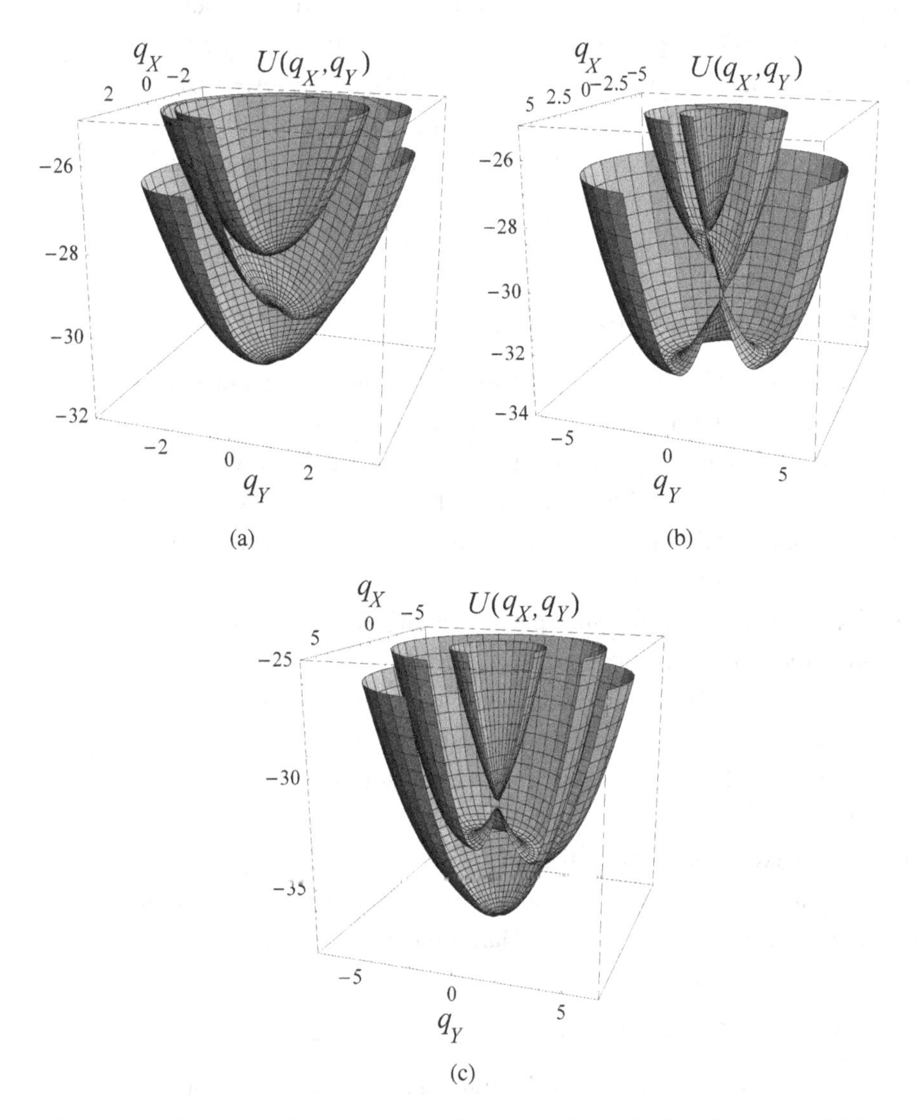

Fig. 30. Adiabatic surfaces of a triangular system for $j_0 = 10$ and $\gamma_l = \gamma_t = 0.1\,j_0$: (a) weak spin-vibronic coupling with $\upsilon = j_0$, $\xi = 3\,j_0$; (b) strong spin-vibronic coupling with $\upsilon = 4\,j_0$, $\xi = 3\,j_0$; (c) strong magnetic field with $\upsilon = 4\,j_0$, $\xi = 3.5\,j_0$.

AS exchange, meanwhile in the case of a relatively strong JT coupling and/or weak AS exchange the symmetric configuration of the magnetic sites becomes unstable and the minima of the adiabatic surface are disposed at the ring of the trough (Fig. 30(b)). Finally, in a high field (beyond the anticrossing region) when the degeneracy is removed the symmetric system becomes stable again (Fig. 30(c)). As distinguished from the static distortion the JT conformations are dynamical in the sense that the motion along the ring of the trough corresponds to a series of isosceles configurations in which the spin frustration is eliminated.

10. NMR Experiments

The low temperature NMR measurements on ^{51}V ions in V_{15} are reported in Refs. Furukawa *et al.*, 2007; Furukawa *et al.*, 2007. The resonant frequency depends on the field according to the conventional relation (Furukawa *et al.*, 2007)

$$f = \gamma_N (H_0 + H_{\text{int}}) \tag{48}$$

where $\gamma_N / 2\pi = 11.825$ MHz/*Tesla* is the gyromagnetic ratio for the ^{51}V nucleus and H_0 is the applied field while H_{int} is the local internal field produced by the electronic spin density on the ^{51}V nuclei. The internal field is proportional to the local spin density.

Four peaks of NMR (Furukawa *et al.*, 2007) were observed whose positions provide information about the average spin magnetic moments of the electronic subsystem formed by the vanadium ions of the inner triangle. These average magnetic moments are found to be: 0, β, $-\beta/3$, $2\beta/3$. The conclusion was made that the two Kramers doublets with $S = 1/2$ of V_{15} are described by the eigenfunctions

$$\psi_a = (1/\sqrt{2})(|\downarrow\downarrow\uparrow\rangle - |\downarrow\uparrow\downarrow\rangle) \tag{49}$$

that correspond to the average spin magnetic moments of 0, 0, β, and

$$\psi_b = (1/\sqrt{6})(2\,|\uparrow\downarrow\downarrow\rangle - |\downarrow\uparrow\downarrow\rangle - |\downarrow\downarrow\uparrow\rangle) \tag{50}$$

with the magnetic moments $-\beta/3$, $2\beta/3$ and $2\beta/3$. The spin densities corresponding to these eigenfunctions are symbolically depicted in Fig. 31. In

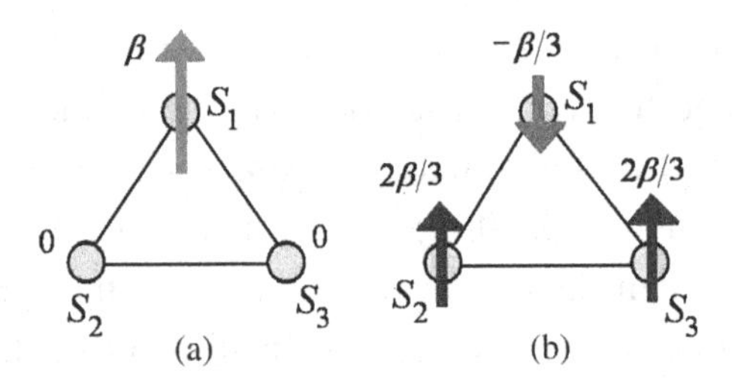

Fig. 31. (modified from Ref. Furukawa *et al.*, 2007) Schematic view of the expectation values for the spin moments of the sites 1, 2, and 3 (modified from Ref. Furukawa *et al.*, 2007).

fact, the peak corresponding to ion 1 (Fig. 31(a), average magnetic moment β) is stronger than the peak corresponding to the same ion in the Kramers doublet ψ_b (magnetic moment $-\beta/3$). Therefore the doublet ψ_α is recognized as the ground state. This implies the following interrelation between the exchange parameters: $J_{12} > J_{23} \approx J_{31}$, so the parameter η proves to be positive.

One can conclude that the two spin sites in the inner triangle are approximately identical to each other while the third one stands distinctively apart from them. These observations indicate the presence of a structural deformation and show that the triangle in V_{15} is approximately isosceles in agreement with the INS experiments (Chaboussant *et al.*, 2004; Chaboussant *et al.*, 2004) (Sec. 9.2).

The ^{51}V-NMR measurements performed at high magnetic fields beyond the crossover point ($S = 3/2$) are reported in Refs. Furukawa *et al.*, 2007; Furukawa *et al.*, 2007. One of the three observed peaks was associated with the vanadium ions having spin magnetic moment of about β (V3 ions in Fig. 2) while two other signals come from the vanadium ions V1 and V2 (Fig. 2) whose magnetic moments are slightly different from each other but having different signs and both close to zero in magnitude. The V_3 ions are identified with ions of the central triangle while V1 and V2 are recognized as vanadium ions in the hexagons (Fig. 2). So this experimental data seem to be a direct evidence of the fact that at low temperatures the hexagons are inactive and the cluster's properties are entirely defined by the spins of the inner triangle.

NMR experiments on the hydrogen nuclei which belong to eight solvent water molecules adjacent to V_{15} and to water molecules located in the center of the cluster are reported in Refs. Kumagai *et al.*, 2005; Furukawa *et al.*, 2007; Furukawa *et al.*, 2005; Procissi *et al.*, 2003, the results on the temperature and field dependencies of the nuclear relaxation time T_1 measured by a saturation method and the relaxation time T_2 that was determined by spin-echo delay curve are presented in Refs. Kumagai *et al.*, 2005; Furukawa *et al.*, 2007; Furukawa *et al.*, 2005.

At high magnetic fields $H > H_{cr}(H_{cr} \approx 2.7\,Tesla$ is a critical field of the crossover $S = 1/2 \rightarrow S = 3/2$) the time T_1, as expected, is well described by the model of nuclear spin-phonon relaxation but a discrepancy with this model is observed at low fields where the time T_1 becomes significantly shorter than that prescribed by the spin-lattice mechanism. The interaction of the proton spin with the electronic spins of vanadium ions in the central triangle was proposed as the main decoherence factor. These three ionic spins are supposed to fluctuate due to frustration of the $S = 1/2$ ground state. The same factor was pinpointed as a cause for decoherence of the electronic spins in Refs. Bertaina *et al.*, 2008; Chiorescu *et al.*, 2003; Barbara *et al.*, 2002.

Another interesting finding is that the T_1 time is temperature independent at low (less than 0.5 *Tesla*) fields. This phenomenon remarkably recalls the temperature-independent persistent spin dynamics observed in the muon-scattering experiment (Sec. 7.2, Ref. Salman *et al.*, 2008). These similar observations made in different experiments (NMR and μSR) have probably a common origin.

Relaxation time T_2 cannot also be explained entirely in terms of the nuclear spin-spin mechanism and the additional factors arise again from the fluctuating field of vanadium ions in the cluster. Of course, it is not accidentally that $1/T_2$ *vs.* H has a characteristic deep minimum just next to the crossover point H_{cr} (Kumagai *et al.*, 2005).

By analyzing NMR peaks' behavior in the alternating field (Kumagai *et al.*, 2005) it was possible to distinguish between the signals coming from the adjacent solvent molecules and the peaks associated with a water molecule located inside the cluster. In fact, in solvent molecules the local magnetic field at protons originates from the dipolar interaction with vanadium ions while in the central water molecule the field arises from the contact hyperfine interaction with vanadium sites in the triangle. From the observation of a few different NMR peaks associated with the central water molecule it was concluded that this molecule has more than one stable position inside the cluster.

A strong difference between the NMR properties of the central triangle of V_{15} and triangles inside the $[CN_3H_6]_4Na_2[H_{42}P_4O_4]\cdot14H_2O$ complex known as V_6 was found and interpreted in Ref. Procissi *et al.*, 2003.

11. Conclusions and Outlook

We have given a comprehensive analysis of the properties of the unique spin frustrated V_{15} magnetic polyoxometalate. This system attracted continuous interest during two decades. Last but not least the interest in the V_{15} increased again after it had been proposed (Dobrovitski *et al.*, 2000) that it could be a candidate for a qubit in the quantum computation systems due to the small $S = 1/2$ spin in a comparably large cluster thereby reducing the dipolar interactions. Recently Rabi oscillations in this system have been observed and interpreted in the framework of the triangle model with due account for the AS exchange (Bertaina *et al.*, 2008).

In this article the energy levels' behavior and crossover as well as the low-temperature magnetization are described with the use of the pseudo-angular momentum representations for the many-spin states within the triangle model. A special physical role of the AS exchange in spin frustrated triangular systems has been emphasized. In particular, it was underlined that in-plane and normal components of the AS exchange have different effects on magnetization as a function of field. Afterwards using pseudo-angular quantum numbers we have formulated the selection rules for the EPR transitions and have given an analysis of the EPR spectra at different frequencies. In this context we have recognized the high-frequency temperature dependence of the spectral bandwidth as a problem which can not be explained within the foregoing model and thus addresses the study of additional phenomena like relaxation processes and the hyperfine structure in vanadium clusters. Also the influence of structural deformations on the magnetic properties, anisotropy of the system and EPR transitions are described.

JT instability is shown to eliminate spin frustration due to removal of the 'accidental' degeneracy in course of the dynamical structural distortions. The influence of the vibronic interaction on the magnetization is revealed with the aid of the semiclassical adiabatic approach that provides qualitatively transparent results and gives numerical results with a good accuracy. The first and second Van Vleck coefficients in the Zeeman energies are deduced as the functions of the direction of the field, AS exchange and vibronic coupling. The JT coupling is shown to be competitive to the AS exchange so that the increase of the vibronic coupling decreases the magnetic anisotropy of the system. On the other hand sufficiently strong AS exchange tends to suppress the JTE. This is demonstrated by the theoretical modeling of the field dependence of the magnetization that clearly exhibits the crucial role of the pseudo JT coupling in spin-frustrated systems. Furthermore we can conclude that the results so far described allow a deeper understanding of the physical contents of the concept of spin frustration as well as about orbitally degenerate frustrated units integrated in complex systems.

Also the dynamical properties of V_{15} and relaxation mechanisms responsible for the hysteresis loop are discussed in detail. Special attention is paid to the INS spectra and muon scattering experiments.

The study of V_{15} is considered to be part of a more extended investigation regarding multicenter JTE and structure-related magnetism in nanoscopic polyoxometalates/vanadates showing frustration from one-triangle via two- (in the case of M_6Mo_{57} type clusters see Ref. Gatteschi *et al.*, 1996 and papers cited therein) to 20-triangle spin-arrays (in case of the clusters $M_{30}Mo_{72}$; $M = V^{IV}, Fe^{III}, Cr^{III}$, see Ref. Müller *et al.*, 2005 and papers cited therein). This has been discussed in a review (Kögerler *et al.*, 2010) along with the related questions. The study of the interplay of spin frustration and the JTE in a simple case of a triangular system seems to be a good background for the understanding of the properties of solids like Kagomé antiferromagnets based on the triangular lattices (Matan *et al.*, 2006; Anderson, 1987; Shores *et al.*, 2005; Bulaevskii *et al.*, 2008; Shores *et al.*, 2005). The model of the vibronic interaction so far developed is a good background for the consideration of spin-phonon relaxation in molecular magnets, in particular for the discussion of the decoherence in V_{15} (Tarantul and Tsukerblat, 2010). Finally to repeat it, the prediction (Dobrovitski *et al.*, 2000) and realization (Bertaina *et al.*, 2008) of the long living coherent oscillations in V_{15} will undoubtedly wield far reaching influence on the use of molecular magnets as qubits in quantum computing.

Acknowledgements

Financial support from the Israel Science Foundation, ISF (grant no.168/09) is gratefully acknowledged. We thank Professor Achim Müller for careful reading the manuscript and many valuable comments.

References

Abragam, A. and Bleaney, B. (1970). *Electronic Paramagnetic Resonance of Transition Ions*, Clarendon Press, Oxford.

Affronte, M., Troiani, F., Ghirri, A., Candini, A., Evangelisti, M., Corradini, V., Carretta, S., Santini, P., Amoretti, G., Tuna, F., Timco, G. and Winpenny, R.E.P. (2007). *Phys. J. D: Appl. Phys.* **40**, 2999.

Affronte, M., Troiani, F., Ghirri, A., Carretta, S., Santini, P., Corradini, V., Schuecker, R., Muryn, C., Timco, G. and Winpenny, R.E.P. (2006). *Dalton Trans.* 2810.

Ajiro, Y., Itoh, H., Inagaki, Y., Asano, T., Narumi, Y., Kindo, K., Sakon, T., Motokawa, M., Cornia, A., Gatteschi, D., Müller, A. and Barbara, B. (2001). French — Japanese Symposium, Fukuoka, Japan.

Anderson, P.W. (1987). *Science* **235**, 1196.

Ardavan, A., Rival, O., Morton, J.J.L., Blundell, S., Tyryshkin, A.M., Timco, G.A. and Winpenny, R.E.P. (2007). *Phys. Rev. Lett.* **98**, 057201.

Barbara, B. (2003). *J. Mol. Struct.* **656**, 135.

Barbara, B., Chiorescu, I., Giraud, R., Jansen, A.G.M. and Ganeschi, A. (2000). *J. Phys. Soc. Japan* **69**, 383.

Barbara, B., Chiorescu, I., Wernsdorfer, W., Bögge, H. and Müller, A. (2002). *Progress of Theoretical Physics Supplements* **145**, 357.

Barbour, A., Luttrell, R.D., Choi, J., Musfeldt, J.L., Zipse, D., Dalal, N.S., Boukhvalov, D.W., Dobrovitski, V.V., Katsnelson, M.I., Lichtenstein, A.I., Harmon, B.N. and Kögerler, P. (2006). *Phys. Rev. B* **74**, 014411.

Barra, A.-L., Gatteschi, D., Pardi, L., Müller, A. and Döring, J. (1992). *J. Am. Chem. Soc.* **114**, 8509.

Bartlett, B.M. and Nocera, D.G. (2005). *J. Am. Chem. Soc.* **127**, 8985.

Bartlett, B.M., Grohol, D., Papoutsakis, D., Shores, M.P. and Nocera, D.G. (2004). *Chem. Eur. J.* **10**, 3850.

Bates, C.A. and Jasper, R.F. (1971). *J. Phys. C: Sol. State Phys.* **4**, 2341.

Belinskii, M.I., Tsukerblat, B.S. and Ablov, A.V. (1972). *Phys. Stat. Solidi* K71.

Belinskii, M.I., Tsukerblat, B.S. and Ablov, A.V. (1973). *Fizika. Tverdogo Tela* (*Rus*) **15**, 29.

Belinskii, M.I., Tsukerblat, B.S. and Ablov, A.V. (1974). *Mol. Phys.* **28**, 283.
Belinsky, M.I. (2008). *Inorg. Chem.* **47**, 3521.
Belinsky, M.I. (2008). *Inorg. Chem.* **47**, 3532.
Belinsky, M.I. (2009). *Chem. Phys. Lett.* **361**, 137.
Bersuker, I.B. (2006). *The Jahn-Teller effect*, University Press, Cambridge.
Bersuker, I.B. and Polinger, V.Z. (1989). *Vibronic Interactions in Molecules and Crystals*, Springer Series in Chemical Physics, 49, Springer-Verlag.
Bertaina, S., Gambarelli, S., Mitra, T., Tsukerblat, B., Müller, A. and Barbara, B. (2008). *Nature* **453**, 20.
Bogani, L., Cavigli, L., Gurioli, M., Novak, R.L., Mannini, M., Caneschi, A., Pineider, F., Sessoli, R., Clemente-Le6n, M., Coronado, E., Cornia, A. and Gatteschi, D. (2007). *Adv. Mater.* **19**, 3906.
Borras-Almenar, J.J., Clemente-Juan, J.M., Coronado, E. and Tsukerblat, B.S. (2001). *J. of Comp. Chem.* **22**, 985.
Boukhvalov, D.W., Dobrovitski, V.V., Katsnelson, M.I., Lichtenstein, A.I., Harmon, B.N. and Kögerler, P. (2004). *Phys. Rev. B* **70**, 054417.
Boukhvalov, D.W., Kurmaev, E.Z., Moiwes, A., Zatsepin, D.A., Cherkashenko, V.M., Nemnonov, S.N., Finkelstein, L.D., Yarmoshenko, Yu.M., Neumann, M., Dobrovitski, V.V., Katsnelson, M.I., Lichtenstein, A.I., Harmon, B.N. and Koegerler, P. (2003). *Phys. Rev. B* **67**, 134408.
Bulaevskii, L.N., Batista, C.D., Mostovoy, M.V. and Khomskii, D.I. (2008). *Phys. Rev. B* **78**, 024402.
Carretta, S., Santini, P., Amoretti, G., Troiani, F. and Affronte, M. (2007). *Phys. Rev. B* **76**, 024408.
Chaboussant, G., Basler, R., Sieber, A., Ochsenbein, S.T. and Güdel, H.-U. (2004). *Physica B* **350**, e51.
Chaboussant, G., Basler, R., Sieber, A., Ochsenbein, S.T., Desmedt, A., Lechner, R.E., Telling, M.T.F., Kögerler, P., Müller, A. and Güdel, H.-U. (2002). *Europhys. Lett.* **59**(2), 291.
Chaboussant, G., Ochsenbein, S.T., Sieber, A., Güdel, H.-U., Barbara, B., Müller, A. and Mutka, H. (2004). *Phys. Stat. Sol. (c)* **1**(12), 3399.
Chaboussant, G., Ochsenbein, S.T., Sieber, A., Güdel, H.-U., Mutka, H., Müller, A. and Barbara, B. (2004). *Europhys. Lett.* **66**(3), 423.
Chiorescu, I., Wernsdorfer, W. and Müller, A. (2003). *Phys. Rev. B* **67**, 020402(R).
Chiorescu, I., Wernsdorfer, W., Müller, A. and Bögge, H. (2000). *Phys. Rev. Lett.* **84**, 3454.
Chiorescu, I., Wernsdorfer, W., Müller, A., Bögge, H. and Barbara, B. (2000). *J. Magn. Magn. Mater.* **221**, 103.
Choi, J., Sanderson, L.A.W., Musfeldt, J.L., Ellern, A. and Kögerler, P. (2003). *Phys. Rev. B* **68**, 064412.
Clemente-Juan, J.M. and Coronado, E. (1999). *Coord. Chem. Rev.* 193–195.
Condorelli, G.G., Motta, A., Fragalà, I.L., Giannazzo, F., Raineri, V., Caneschi, A. and Gatteschi, D. (2004). *Angew. Chem. Int. Ed.* **43**, 4081.
Cornia, A., Fabretti, A.C., Pacchioni, M., Zobbi, L., Bonacchi, D., Caneschi, A., Gatteschi, D., Biagi, R., del Pennino, U., DeRenzi, V., Gurevich, L., van der Zant, H.S.J. (2003). *Angew. Chem. Int. Ed.* **42**, 1645.
Coronado, E., Forment-Aliaga, A., Romero, F.M., Corradini, V., Biagi, R., DeRenzi, V., Gambardella, A. and del Pennino, U. *Inorg. Chem.* (22) 7693.

Corradini, V., Biagi, R., del Pennino, U., DeRenzi, V., Gambardella, A., Affronte, M., Muryn, C.A., Timco, G.A. and Winpenny, R.E.P. *Inorg. Chem.* **46**(12), 4937.

Döring, J. (1990). Reduzierte Polyoxovanadate eine Neue Klasse von Metall-Sauerstoff-Clustern. University of Bielefeld.

De Raedt, H., Miyashita, S. and Michielsen, K. (2003). *arXiv:cond-mat/0306275v1.*

De Raedt, H., Miyashita, S. and Michielsen, K. (2004). *Phys. Stat. Solidi* (*b*) **241**, 1180.

De Raedt, H., Miyashita, S., Michielsen, K. and Machida, M. (2003). *arXiv:cond-mat/0312581v1.*

De Raedt, H., Miyashita, S., Michielsen, K. and Machida, M. (2004). *Phys. Rev. B* **70**, 064401.

Dobrovitski, V.V., Katsnelson, M.I. and Harmon, B.N. (2000). *Phys. Rev. Lett.* **84**, 3458.

Dzyaloshinsky, I.E. (1957). *Zh. Exp. Teor. Fiz.* **32**, 1547.

Elhajal, M., Canals, B. and Lacroix, C. (2002). *Phys. Rev. B* **66**, 014422/1.

Englman, R. (1972). *The Jahn-Teller Effect in Molecules and Crystals*, Wiley Chemical Monographs in Physics WILEY-INTERSCIENCE, a division of John Wiley and Sons.

Fainzilberg, V.E., Belinskii, M.I. and Tsukerblat, B.S. (1980). *Solid State Comm.* **36**, 639.

Fainzilberg, V.E., Belinskii, M.I. and Tsukerblat, B.S. (1981). *Mol. Phys.* **44**, 1177.

Fainzilberg, V.E., Belinskii, M.I. and Tsukerblat, B.S. (1981). *Mol. Phys.* **44**, 1195.

Fainzilberg, V.E., Belinskii, M.I. and Tsukerblat, B.S. (1982). *Mol. Phys.* **45**, 807.

Fainzilberg, V.E., Belinskii, M.I., Kuyavskaya, B.Ya. and Tsukerblat, B.S. (1985). *Mol. Phys.* **54**, 799.

Fleury, B., Catala, L., Huc, V., David, C., Zhong, W.Z., Jegou, P., Baraton, L., Albouy, P.-A. and Mallah, T. (2005). *Chem. Commun.* 2020.

Furukawa, Y. *et al.* (2005). *Polyhedron* **24**, 2737.

Furukawa, Y., Nishisaka, Y., Kumagai, K. and Kögerler, P. (2007). *J. Mag. Mag. Mater.* **310**, 1429.

Furukawa, Y., Nishisaka, Y., Kumagai, K., Kögerler, P. and Borsa, F. (2007). *Hyperfine Interactions* **176**, 65.

Furukawa, Y., Nishisaka, Y., Kumagai, K., Kögerler, P. and Borsa, F. (2007). *Phys. Rev. B* **75**, 220402(R).

Gatteschi, D., Barra, A.L., Ganeschi, A., Cornia, A., Sessoli, R. and Sorace, L. (2006). *Coord. Chem. Rev.* **250**, 1514.

Gatteschi, D., Pardi, L., Barra, A.-L. and Müller, A. (1993). *Mol. Eng.* **3**, 157.

Gatteschi, D., Pardi, L., Barra, A.-L., Müller, A. and Döring, J. (1991). *Nature* **354**, 465.

Gatteschi, D., Pardi, L., Barra, A.L. and Müller, A. (1994). In *Polyoxometalates: From Platonic Solids to Anti-Retroviral Activity*, Pope, M.T. and Müller, A., Eds.; Kluwer, Dordrecht, p. 219.

Gatteschi, D., Sessoli, R. and Villain, J. (2006). *Molecular Nanomagnets*, Oxford University Press, Oxford.

Gatteschi, D., Sessoli, R., Müller, A. and Kögerler, P. (2001). In *Polyoxometalate Chemistry: From Topology via Self-Assembly to Applications*, Pope, M.T., Müller, A., Peters, F. and Gatteschi, D., Eds.; Kluwer, Dordrecht, 319.

Gatteschi, D., Sessoli, R., Plass, W., Müller, A., Krickemeyer, E., Meyer, J., Sölter, D. and Adler, P. (1926). *Inorg. Chem.* **35**, 1926.

Grohol, D., Matan, K., Cho, J.-H., Lee, S.-H., Lynn, J.W., Nocera, D.G. and Lee, Y.S. (2005). *Nature Mater.* **4**, 323.

Inami, T., Nishiyama, M., Maegawa, S. and Oka, Y. (2000). *Phys. Rev. B* **61**, 12181.

Kahn, O. (1993). *Molecular Magnetism*, VCH, NY.
Kajiyoshi, K., Kambe, T., Mino, M., Nojiri, H., Koegerler, P. and Luban, M. (2007). *J. Magn. Magn. Mater.* **310**, 1203.
Kane, B.E. (1998). *Nature* **393**, 133.
Kögerler, P., Tsukerblat, B. and Müller, A. (2010). *Dalton. Trans.* **39**, 1.
Konstantinidis, N.P. and Coffey, D. (2002). *Phys. Rev. B* **66**, 174426.
Kortus, J., Hellberg, C.S. and Pederson, M.R. (2001). *Phys. Rev. Lett.* **86**, 3400.
Kortus, J., Pederson, M.R., Hellberg, C.S. and Khanna, S.N. (2001). *Eur. Phys. J.* **D16**, 177.
Kostyuchenko, V.V. and Popov, A.I. (2008). *J. Exp. Theor. Phys.* **107**(4), 595.
Kumagai, K., Fujiyoshi, Y., Furukawa, Y. and Kögerler, P. (2005). *J. Mag. Mag. Mater.* **294**, 141.
Landau, L.D. and Lifshitz, E.M. (1966). *Quantum Mechanics*, MIR, Moscow.
Lehmann, J., Gaita-Arino, A., Coronado, E. and Loss, D. (2007). *Nature Nanotechnology* **2**, 312.
Leibfried, D., DeMarco, B., Meyer, V., Lucas, D., Barrett, M., Britton, J., Itano, W.M., Jelenković, B., Langer, C., Rosenband, T. and Wineland, D.J. (2003). *Nature* **422**, 412.
Lever, A.B.P. (1984). *Inorganic Electronic Spectroscopy, Studies in Physical and Theoretical Chemistry*, 2 ed.; Elsevier, Vol. 33.
Loss, D. and DiVincenzo, D.P. (1998). *Phys. Rev. A* **57**, 120.
Loss, D. and Leuenberger, M.N. (2001). *Nature* **410**, 789.
Machida, M. and Miyashita, S. (2005). *Physica E* **29**, 538.
Machida, M., Iitaka, T. and Miyashita, S. (2005). *arXiv: cond-mat/0501439 v2.*
Matan, K., Grohol, D., Nocera, D.G., Yildirim, T., Harris, A.B., Lee, S.H., Nagler, S.E. and Lee, Y.S. (2006). *Phys. Rev. Lett.* **96**, 247201/1.
Mischenko, A., Zvezdin, A. and Barbara, B. (2003). *J. Magn. Magn. Mater.* **258**, 352.
Mitrikas, G., Sanakis, Y., Raptopoulou, C.P., Kordas, G. and Papavassiliou, G. (2008). *Phys. Chem. Chem. Phys.* **10**, 743.
Miyashita, S. (1995). *J. Phys. Soc. Japan* **64**, 3207.
Miyashita, S. (1996). *J. Phys. Soc. Japan* **65**, 2734.
Miyashita, S. and Nagaosa, N. (2001). *Prog. Theor. Phys.* **106**, 533.
Miyashita, S., De Raedt, H. and Michielsen, K. (2003). *Prog. Theor. Phys.* **110**, 889.
Moriya, T. (1960). *Phys. Rev. B* **120**, 91–98.
Morton, J.J.L. (2006). *Nature Phys.* **2**, 365.
Müller, A. and Döring, J. (1988). *Angew. Chem. Int. Ed. Eng.* **27**, 1719.
Müller, A., Peters, F., Pope, M.T. and Gatteschi, D. (1998). *Chem. Rev.* **98**, 239.
Müller, A., Todea, A.M., van Slageren, J., Dressel, M., Bögge, H., Schmidtmann, M., Luban, M., Engelhardt, L. and Rusu, M. (2005). *Angew. Chem. Int. Ed.* **44**, 3857.
Nielsen, M.A. and Chuang, I.L. (2000). *Quantum Computation and Quantum Information*, Cambridge University Press, Cambridge.
Nojiria, H., Taniguchia, T., Ajiro, Y., Müller, A. and Barbara, B. (2004). *Physica B* **216**, 346–347.
Park, K. and Pederson, M.R. (2005). *Users Group Conference* http://doi.ieeecomputer-society.org/10.1109/DODUGC.2005.22.
Platonov, V.V., Tatsenko, O.M., Plis, V.I., Zvezdin, A.K. and Barbara, B. (2002). *Solid State Physics* (*Rus. Fizika Tverdogo Tela*), **44**, 2010.
Pope, M.T. and Müller, A. (1991). *Angew. Chem. Int. Ed. Engl.* **30**, 34.

Popkov, A.F., Kulagin, N.E., Mukhanova, A.I., Popov, A.I. and Zvezdin, A.K. (2005). *Phys. Rev. B* **72**, 104410.

Popov, A.I., Plis, V.I., Popkov, A.F. and Zvezdin, A.K. (2004). *Phys. Rev. B* **69**, 104418.

Procissi, D., Suh, B.J., Jung, J.K., Kögerler, P., Vincent, R. and Borsa, F. (2003). *Journal of Applied Physics* **93**, 7810.

Raghu, C., Rudra, I., Sen, D. and Ramasesha, S. (2003). *Phys. Rev. B* **68**, 029902.

Rudra, I., Ramasesha, S. and Sen, D. (2001). *J. Phys.: Condens. Matter* **13**, 11717.

Saito, K. and Miyashita, S. (2001). *J. Phys. Soc. Japan* **70**, 3385.

Sakon, T., Koyama, K., Müller, M.A., Barbara, B. and Ajiro, Y. (2005). *Progress of Theoreticla Physics Supplement* No.159.

Sakon, T., Koyama, K., Motokawa, M., Ajiro, Y., Müller, A. and Barbara, B. (2004). *Physica B* **206**, 346–347.

Salman, Z., Chow, K.H., Miller, R.I., Morello, A., Parolin, T.J., Hossain, M.D., Keeler, T.A., Levy, C.D.P., MacFarlane, W.A., Morris, G.D., Saadaoui, H., Wang, D., Sessoli, R., Condorelli, G.G. and Kiefl, R.F. (2007). *Nano Lett.* **7**, 1551.

Salman, Z., Kiefl, R.F., Chow, K.H., Mac Farlane, W.A., Keeler, T., Parolin, T., Tabbara, S. and Wang, D. (2008). *Phys. Rev. B* **72**, 214415.

Schlegel, C., van Slageren, J., Manoli, M., Brechin, E.K. and Dressel, M. (2008). *Phys. Rev. Lett.* **101**, 147203.

Schmidt-Kaler, F., Häffner, H., Riebe, M., Gulde, S., Lancaster, G.P.T., Deuschle, T., Becher, C., Roos, C.F., Eschner, J. and Blatt, R. (2003). *Nature* **422**, 408.

Schnack, J. (2006). *J. Low Temp. Phys.* **142**, 279.

Schnack, J. (2007). *Comptes Rendus Chimie* **10**, 15.

Schnack, J. (2010). *Dalton Transactions* **39**, 4677.

Schnalle, R. and Schnack, J. (2009). *Polyhedron* **28**, 1620.

Shores, M.P., Bartlett, B.M. and Nocera, D.G. (2005). *J. Am. Chem. Soc.* **127**, 17986.

Shores, M.P., Nytko, E.A., Bartlett, B.M. and Nocera, D.G. (2005). *J. Am. Chem. Soc.* **127**, 13462.

Stamp, P.C.E. (2008). *Nature* **453**, 167.

Stepanenko, D., Trif, M. and Loss, D. (2008). *Inorg. Chim. Acta* **361**, 3740.

Stoneham, A.M., Fisher, A.J. and Greenland, P.T.J. (2003). *J. Phys. Cond. Matter* **15**, L447.

Tarantul, A. and Tsukerblat, B. (2010). *Inorg. Chim. Acta*, **363**, 4361.

Tarantul, A., Tsukerblat, B. and Müller, A (2008). *J. Mol. Struct.* **890**, 170–177.

Tarantul, A., Tsukerblat, B. and Müller, A. (2006). *Chem. Phys. Lett.* **428**, 361–366.

Tarantul, A., Tsukerblat, B. and Müller, A. (2007). *Inorg. Chem.* **46**(1), 161.

Tarantul, A., Tsukerblat, B. and Müller, A. (2008). *Solid State Sciences* **10**, 1814.

Timco, G.A., Carretta, S., Troiani, F., Tuna, F., Pritchard, R.J., Muryn, C.A., McInnes, E.J.L., Ghirri, A., Candina, A., Santini, P., Amoretti, G., Affronte, M. and Winpenny, R.E.P. (2009). *Nature Nanotechnology*, **4**, 173.

Troiani, F., Ghirri, A., Affronte, M., Carretta, S., Santini, P., Amoretti, G., Piligkos, S., Timco, G. and Winpenny, R.E.P. (2005). *Phys. Rev. Lett.* **94**, 207208–1.

Tsukerblat, B., Tarantul, A. and Müller, A. (2006). *Chem. J. Phys.* **125**, 0547141.

Tsukerblat, B., Tarantul, A. and Müller, A. (2006). *Phys. Lett. A* **353**, 48–59.

Tsukerblat, B., Tarantul, A. and Müller, A. (2007). *J. Mol. Struct.* **838**, 124.

Tsukerblat, B.S. (2006). *Group Theory in Chemistry and Spectroscopy: A Simple Guide to Advanced Usage*, Dover Pub., Mineola, NY.

Tsukerblat, B.S. and Belinskii, M.I. (1983). *Magnetochemistry and Radiospectroscopy of Exchange Clusters (Rus)*, Stiintsa, Kishinev.

Tsukerblat, B.S., Belinskii, M.I. and Ablov, A.V. (1974). *Fizika. Tverdogo Tela (Rus)* **16**, 989.

Tsukerblat, B.S., Belinskii, M.I. and Fainzilberg, V.E. (1987). *Magnetochemistry and Spectroscopy of Transition Metal Exchange Clusters*, In *Soviet Sci. Rev. B*, M, Ed.; Harwood Acad. Pub, New York, 1987, Vol. 9, p. 337.

Tsukerblat, B.S., Botsan, I.G., Belinskii, M.I. and Faiinzilberg, V.E. (1985). *Mol. Phys.* **54**, 813.

Tsukerblat, B.S., Fainzilberg, V.E., Belinskii, M.I. and Kuyavskaya, B.Ya. (1983). *Chem. Phys. Lett.* **98**, 149.

Tsukerblat, B.S., Kuavskaya, B.Ya., Belinskii, M.I., Ablov, A.V., Novotortsev, V.M. and Kalinnikov, V.T. (1975). *Theoretica Chimica Acta.* **38**, 131.

Tsukerblat, B.S., Kuyavskaya, B.Ya., Fainzilberg, V.E. and Belinskii, M.I. (1984). *Chem. Phys. Lett.* **90**, 361.

Tsukerblat, B.S., Kuyavskaya, B.Ya., Fainzilberg, V.E. and Belinskii, M.I. (1984). *Chem. Phys. Lett.* **90**, 373.

Varshalovich, D.A., Moskalev, A.N. and Khersonskii, V.K. (1988). *Quantum Theory of Angular Momentum*, World Scientific, Singapore.

Vongtragool, S., Gorshunov, B., Mukhin, A.A., van Slageren, J., Dressel, M. and Müller, A. (2003). *Phys. Chem. Chem. Phys.* **5**, 2778–2782.

Waldmann, O. (2007). *Phys. Rev. B* **75**, 174440.

Waldmann, O., Dobe, C., Ochsenbein, S.T., Güdel, H.-U. and Sheikin, I. (2006). *Phys. Rev. Lett.* **96**, 027206.

Wernsdorfer, W., Müller, A., Mailly, D. and Barbara, B. (2004). *arXiv:cond-mat/0404410v1*.

White, R.M. and Geballe, T.H. (1979). *Solid State Physics Advances in Research: Long Range Order in Solids (Solid state physics: Supplement)*, Academic Press.

Winpenny, R.E.P. (2008). *Angew. Chem. Int. Ed.* **47**, 2.

Yoon, J., Mirica, L.M., Stack, T.D.P. and Solomon, E.I. (2004). *J. Am. Chem. Soc.* **126**, 12586.

Zipse, D., Dalal, N.S., Vasic, R., Brooks, J.S. and Kögerler, P. (2005). *Phys. Rev. B* **71**, 064417.

Zobbi, L., Mannini, M., Pacchioni, M., Chastanet, G., Bonacchi, D., Zanardi, C., Biagi, R., del Pennino, U., Gatteschi, D., Cornia, A. and Sessoli, R. (2005). *Chem. Commun.* 1640.

Zvezdin, K., Plis, V.I., Popov, A.I. and Barbara, B. (2001). *Physics of the Solid State* **43**(1), 185.

Chapter 4

NEUTRON SPECTROSCOPY OF MOLECULAR NANOMAGNETS

TATIANA GUIDI
ISIS Facility Rutherford Appleton Laboratory
Chilton, Didcot OX11 0QX, United Kingdom

1. Introduction

The interaction between neutrons and condensed matter provides unique information about the spatial and temporal correlations and the magnetic properties (magnetization density, magnetic excitations, magnetic order) of materials. Neutron scattering is a widely used technique for the study of spin excitations and static correlations in one and two-dimensional quantum spin systems, and it has been naturally extended to the field of zero-dimensional magnetism. In particular, *neutron spectroscopy* is a very powerful technique because important information can be obtained from the energy and the **Q** dependent intensity of the inelastic peaks. It provides direct access to the energies and wavefunctions of the cluster spin states and gives exact information on quantities that can be calculated from first principles. This technique requires the use of large samples, the preparation of which implies a considerable chemical effort especially if fully deuterated materials are necessary. The chemical effort is rewarded by the possibility of performing measurements without having to apply an external magnetic field, which is an unavoidable perturbation when using standard spectroscopic techniques, like Electron Paramagnetic Resonance (EPR) spectroscopy. Moreover, the number of spin multiplets accessible by Inelastic Neutron Scattering (INS) is not limited by the relaxation time as it is for EPR, and the relative energy for several spin levels can be determined. In contrast with extended systems, the excitations for isolated magnetic clusters are localized and dispersionless. Nevertheless, the dependence of the excitation intensities on the transferred wave vector **Q** provides information useful to label the spin levels according to the irreducible representations of the molecular symmetry group and gives information on the local structure of the clusters.

This chapter is intended as a review of the most significant experiments that have been performed on molecular nanomagnets (MNMs) using neutron spectroscopy techniques.

2. Neutron Scattering: Basic Principles

Neutrons are particularly suited for magnetic and structural investigations in MNM systems. The neutron's wavelength ($\lambda = 1.8$ Å for thermal neutrons) is comparable to interatomic distances and thus suitable for diffraction studies. The neutron is uncharged, therefore it interacts directly with the nucleus and not with the electrons in the atoms. This means that, contrary to x-rays, certain atoms with a small number of electrons (e.g. hydrogen) may have a strong neutron scattering cross section. The neutron has a magnetic moment ($s = 1/2$) that couples with unpaired electrons in magnetic materials. This magnetic interaction provides information on the density distribution and correlations of the unpaired electrons in elastic scattering whilst in inelastic scattering it provides the energy of the magnetic excitations. The energies of thermal neutrons (5–100 meV) are comparable to the energies of collective excitations in condensed matter. Cold neutrons (0.1–10 meV) are particularly suitable for the study of magnetic excitations in MNMs.

2.1. *Neutron scattering cross section*

The interaction of the neutrons with matter involves three principal mechanisms:

(1) scattering by the nuclei;
(2) absorption by the nuclei;

(3) scattering through the magnetic interaction between the neutron and the atomic magnetic moment.

The first mechanism is responsible for the nuclear elastic scattering (e.g. Bragg scattering) and phonon scattering. The second mechanism leads to a loss of intensity, since the neutron is captured by the nucleus to form a compound nucleus (Cd and Gd are examples of strongly absorbing nuclei). The third mechanism is the one we are mainly interested in and we therefore want to minimize the other two. In a scattering process the quantity that is experimentally measured is the *cross-section*. The *partial differential cross-section* is defined as (Squires, 1978):

$$\frac{d^2\sigma}{d\Omega dE} = \frac{\begin{array}{c}\text{neutrons scattered per sec into } d\Omega(\theta,\phi)\\ \text{with final energy between } E \text{ and } E+dE\end{array}}{\Phi d\Omega dE}$$

where Φ is the flux of incident neutrons, i.e. the number of incoming neutrons per unit area and per second and $d\Omega$ is the elementary solid angle around the direction specified by the angles θ and ϕ. The incident neutrons have an initial energy E_i and initial momentum $\mathbf{k}_i$, whilst the sample is characterized by an initial state ν with energy E_ν. After interacting with the sample, the neutrons with final energy E_f and final momentum $\mathbf{k}_f$ are collected by a detector. The final state of the scattering system, ν', has energy $E_{\nu'}$. From the conservation of the energy and momentum of the overall system (neutron plus sample) the following equations must be satisfied:

$$E_i + E_\nu = E_f + E_{\nu'} \Rightarrow \hbar\omega = E_{\nu'} - E_\nu = E_i - E_f = \frac{\hbar^2}{2m_N}(k_i^2 - k_f^2)$$

$$\mathbf{Q} = \mathbf{k}_i - \mathbf{k}_f$$

where $\hbar\omega$ is the *energy transfer* and $\mathbf{Q}$ is the *scattering vector*. The general expression for the scattering cross-section for a transition from an initial k_i quantum state of the neutron to a final k_f state, ignoring the spin of the neutron, can be derived using Fermi's golden rule (Squires, 1978; Lovesey, 1987):

$$\left(\frac{d^2\sigma}{d\Omega dE}\right)_{\nu\to\nu'} = \left(\frac{m_N}{2\pi\hbar^2}\right)^2 \frac{k_f}{k_i}|\langle \mathbf{k}_f\nu'|V|\mathbf{k}_i\nu\rangle|^2\delta(E_i+E_\nu-E_f-E_{\nu'}) \quad (1)$$

where m_N is the neutron mass and V is the interaction potential between the neutron and the sample relative to the kind of scattering process we want to analyze and can be treated as a small perturbation. We can define the *scattering function* $S(\mathbf{Q}, \omega)$ as:

$$S(\mathbf{Q}, \omega) \propto \frac{k_i}{k_f} \frac{d^2\sigma}{d\Omega dE}$$

The scattering profile can be decomposed into two contributions:

$$S(\mathbf{Q}, \omega) = S_{\text{nuclear}}(\mathbf{Q}, \omega) + S_{\text{magn}}(\mathbf{Q}, \omega)$$

that will be discussed in the following sections.

2.1.1. *Nuclear scattering*

The nuclear scattering can be calculated using for the interaction potential $V(\mathbf{r},t)$ between a neutron (with position vector $\mathbf{r}$) and a nucleus ($\mathbf{R}_i$), the so called *Fermi pseudopotential*:

$$V^i_{\text{nuc}}(\mathbf{r}) = \frac{2\pi\hbar^2}{m_N} b_i \delta(\mathbf{r} - \mathbf{R}_i) \tag{2}$$

where b_i is the scattering length of the ith nucleus. The value of the scattering length depends on the particular nuclide and on the combined spin of the nucleus-neutron system. This leads to two different terms in the nuclear scattering: *coherent scattering* $S_{\text{coh}}(\mathbf{Q}, \omega)$ and *incoherent scattering* $S_{\text{inc}}(\mathbf{Q}, \omega)$. The coherent scattering depends on the correlation between the positions of the nuclei at the same or different times and therefore gives interference effects (Bragg scattering, phonon excitations, etc.). The incoherent cross-section arises from the random distribution of the deviations of the scattering lengths from their mean value. Incoherent scattering is isotropic and an inconvenient source of background. Molecular magnets usually have a large number of hydrogen atoms (in the organic ligand part of the molecule) and the high incoherent cross section of ^{1}H isotope is the main source of background in these systems. For this reason the samples used for neutron scattering experiments are usually deuterated, since the incoherent cross-section for the ^{2}H (D) isotope is about 40 times smaller than for ^{1}H.

2.1.2. *Magnetic scattering*

The scattering cross section due to magnetic interaction can be calculated using Eq. 1 and substituting for $V(\mathbf{r},t)$ the expression for the magnetic interaction potential between the neutron and the electrons of the system:

$$V_{\text{mag}} = -\boldsymbol{\mu}_{\mathbf{n}} \cdot \mathbf{B} \tag{3}$$

where $\boldsymbol{\mu}_{\mathbf{n}}$ is the neutron magnetic moment operator and $\mathbf{B}$ is the magnetic field arising from the electron spin and its orbital motion around the nucleus. If we consider the particular case of a molecular cluster with spin only moments in which the electrons are localized around each ith nucleus with position vector $\mathbf{R}_i$ (Heitler-London model), the scattering cross section becomes:

$$\frac{d^2\sigma}{d\Omega dE} = \frac{A}{2\pi\hbar}\frac{k_f}{k_i}\frac{1}{4\mu_B^2}\sum_{\alpha,\beta}\left(\delta_{\alpha\beta} - \frac{Q_\alpha Q_\beta}{Q^2}\right)\sum_{ij} F_i(\mathbf{Q})F_j(\mathbf{Q})$$
$$\times \int \langle e^{-i\mathbf{Q}\cdot\mathbf{R}_i(0)} e^{i\mathbf{Q}\cdot\mathbf{R}_j(t)}\rangle \langle s_\alpha^\dagger(i,0)s_\beta(j,t)\rangle e^{-i\omega t}dt \tag{4}$$

where $A = 0.29$ *barn* and $F_i(\mathbf{Q})$ is the Fourier transform of the normalized density of the unpaired electrons of the ith ion. It is known as the *magnetic form factor* and falls off rapidly with $\mathbf{Q}$. The quantity $s_\alpha(i,t)$ is the α component of the spin operator $s(i,t)$ at position $\mathbf{R}_i$. The factor $(\delta_{\alpha\beta} - \frac{Q_\alpha Q_\beta}{Q^2})$ means that the neutron can only couple to components of the magnetic moment that are perpendicular to the wave-vector $\mathbf{Q}$. The second line in the preceding equation is the space and time Fourier transform of the time-dependent spin-spin correlation function and contains all the information about the magnetic structure and spin dynamics. Equation 4 can be written more explicitly as (Lovesey, 1987):

$$\frac{d^2\sigma}{d\Omega dE} = \frac{A}{Z}\frac{k_f}{k_i}\sum_{\alpha,\beta}\left(\delta_{\alpha\beta} - \frac{Q_\alpha Q_\beta}{Q^2}\right)\sum_{\nu,\nu'}\exp\left(-\frac{E_\nu}{k_B T}\right)$$
$$\times \sum_{ij} F_i(\mathbf{Q})F_j(\mathbf{Q})\exp[i\mathbf{Q}\cdot(\mathbf{R}_i - \mathbf{R}_j)]\langle\nu|s_\alpha^\dagger(i)|\nu'\rangle$$
$$\times \langle\nu'|s_\beta(j)|\nu\rangle\delta(E_\nu - E_{\nu'} + \hbar\omega) \tag{5}$$

where Z is the partition function. In the above formula, we can write the spin states $|\nu\rangle$ as a linear superposition of basis states $|SM\rangle$ and express the spin operator components $s_\alpha(i)$ in terms of spherical tensor operators:

$$s_x = \frac{(s_{-1} - s_{+1})}{\sqrt{2}}$$
$$s_y = i\frac{(s_{-1} + s_{+1})}{\sqrt{2}}$$
$$s_z = s_0 \tag{6}$$

The matrix elements $\langle SM|s_\alpha(i)|S'M'\rangle$ can then be calculated using the Wigner-Eckart theorem:

$$\langle SM|s_q(i)|S'M'\rangle = (-1)^{S-M}\begin{pmatrix} S & 1 & S' \\ -M & q & M' \end{pmatrix}\langle S||\mathbf{s}(i)||S'\rangle \tag{7}$$

where q $= -1, 0, +1$. The properties of the 3-j symbol imply the following selection rules for the neutron scattering cross-section (Furrer and Güdel, 1979):

$$\Delta S = 0, \pm 1 \quad \text{and} \quad \Delta M = 0, \pm 1$$

INS experiments are usually performed on polycrystalline samples, thus the magnetic scattering cross section of Eq. 5 has to be averaged over all directions of the wave vector transfer **Q**. Borras-Almenar *et al.* (1999) report the following formula for the orientation-averaged INS cross-section:

$$\left(\frac{\partial^2\sigma}{\partial\Omega\partial E_f}\right)_{\nu\to\nu'} = A\frac{N}{Z}\frac{k_f}{k_i}\exp[-2W]\exp\left\{-\frac{E_\nu}{k_B T}\right\}\sum_{i\le j} F_i^\star(Q)F_j(Q)$$

$$\times \sum_{(\tilde{S})SM}\sum_{(\tilde{S}')S'M'}\sum_{(\tilde{S''})S''M''}\sum_{(\tilde{S'''})S'''M'''}\langle\nu|(\tilde{S})SM\rangle\langle(\tilde{S}')S'M'|\nu'\rangle$$

$$\times \langle\nu\prime|(\tilde{S''})S''M''\rangle\langle(\tilde{S'''})S'''M'''|\nu\rangle\langle(\tilde{S})S||\hat{\mathbf{s}}(i)||(\tilde{S}')S'\rangle$$

$$\times \frac{\langle(\tilde{S''})S''||\hat{\mathbf{s}}(j)||(\tilde{S'''})S'''\rangle}{\sqrt{(2S'+1)(2S'''+1)}}\left\{\left[\frac{2}{3}\delta_{ij} - 4\left(\frac{\cos(QR_{ij})}{(QR_{ij})^2} - \frac{\sin(QR_{ij})}{(QR_{ij})^3}\right)\right.\right.$$

$$\left.\times (1-\delta_{ij})\right]C^{SM}_{S'M'10}C^{S''M''}_{S'''M'''10} - \left[\frac{2}{3}\delta_{ij} + 2\left(\frac{\sin(QR_{ij})}{(QR_{ij})} + \frac{\cos(QR_{ij})}{(QR_{ij})^2}\right.\right.$$

$$- \left. \frac{\sin(QR_{ij})}{(QR_{ij})^3} \right) (1 - \delta_{ij}) \Big] \left(C^{SM}_{S'M'11} \, C^{S''M''}_{S'''M'''1-1} + C^{SM}_{S'M'1-1} C^{S''M''}_{S'''M'''11} \right) \Big\}$$

$$\times \, \delta[E_\nu - E_{\nu'} + E_i - E_f] \tag{8}$$

where N is the total number of magnetic clusters, $\exp(-2W)$ is the Debye-Waller factor, $C^{SM}_{S'M'1q}$ are Cebsch-Gordan coefficients and R_{ij} gives the relative position of ions i and j. This formula has been widely used for simulating INS experiments, but it is not of general applicability. In particular, it may fail to describe correctly the scattering from systems with large magnetic anisotropy and does not properly take into account intramolecular interference effects. A correct interpretation of all the experimental spectra requires a generalization of formula (8) (Waldmann, 2003; Caciuffo *et al.*, 2005):

$$\frac{\partial^2 \sigma}{\partial \Omega \partial \omega} = A \frac{k_f}{k_i} e^{-2W} \sum_{\nu,\nu'} \frac{e^{-\beta E_\nu}}{Z} I_{\nu\nu'}(Q) \delta(\hbar\omega - E_{\nu'} + E_\nu) \tag{9}$$

where:

$$\begin{aligned} I_{\nu\nu'}(Q) = \sum_{i,j} F_i^*(Q) F_j(Q) \times \Big\{ & \frac{2}{3} [j_0(QR_{ij}) + C_0^2(\mathbf{R}_{ij}) j_2(QR_{ij})] \\ & \times \tilde{s}_z(i)\tilde{s}_z(j) + \frac{2}{3} \left[j_0(QR_{ij}) - \frac{1}{2} C_0^2(\mathbf{R}_{ij}) \right] (\tilde{s}_x(i)\tilde{s}_x(j) \\ & + \tilde{s}_y(i)\tilde{s}_y(j)) + \frac{1}{2} j_2(QR_{ij}) \big[C_2^2 (\tilde{s}_x(i)\tilde{s}_x(j) - \tilde{s}_y(i)\tilde{s}_y(j)) \\ & + C_{-2}^2 (\tilde{s}_x(i)\tilde{s}_y(j) + \tilde{s}_y(i)\tilde{s}_x(j)) \big] + j_2(QR_{ij}) \big[C_1^2 (\tilde{s}_z(i)\tilde{s}_x(j) \\ & + \tilde{s}_x(i)\tilde{s}_z(j)) + C_{-1}^2 (\tilde{s}_z(i)\tilde{s}_y(j) + \tilde{s}_y(i)\tilde{s}_z(j)) \big] \Big\}, \end{aligned} \tag{10}$$

$$C_0^2 = \frac{1}{2} \left[3 \left(\frac{R_{ijz}}{R_{ij}} \right)^2 - 1 \right]$$

$$C_2^2 = \frac{R_{ijx}^2 - R_{ijy}^2}{R_{ij}^2}$$

$$C_{-2}^2 = 2 \frac{R_{ijx} R_{ijy}}{R_{ij}^2}$$

$$C_1^2 = \frac{R_{ijx} R_{ijz}}{R_{ij}^2}$$
$$C_{-1}^2 = \frac{R_{ijy} R_{ijz}}{R_{ij}^2} \tag{11}$$

and

$$\tilde{s}_\alpha(i)\tilde{s}_\gamma(j) = \langle \nu | s_\alpha(i) | \nu' \rangle \langle \nu' | s_\gamma(j) | \nu \rangle \quad (\alpha, \gamma = x, y, z). \tag{12}$$

Besides being valid whatever the symmetry of the magnetic anisotropy, Eq. 9 is also considerably less time-consuming when calculated numerically than the formula (8). The intrinsic linewidth of the excitations and the finite instrumental resolution are taken into account by replacing the δ function in Eq. 9 by an appropriate spectral weight function, usually a Gaussian or a Lorentzian line-shape.

In the strong exchange limit, Eq. 9 can be replaced by a much simpler expression involving the matrix elements of the cartesian components of the total spin S (Caciuffo *et al.*, 1998):

$$\frac{\partial^2 \sigma}{\partial \Omega \partial \omega} \propto \sum_{\nu, \nu'} \frac{e^{-\beta E_\nu}}{Z} \sum_{\alpha = x, y, z} |\langle \nu | S_\alpha | \nu' \rangle|^2 \delta(\hbar\omega - E_{\nu'} + E_\nu), \tag{13}$$

Its validity is limited to cases where the giant-spin Hamiltonian model can be used and for $Q \simeq 0$. Indeed, it does not account for the interference effects between various spin centers that give rise to a characteristic oscillatory Q dependence (Güdel and Furrer, 1977; Waldmann and Güdel, 2005).

2.2. *The time-of-flight technique*

The most efficient way of measuring the energy and Q dependence of the magnetic scattering intensity of MNMs is to use the time-of-flight (TOF) technique. In a TOF neutron spectrometer the energy analysis of the scattered beam is performed by measuring the time taken by the neutrons to travel a known distance, with the incident (direct geometry) or the scattered (indirect geometry) neutron energy fixed at a particular value. A schematic layout of a direct TOF instrument is shown in Fig. 1. In a direct geometry TOF spectrometer pulses of monochromatic neutrons are produced at the sample by phasing a Fermi chopper (i.e. a package of neutron absorbing/transparent alternating slits spinning at a frequency in the range of

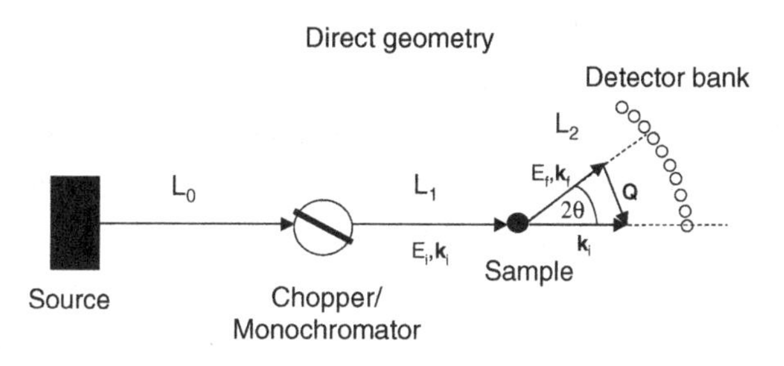

Fig. 1. Schematic representation of a time-of-flight direct geometry neutron scattering spectrometer. A white pulsed neutron beam comes from the source and a chopper/monochromator selects the desired incident neutron energy E_i with a defined k_i. After the sample, the neutrons are collected by the time-resolved array of detectors. The final energy E_f is determined by the measured time-of-flight between the sample and the detector.

50–600 Hz), or a series of disk choppers, to the pulse of neutrons produced by a time-structured source (for instance a proton-spallation source, or the continuous beam produced by a reactor pulsed by another chopper). The scattered neutrons are detected by an array of detectors covering an extended scattering angle range, and the signals from the detectors are sorted according to TOF and scattering angle, so that the cross section as a function of Q and ω may be derived. Therefore, a scan over the total time of flight corresponds to a scan on the energy transfer at constant scattering angle 2θ. To correctly extract the Q dependence of the inelastic features of the spectra an appropriate calibration of the neutron intensity has to be done. The scattering from vanadium is incoherent, thus reflects the incident spectrum and the efficiency of the detectors. Normalization corrections are based on the elastic peak intensity measured for a vanadium foil identical in size to the sample. Empty can and background counting are usually subtracted from the measured intensity of the sample before the time-of-flight to energy conversion.

3. Exchange Interaction: A Spectroscopic Measurement

The great advantage of INS with respect to other spectroscopic techniques like infrared spectroscopy or EPR, is the possibility it offers to probe magnetic transitions between states that have different total spin. The direct access to the energy difference between different spin states gives unique information about the exchange interaction. For example, in the simplest case of a s $= 1/2$ Heisenberg spin dimer, the transition energy between the singlet and triplet states provides the direct microscopic measurement of the isotropic exchange parameter J, since $E_{triplet} - E_{singlet} = J$.

The INS transition spectrum provides also additional information through its Q dependence. The intensity for the transition between the singlet and the triplet takes for the dimer the very simple form:

$$I(\mathbf{Q}) \propto F^2(\mathbf{Q})[1 - \cos(\mathbf{Q} \cdot \mathbf{R})] \tag{14}$$

The angular average for a powder sample gives (Furrer and Güdel, 1977):

$$I(Q) \propto F^2(Q)\left[1 - \frac{\sin(QR)}{QR}\right] \tag{15}$$

where R is the distance between the two magnetic ions. The Q dependence therefore contains information on the geometry of the dimer. The first INS measurement that allowed the determination of the exchange integral on a metal-organic complex was performed by Güdel and Furrer (see for example Güdel and Furrer (1977) or Basler *et al.* (2003) for a review) on a deuterated powder sample of acid rhodo chromium chloride, $[(ND_3)_5Cr(OD)Cr(ND_3)_5Cl_5] \cdot D_2O$. The system is composed of two antiferromagnetic (AF) exchange coupled Cr(III) ions ($s = 3/2$), and the transitions between the four different spin states ($S = 0, 1, 2, 3$) were measured using a triple axis spectrometer. The INS spectra (Fig. 2) give a textbook example of the potential of the technique. As a consequence of the INS selection rules, all the energy levels can be probed. The position and the temperature dependence of the observed excitations allows a straightforward identification of the INS transitions. Furthermore, the intensity as a function of scattering vector Q (Fig. 3) gives experimental evidence of the interference term $[1 - \sin(QR)/(QR)]$ in Eq. 15 and allows the experimental determination of the magnetic form factor of the Cr(III) ion.

This first experiment was followed by a series of INS experiments on di- tri- and tetra-nuclear clusters. As the nuclearity of the cluster increases, more exchange pathways and exchange constants are involved and more interference terms appear in the Q dependence (see for example the calculations

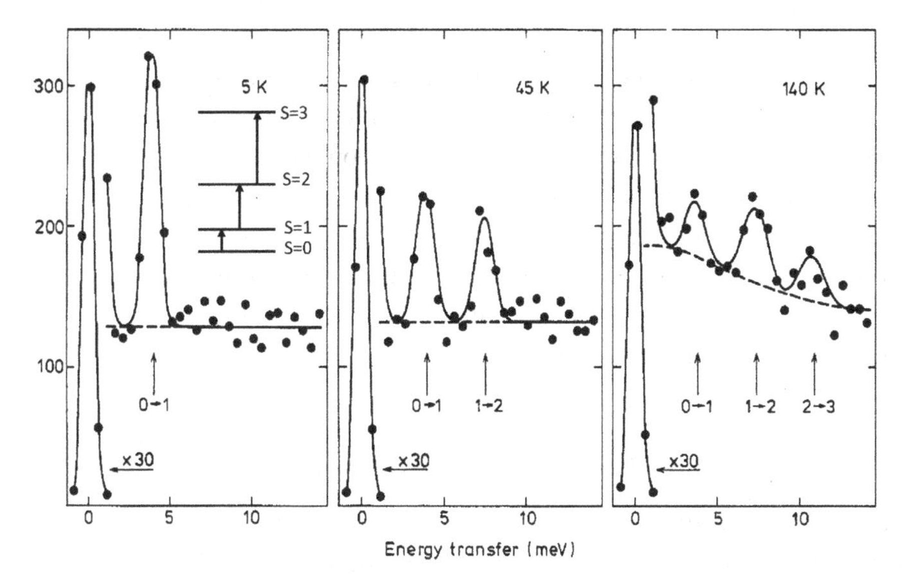

Fig. 2. INS spectra of rhodo chromium chloride collected on a triple axis spectrometer from Güdel and Furrer (1977). The corresponding energy level diagram is shown in the inset.

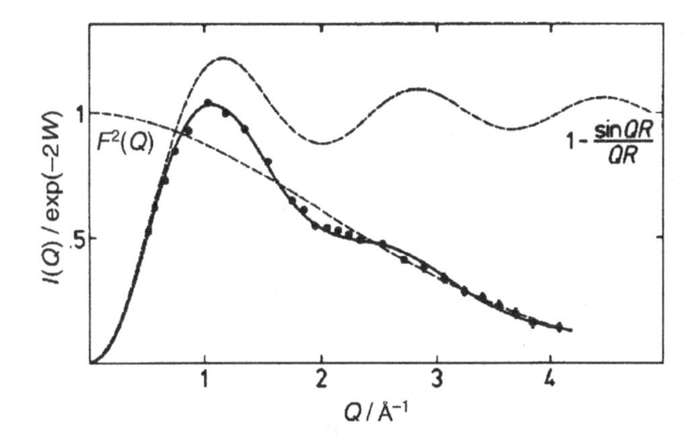

Fig. 3. Q dependence of the first INS transition in Fig. 2 from Furrer and Güdel (1977).

by Waldmann (2003) and Haraldsen *et al.* (2005). Magnetic susceptibility and heat capacity measurements are also sensitive to the exchange interaction. INS is a very useful complementary technique to restrict the number of possible solutions that simultaneously fit all the available experimental data. However, a consistent number of inter-multiplet transitions has to be measured to get a unique set of exchange parameters. The Q dependence of the transitions can be used to assign the origin of different transitions. It is also used to discriminate magnetic excitations from vibrational ones. The intensity of a phonon excitation increases as Q^2 whereas magnetic excitations drop down as Q increases, due to the magnetic form factor.

A good example of the extensive use of INS to investigate the exchange interaction and wavefunction symmetries in magnetic clusters is given by the experiments performed on AF rings.

3.1. *Spin dynamics in antiferromagnetic molecular rings*

Ring-shaped homometallic molecular clusters are a subclass of molecular magnets in which the metal ions are arranged at the vertices of regular polygons thus forming perfect ring structures (Taft and Lippard, 1990; van Slageren *et al.*, 2002; Larsen *et al.*, 2003a; Affronte *et al.*, 2007). With an even number of ions and a dominant AF nearest-neighbor exchange coupling, the spin cancellation in these cyclic clusters is complete, and the spin ground state in zero field is $S=0$. The spectrum of the excited states is characterized by rotational and spin-wave like bands (Schnack and Luban, 2001; Waldmann, 2002) that are expected to merge to those of an infinite chain as the number of spin centers increases, therefore they can also serve as model systems for one-dimensional AF chains. Molecular rings have also been proposed as candidates for the observation of the quantum tunneling of the Néel vector through the anisotropy barrier (Chiolero and Loss, 1998), although the conditions to be met in real samples are still a matter of debate.

The first example of a molecular ring that has raised attention is the Fe_{10} molecule, the so called "ferric wheel", first synthesized by Taft and Lippard (1990). The dominant antiferromagnetic coupling between the ten spins $s=5/2$ leads to a total spin ground state $S=0$. Evidence of a discrete structure of the energy spectrum was given by high field DC and pulsed magnetization measurements (Taft *et al.*, 1994). The presence of well defined steps at evenly spaced magnetic field values $H_{S,S+1} = [E(S+1) - E(S)]/g\mu_B$ was associated with the occurrence of level-crossing conditions. To describe the spin dynamics of AF rings the following microscopic spin Hamiltonian is generally used:

$$H = \sum_i J_i \mathbf{s}(i) \cdot \mathbf{s}(i+1) + \sum_i d_i[s_z^2(i) - s_i(s_i+1)/3] + \sum_{i>j} \mathbf{s}(i) \cdot \mathbf{D}_{ij} \cdot \mathbf{s}(j) + \mu_B \sum_i g_i \mathbf{B} \cdot \mathbf{s}(i) \quad (16)$$

where $\mathbf{s}(i)$ is the spin operator for the ith ion in the molecule ($\mathbf{s}(N) \equiv \mathbf{s}(1)$). The first term is the dominant nearest neighbor (NN) isotropic Heisenberg exchange interaction. The second and third terms describe uniaxial local crystal fields and anisotropic intra-cluster spin-spin interactions (the z axis is perpendicular to the ring plane). The last term is the Zeeman coupling with an external field $\mathbf{B}$. The isotropic exchange interaction is usually the dominant term in the spin Hamiltonian and S is a good quantum number. INS has been extensively used to investigate molecular rings. The large number of magnetic excitations that can usually be measured offer the opportunity to accurately determine NN and NNN (next-nearest neighbor) exchange parameters, to closely investigate wavefunction symmetries and spin mixing effects. Also from the Q dependence of the magnetic excitations detailed information on the nature of the spin states can be extracted.

3.1.1. *Elementary excitations in antiferromagnetic rings*

The first molecular wheel that has been investigated by INS is the Fe_6 AF ring (Waldmann *et al.*, 1999). The combination of INS experiments with magnetic susceptibility and high-field torque magnetometry measurements allowed the coupling constants and the anisotropy parameters to be determined. However, the analysis of the Q dependence of the observed INS excitations was found to be not fully reliable due to instrumental artifacts, therefore not all the relevant information could be extracted from the INS data. An exhaustive study of the elementary spin excitation spectrum of AF rings has been obtained later on by Carretta *et al.* (2003b) with accurate INS measurements performed on the octanuclear Cr_8 ring. This system has become a prototype for the study of elementary excitations in AF rings (Waldmann *et al.*, 2003). The magnetic characterization of Cr_8 by susceptibility, torque magnetometry and high-frequency electron paramagnetic resonance was performed by van Slageren *et al.* (2002). The data were modeled with J = 1.5 meV for the isotropic exchange and D_1 = 0.2 meV and D_2 = 0.05 meV for the anisotropic parameter of the first two excited states ($S = 1$ and $S = 2$). INS was subsequently used to explore higher energy levels and the symmetry of the spin wavefunctions. Measurements were performed on $Cr_8F_8(Piv_2D_9)_{16}$ (Cr_8D), which is the fully deuterated version of Cr_8, using IN6 at the Institute Laue Langevin (ILL, France) and MARI time of flight spectrometers at the ISIS spallation source (UK) (Carretta *et al.*, 2003b). A best fit of the INS spectra was obtained with a single value for the isotropic exchange parameters $J = 1.46 \pm 0.04$ meV, $D_1 = 0.188$ meV and $D_2 = 0.044$ meV, $E_1 = 0.02$ meV and $E_2 = 0.005$ meV for the axial and rhombic anisotropic parameters, respectively. The inclusion of a rhombic term was necessary to improve the quality of the fit, especially for the $|S = 0\rangle \rightarrow |S = 1\rangle$ transition. These values are comparable with the ones previously obtained, even though the rhombic parameter was found to be about 5 times larger than that determined before. This discrepancy has been attributed to the substantial mixing between the $|S = 0\rangle$ and $|S = 2\rangle$ states and $|S = 1\rangle$ and $|S = 3\rangle$ states that was neglected in the analysis of the EPR data.

The INS results have confirmed the theoretical predictions based on models that exploit the underlying bipartite structure and high symmetry of the rings to describe the spin level structure and spin dynamics. As observed by (Taft *et al.*, 1994) and predicted theoretically by Schnack and Luban (2001), the lowest lying energy levels of Cr_8 are found to follow the Landé rule:

$$E_{S,\min} = -\frac{4J}{N}\left[S(S+1) - \frac{2Ns_i}{2}\left(\frac{Ns_i}{2}+1\right)\right], (N = 8, s_i = 3/2) \tag{17}$$

and thus to belong to the so called rotational (L) band. The excitations observed at higher energy transfers confirm the presence of a set of parallel rotational bands (E-band) and above that a quasicontinuum of states, as predicted by Waldmann (2001). Physically, the L-band is associated with the mutual rotations of the two sublattices, whilst the E-band corresponds to the (discrete) antiferromagnetic spin-wave excitations. The INS data collected on the Cr_8 have nicely confirmed the above picture, as evidenced in Fig. 4.

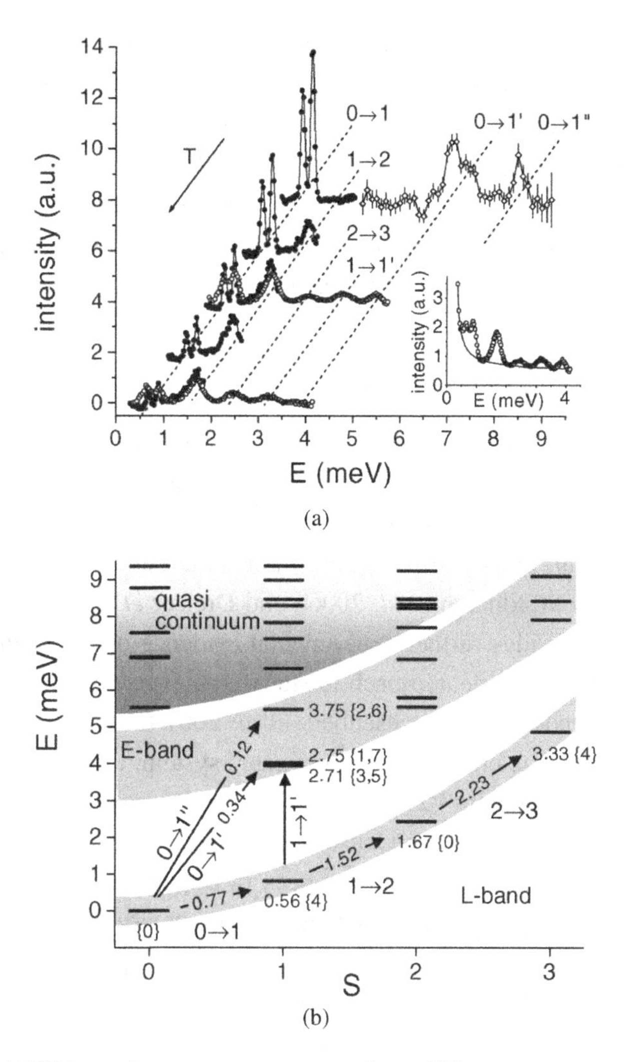

Fig. 4. (a) INS intensity versus energy transfer at different temperatures for Cr_8. Data collected on IN6 at different energies and MARI are combined together. (b) Isotropic exchange calculated energy level diagram for Cr_8 as a function of the total spin quantum number S. Arrows indicate the observed transitions. Values at arrows give the matrix element $\langle m|s_i^z|n\rangle^2$. From Waldmann *et al.* (2003).

Additional information is provided by the intensity of the inelastic peaks. Indeed, the calculations reported in Waldmann (2001) predict that the matrix element between different spin states in the L band should increase linearly as $(S+1)$ (see Waldmann (2001) for details). INS experiments have indeed confirmed the predicted behavior through the analysis of the intensity of the inelastic peaks, which is directly proportional to the transition matrix elements. On the other hand, the transitions to the quasicontinuum have been found to have negligible intensity as expected. Furthermore, the theory predicts that, because of the different spatial symmetry properties of the L and E bands, the transitions within the L band have a different Q dependence with respect to transitions to the E band. This has also been confirmed by the experimental data.

The study initiated with Cr_8 was then extended to other members of the family of AF homonuclear rings, and further experimental evidence of the rotational-band model (RBM) was found for Fe_{10} (Santini *et al.*, 2005), $CsFe_8$ (Waldmann *et al.*, 2005, 2006b), Fe_{18} (Waldmann *et al.*, 2009) as well as for frustrated molecules like the Fe_{30} magnetic Keplerate (Garlea *et al.*, 2006; Waldmann, 2007). In particular, the detailed and exhaustive INS study reported by Waldmann *et al.* 2006b and Dreiser *et al.* 2010 on the $CsFe_8$ molecule provides further experimental evidence of the nature of the E-band. Another possible approach to study elementary excitations in AF rings has been adopted by Ochsenbein *et al.* 2007 to describe the spin excitations in AF finite chains. The investigated system, the so called "horseshoe", consists of 6 spin-3/2 Cr^{3+} ions arranged on a ring with a missing bond (Larsen *et al.*, 2003b; Ochsenbein *et al.*, 2008). The open boundary condition introduces some differences in the energy spectrum if compared to that of the parent ring with periodic boundary conditions. These effects have been investigated using INS experiments performed at ILL (IN5) and at ISIS (IRIS) (Ochsenbein *et al.*, 2007). Several transitions within the Landé band and toward the E-band were detected, revealing the splitting of states in the E-band otherwise degenerate in the closed ring. Furthermore, the energy gap between the ground state $S = 0$ and the lowest triplet $S = 1$ has been found to be smaller than in the parent compound and the concept of "standing" spin waves, as opposed to "running" spin waves, was introduced. Several spin wave theories (SWT), standard for an extended system, have been applied by Ochsenbein *et al.* (2007) to interpret the excitation energies in the framework of spin wave dispersions. The theory that better reproduces the experimental data was found to be the spin-level SWT (Ochsenbein *et al.*, 2007).

An alternative method to create a finite chain is to introduce a diamagnetic (Zn, Cd) impurity in the homonuclear ring (Larsen *et al.*, 2003a,b). Effectively opened AF rings have been obtained by inserting Zn^{2+} or Cd^{2+} ions in the parent Cr_8 compound. The resulting Cr_8Cd and Cr_8Zn compounds have been extensively studied using magnetization and heat-capacity measurements that revealed deviations from the Landé rule (Ghirri *et al.*, 2007; Furukawa *et al.*, 2008).

The effects induced by the breaking of the ideal ring symmetry on the spin wavefunctions have been investigated in more detail by Bianchi *et al.* (2009) using inelastic neutron-scattering measurements. Data were collected on a deuterated powder sample of Cr_8Zn using the time-of flight Disk Chopper Spectrometer DCS at the National Institute of Standards and Technology NIST Center for Neutron Research, USA (see Fig. 5) [Copley, J.R.D. and Cook, J.C. (2003). Chem. Phys. **292**, 477]. Qualitative differences with respect to the parent compound were found in the internal structure of the spin eigenstates, probed through the wave-vector transfer dependence of the neutron cross section. The usual rotational bands picture has been found to be no more valid because of the occurrence of disjoint quantum fluctuations of the total-spin length of the two sublattices. This effectively removes the sharp separation of the low-lying states in two distinct rotational bands, as shown in Fig. 5.

4. Probing Quantum Coherence

Quantum tunneling is a phenomenon that has been widely studied in molecular nanomagnets. In high-spin magnetic clusters the tunneling generally involves fluctuations between the $+M_S$ and $-M_S$ components of a well defined S multiplet, but the tunnel splitting is usually too small to be detected through spectroscopic techniques. This also means that the tunnel rate is generally smaller than the decoherence time and coherent tunneling is difficult to observe. An opportunity to observe quantum coherent phenomena is offered by AF rings, where the tunnel splitting is expected to be considerably larger (Barbara and Chudnovsky, 1990). In the following sections the results of INS experiments that have allowed to make steps forward in establishing quantum coherent scenarios are reported.

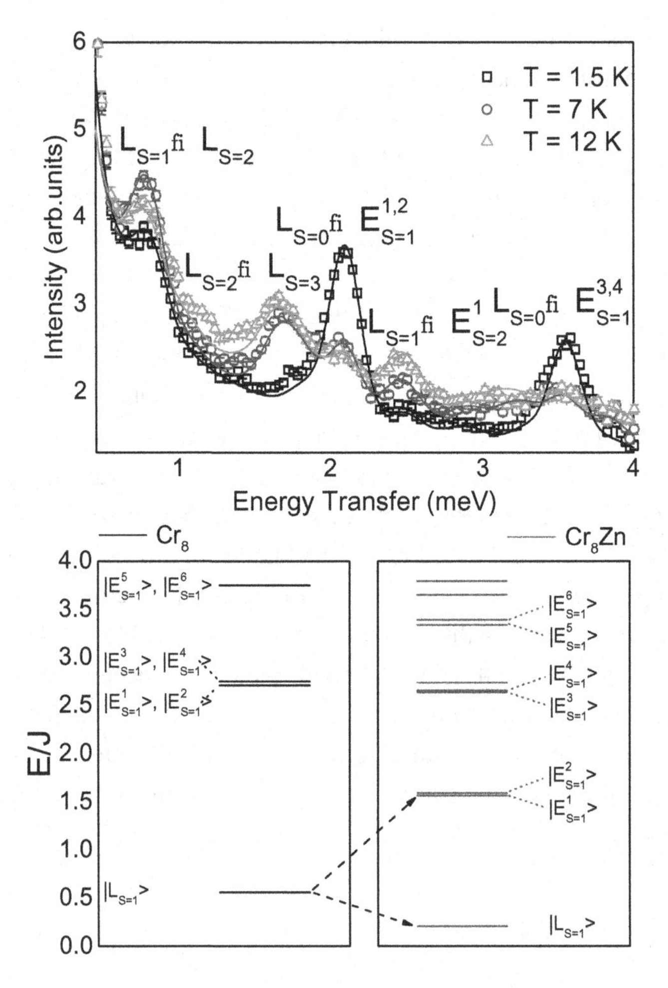

Fig. 5. (top) INS spectra collected on DCS on Cr_8Zn compound as a function of temperature. (bottom) Energy of the low-lying triplet eigenstates for Cr_8 and Cr_8Zn. For the latter, the labelling of states as L or E refers to the dominant component in the state. From Bianchi *et al.* (2009).

4.1. *Tunneling of the Néel vector*

The spins of an AF ring can be subdivided into two ferromagnetic sublattices and the Néel vector can be defined as:

$$\mathbf{n} = (\mathbf{S_A} - \mathbf{S_B})/N \tag{18}$$

where $\mathbf{S}_A$ and $\mathbf{S}_B$ are the total spin of each sublattice. Due to the symmetry of the two sublattices, the molecule possesses two energetically equivalent states, one corresponding to the Néel vector **n** pointing along the $+\mathbf{z}$ direction (perpendicular to the ring plane) and the other one with **n** pointing along $-\mathbf{z}$. In the presence of strong anisotropy the two states are separated by an energy barrier and in analogy to the Quantum Tunneling of the Magnetization (QTM) in high spin molecules, tunneling of the Néel vector (NVT) can occur.

Macroscopic quantum tunneling of the Néel vector in antiferromagnets was introduced by Barbara and Chudnovsky (1990) and the concept was extended to ringlike structures by Chiolero and Loss (1998). In tunneling conditions, the system is expected to coherently oscillate between the two opposite configurations with a frequency that has to be much larger than the decoherence rate. The coherent quantum oscillations of **n** will leave the total spin invariant, hence to experimentally demonstrate this phenomenon a local probe that couples with the spin of one sublattice is needed. Alternatively, the inelastic neutron scattering technique can be used to verify if the conditions for tunneling are fulfilled. The dynamics of **n** is described by the time correlation function given by Honecker *et al.* (2002):

$$\langle n_\alpha(t) n_\alpha(0)\rangle = \sum_\nu e^{-i\Delta_\nu t/\hbar} |\langle \nu | n_\alpha | 0\rangle|^2 \qquad (19)$$

where $\alpha = x, y, z$, ν is the spin state and $\Delta_\nu = E(\nu) - E(0)$. For the tunneling to occur the condition: $|\langle 1 | n_\alpha | 0\rangle|^2 = s^2$ has to be fulfilled.

INS provides all the information relevant to verify this. Indeed, the intensity of the inelastic transition within the tunnel-split doublet is proportional to the value of the spectral weight $|\langle 1 | n_z | 0\rangle|^2$. To extract this value from the measured spectra, it is necessary to perform an absolute units measurement of the inelastic transition, but the accuracy of these measurements is generally not enough to get a conclusive answer. On the other hand, the high accuracy in determining the spin wave functions from INS data allows a precise calculation of the spectral weight. Also, the position of the INS peaks directly gives the energy gap $\Delta_1 = E(1) - E(0)$, proportional to the tunnel frequency. INS has therefore been used to study to what extent the tunneling of the Néel vector characterizes the low-temperature spin dynamics of some of the molecular AF rings that are expected to show this phenomenon. The first investigated molecule was $CsFe_8$ (Waldmann *et al.* 2005; Waldmann *et al.* 2006b). This molecule has been found to fulfill one of the prerequisites for the tunneling picture, i.e. the tunneling action $S_0/\hbar$ is greater than 4, where $S_0/\hbar = \mathrm{N}s\sqrt{-2d/J}$ (with d the effective uniaxial anisotropy) (Chiolero and Loss, 1998; Meier and Loss, 2001; Waldmann *et al.*, 2005). This means that the axial anisotropy is of the same order of magnitude as the isotropic exchange. In this condition, the energy gap Δ_1 between the singlet ground state and the triplet excited state becomes smaller than the splitting Δ_2 of the $S = 1$ multiplet. INS measurements performed by Waldmann *et al.* (2005) on a non deuterated powder sample of $CsFe_8$ using the IN5 spectrometer at the ILL have indeed revealed that $\Delta_1 = 0.51$ meV and $\Delta_2 = 0.80$ meV. From these measurements values of $J = 1.78$ meV and $d = 0.048$ meV were determined, giving $S_0/\hbar = 4.6$. Therefore the semiclassical criterion for the tunneling scenario is fulfilled. However, a calculation of the spectral weight $|\langle 1|n_z|0\rangle|^2$ as a function of d/J reported by Santini *et al.* (2005) for a ring of 8 spins 5/2 shows that $|\langle 1|n_z|0\rangle|^2$ approaches the value of $s^2 = 6.25$ only for $d/J \gg 0.1$. Therefore the experimental value $d/J \simeq 0.026$ found for $CsFe_8$ allows one to conclude that in this molecule the tunneling scenario is only approximately valid. After the $CsFe_8$ study, other AF rings have been characterized. In particular, the Fe_{10} ferric wheel, being originally identified as a promising candidate for the NVT by Chiolero and Loss (1998) and Meier and Loss (2001), has been accurately investigated using INS. The experiments were carried out using the DCS time-of-flight spectrometer at NIST (Santini *et al.*, 2005). The measurements of several intermultiplet excitations allowed an accurate determination of exchange integrals and single-ion anisotropy parameters. The huge dimension of the Hilbert space prevents an exact diagonalization of the microscopic spin Hamiltonian, but the combination of the Lanczos method and the use of irreducible tensor operators (ITOs), together with symmetry considerations, allowed a good interpretation of the INS spectra. The best fit of the experimental data is obtained with $J = 1.23$ meV, $d = -5 \pm 0.25\ \mu$eV and $e = 3.6 \pm 1.8\ \mu$eV. These values give the first indication that the tunneling action $S_0/\hbar = 4.2$ fulfills the tunneling requirements. The tunnel splitting Δ_1 is found to be about 0.35 meV. The calculation of the time correlation function of the Néel vector has however revealed that the value of $|\langle 1|n_z|0\rangle|^2$ is far below $s^2 = 6.25$, also excluding the possibility for this molecule of achieving the ideal tunneling picture. More recently a combined torque, magnetization and INS study of a new Fe_{18} wheel, performed by (Waldmann *et al.*, 2009), reports the highest value for the tunneling action $S_0/\hbar = 5.9$, representing a favorable condition for the tunneling. Furthermore, the observation of wiggles in the torque measurements at high fields has been associated with the oscillations of NVT splitting. Therefore the low temperature spin dynamics of the Fe_{18} wheel is considered to be accurately described by the NVT scenario.

4.2. *Quantum oscillations of the total spin*

Beside the NVT, another interesting quantum phenomenon that has been observed in a series of AF exchange-coupled clusters is the quantum oscillation of the total spin S. The first manifestation of this phenomenon was reported by Waldmann *et al.* (2004) and Carretta *et al.* (2003a) as the evidence of quantum magneto-oscillations in the Mn-[3×3] grid. High-field torque magnetometry revealed for the first time the presence of peaks at particular values of the applied magnetic field. The combination of the torque measurement and INS experiments (Guidi *et al.*, 2004) was found to be essential to determine the parameters of the effective spin Hamiltonian suitable to describe the mechanism. The details of the high energy transfer INS spectra measured on IN6 (ILL) were satisfactorily reproduced by using two exchange AF constants $J_1 = 0.47$ meV and $J_2 = 0.33$ meV (Guidi *et al.*, 2004). The high resolution spectra collected on IRIS (ISIS) allowed a precise determination of the axial anisotropy parameter $D = -6\ \mu$eV of the $S = 5/2$ ground state. The energy level diagram calculated as a function of the magnetic field reveals the presence of several anticrossings (ACs) between states with different total spin values, namely between $S = 5/2$ and $S = 7/2$, $S = 7/2$ and $S = 9/2$, etc. At the ACs, the total spin of the molecule is expected to oscillate between S and $S + 1$ and the peaks in the torque have been associated with this effect. The repulsion between two spin levels is the manifestation of the so-called S-mixing (see Carretta *et al.* (2004a)), an effect that occurs only when the two crossing states belong to the same irreducible representations of the molecule point group (Carretta *et al.*, 2005). S-mixing induced by the application of an external magnetic field is expected to occur also in AF wheels if the ideal ring symmetry or topology is broken. The synthesis of a family of heterometallic AF rings, the Cr_7M (M = divalent ion) (Larsen *et al.*, 2003a,b), has allowed the experimental realization of this phenomenon. In heterometallic AF rings the ideal ring symmetry is broken by the introduction of an ion with different spin on one site. The details of the spin Hamiltonian for Cr_7M (M = Zn, Ni, Mn) have been obtained by accurate INS measurements, reported in Caciuffo *et al.* (2005). The evidence of the occurrence of mixing between different spin states has been provided by the observations of peaks in the magnetic torque at regular intervals of the applied magnetic field (Carretta *et al.*, 2005), in analogy with what was observed in Mn-[3 $\times$ 3]. However, the torque measurements do not give information on the coherence of the quantum oscillations. A direct measurement of the AC gap and the demonstration of the occurrence of *coherent* oscillations of S was provided by INS measurements on a single crystal of Cr_7Ni by Carretta *et al.* (2007). The experiments, performed on DCS (NIST) using an applied magnetic field up to 11.4 T, have allowed for the first time the direct measurement of the AC gap between the $S = 1/2$ ground state and $S = 3/2$ excited state (Fig. 6). At the AC, the gap was found to be about 0.12 meV, therefore a value of

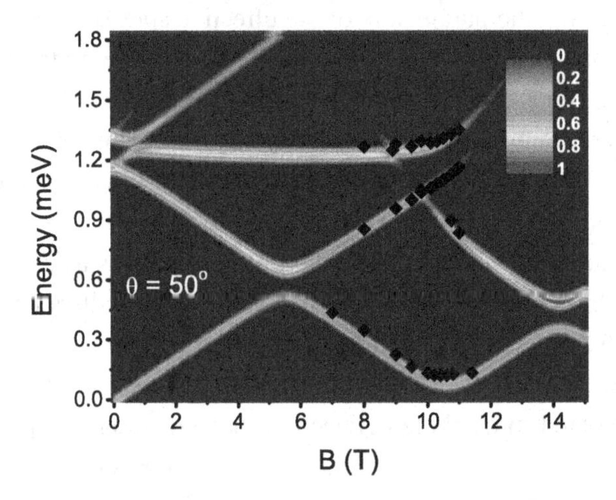

Fig. 6. The color plot shows the intensity of the calculated INS cross section as a function of the magnetic field and energy transfer for T = 66 mK using the best fit spin Hamiltonian parameters for Cr_7Ni. The points represent the position of the observed INS peaks. From Carretta *et al.* (2007).

about 29 GHz for the frequency of the coherent oscillations can be deduced. The inelastic character of the observed AC excitation indicates that the spins of the molecules oscillate coherently between $S = 1/2$ and $S = 3/2$ over a non-negligible number of cycles. Furthermore the measured INS cross section, which strongly depends on the details of the spin wavefunctions, is very well reproduced by the calculation, confirming the validity of the model spin Hamiltonian.

5. Zero-Field Splitting Anisotropy in High Spin Clusters

Whilst INS is a unique technique for directly and accurately determining the exchange integrals in MNMs, its effectiveness in the determination of the anisotropy terms of the spin Hamiltonian is almost comparable to other spectroscopic techniques. The recent introduction of the zero-field version of EPR, the Frequency Domain Magnetic Resonance (FDMR) technique (van Slageren *et al.*, 2003), has indeed overcome the inconvenience of having the magnetic field as unavoidable perturbation in EPR. Nevertheless, the INS technique generally allows the measurement of a larger number of intramultiplet transitions compared to EPR. The main reason for this is that the INS intensity is proportional to the population of the *initial state* (see Eq. 4), and not to the difference in population of the two states involved in the transition, as for EPR. This characteristic is particularly useful when it is necessary to investigate the excitations between states at the top of the anisotropy barrier, where non axial and higher order terms in the spin Hamiltonian show their effect. In the following paragraphs some of the most relevant INS experiments that revealed the details of the anisotropy split ground state are reported.

5.1. *The giant spin approximation and beyond*

The first full characterization of the zero-field split (ZFS) spin Hamiltonian of a MNM using INS was performed by Caciuffo *et al.* (1998) on the Fe_8Br molecule. The high resolution spectra, collected using the IN5 spectrometer (ILL) (see Fig. 7), were modeled using the Stevens operator equivalent formalism and the simplified Eq. 13 for the INS cross section.

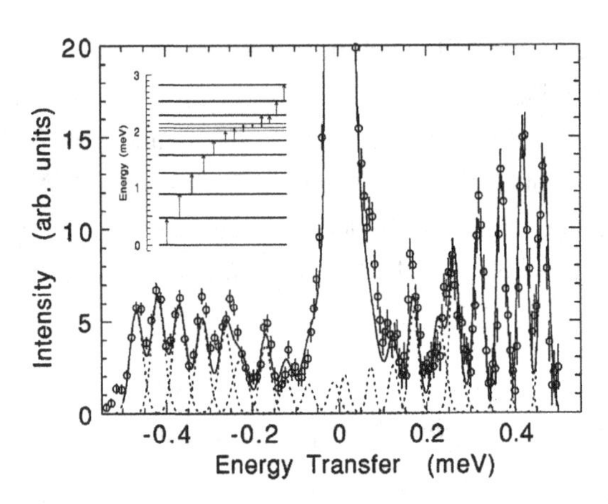

Fig. 7. INS spectrum of Fe_8Br measured at 9.6 K showing the transitions within the ZFS $S = 10$ ground state. The corresponding energy level scheme is reported in the inset. Lines are the theoretical spectrum in the GSA approximation. From Caciuffo *et al.* (1998).

Higher resolution measurements, performed using the IN10 backscattering spectrometer (ILL) (see (Amoretti *et al.*, 2000)), were found to be necessary to better identify the excitations in the low energy region. The INS data were satisfactorily reproduced using the Giant Spin Approximation (GSA), with very similar parameters to the ones deduced by EPR (Hill *et al.*, 2002) and optical spectroscopy (Mukhin *et al.*, 2001). High energy transfer measurements performed on DCS spectrometer by Carretta *et al.* (2006) allowed the determination of the position of the first excited level at about 4 meV. A first excited state well above the ground state justifies the use of the GSA to model the INS data of Fe_8Br, however the mixing between different S states cannot be neglected in understanding the magnetic properties of the molecule. Indeed, using the GSA, the high order parameter B_4^4, which has the greatest effect in determining the value of the tunnel splitting Δ, gives a value of Δ three orders of magnitude lower than what has been experimentally found by Wernsdorfer and Sessoli (1999). This discrepancy could only be resolved taking account of S-mixing, as demonstrated by Carretta *et al.* (2004a). Following the method developed in Liviotti *et al.* (2002), the S-mixing effects were included in the effective spin Hamiltonian and the resulting renormalized Stevens operator parameters allowed a satisfactory interpretation of the observed tunnel splitting.

High resolution INS measurements were also used by Mirebeau *et al.* (1999) and Zhong *et al.* (1999) to resolve the details of the $S = 10$ spin ground state level structure of the extensively studied Mn_{12}-acetate. The measurement performed by Mirebeau *et al.* (1999) on the IN5 spectrometer allowed the accurate determination of the transverse fourth-order terms of the spin Hamiltonian and a slightly lower value of B_4^4 ($\pm 3.0(5) \times 10^{-5}$ cm^{-1}) was found if compared with the EPR results ($\pm 4(1) \times 10^{-5}$ cm^{-1}). The perturbation of a very high external magnetic field (up to 25 T) necessary with EPR to reach the highest levels in the anisotropy split ground multiplet, and the need to additionally fit the g parameter, make the EPR result less precise than the zero-field measurement with INS. Indeed, the INS results were found to reproduce more accurately the relaxation data reported by Thomas *et al.* (1996). However, the symmetry allowed spin Hamiltonian terms are not sufficient to account for all the observed quantum tunneling phenomena and an additional rhombic term ($E(S_x^2 - S_y^2)$) is needed for the interpretation of the observed QTM. The crystallographically determined S_4 symmetry of the molecule does not allow a second order rhombic term. A possible explanation of this discrepancy was found by Cornia *et al.* (2002) through an accurate x-ray diffraction analysis at low temperature which revealed that the solvent disorder gives rise to six different isomers, four of which possess a symmetry lower than S_4. This finding is supported by detailed EPR measurements on a single crystal by Hill *et al.* (2003). Their analysis provided an upper limit of 1.24×10^{-4} meV for the E values for the different variants. Subsequent INS measurements performed by (Bircher *et al.*, 2004) taking advantage of the high flux of the upgraded IN5 spectrometer allowed a better determination of the transverse anisotropy parameters as compared with previous measurements by Mirebeau *et al.* (1999). The INS data were modeled considering the contribution of the six isomers having different anisotropy parameters and the calculated spectra are found to be in good agreement with the experiment, confirming the model proposed by Cornia *et al.* (2002). Higher order terms in the ZFS Hamiltonian that violates the structural symmetry of the molecule were also found by Carretta *et al.* (2004b) for the Fe_4 compound. A best fit of the high resolution INS data required the introduction of a rhombic term forbidden by the D_3 symmetry of the Fe_4 molecule but also crucial to understand the observed QTM.

More recently, Waldmann *et al.* (2006a) succeeded in measuring the relaxation of the magnetization of Mn_{12}-ac using a time-resolved INS experiment on an oriented array of about 500 single crystals. The experiment performed on IN5 allowed the collection of good-statistic spectra in relatively short time intervals. The system was initialized by applying an external magnetic field to populate only the $M = +10$ state. Then the field was quickly inverted to make the $M = +10$ state lie at higher energy and INS data were collected during the time the system relaxed to the more stable $M = -10$ state. The technique allows the direct measurement of the population of the energy levels in time, as nicely shown in the animated movie (EPAPS, 2006). Data collected following an alternative "stop and go" protocol (for details see Waldmann *et al.* (2006a)) are shown as a function of time in Fig. 8. The relaxation time was measured for three different temperatures and a fit to an Arrhenius law gave an activation energy $E_a = 70(2)$ K and $\tau_0 = 1.6$ ns. Similar time-resolved spectroscopic studies on Mn_{12} have been more recently performed by van Slageren *et al.* (2009) using the FDMRS technique.

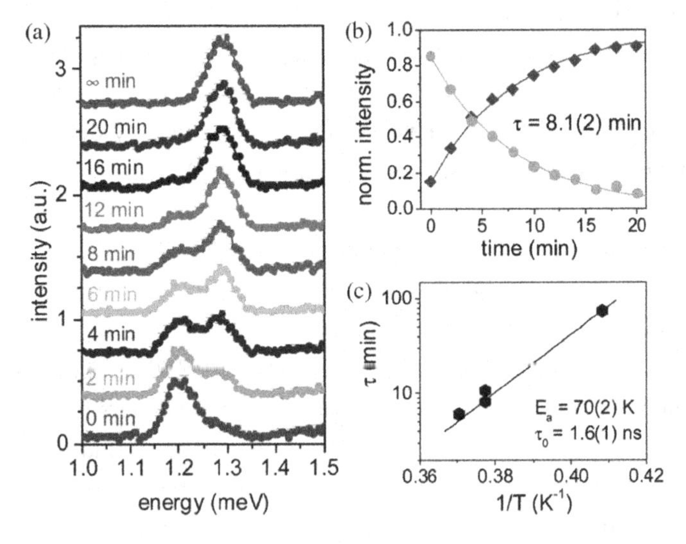

Fig. 8. (a) INS spectra at 2.65 K collected on IN5 following the "stop and go" protocol. (b) Time dependence of the normalized intensities of the transitions $M = -10 \rightarrow -9$ (squares) and $M = +10 \rightarrow +9$ (circles). (c) Arrhenius plot for the relaxation times. Circles are experimental points, the line is a fit to the Arrhenius law with parameters as indicated. From Waldmann *et al.* (2006a).

It is known that some of the Mn_{12}-ac derivatives exhibit two out-of-phase ac magnetic susceptibility signals, one in the 4–7 K region and the other in the 2–3 K region. It has been argued by Aubin *et al.* (2001) that Jahn-Teller isomerism is the origin of this magnetic behavior and a slow relaxing (SR) and a fast relaxing (FR) species were identified. The single crystal INS experiments by Waldmann *et al.* (2006a) have served also as spectroscopic evidence of the presence of these two species. Indeed, the FR and SR can be distinguished in the INS spectra due to their different D values and behavior as a function of the magnetic field. The SR majority species has the largest D parameter (-0.057 meV) whilst the D parameter for the FR species was determined to be $D = -0.036$ meV (Waldmann *et al.*, 2007b). The same result was reported in the spectroscopic study of Mn_{12}-acetate under hydrostatic pressure by (Sieber *et al.*, 2005). This study also revealed that the normal SR molecule partially transforms into the FR species as the pressure increases and that the D parameter increases with increasing pressure.

Beside the details of the anisotropy split ground state, INS has also been used to identify the position of the first excited states in Mn_{12}-ac. Transitions from the $S = 10$ ground state to excited $S = 9$ states were first measured by Hennion *et al.* (1997) using a triple axis spectrometer. These authors reported a series of intermultiplet peaks between 4 and 10 meV that were subsequently confirmed and more deeply investigated by Chaboussant *et al.* (2004). The study of Chaboussant *et al.* (2004) has extended the region of INS measurements to energy transfers up to 45 meV to be able to derive a unique set of exchange constants. The combination of measurements using IN4 (ILL), MARI (ISIS) and FOCUS (Paul Scherrer Institut, Switzerland), together with magnetic susceptibility data, has led to a consistent set of four different J parameters reflecting the possible exchange paths in the molecule. The dominant parameters were found to be the AF exchanges J_1 and J_2 between the inner core Mn^{4+} atoms and the outer ring Mn^{3+} atoms of the molecule (see Chaboussant *et al.* (2004) for details). Several J parameters and exchange coupling paths have been reported in the literature by Sessoli *et al.* (1993), Raghu *et al.* (2001), Regnault *et al.* (2002) and Chaboussant *et al.* (2004). Waldmann *et al.* (2007a) performed a careful calculation of the INS Q dependencies of the excitations for which experimental data are available using different proposed coupling topographies. They have shown how the variations in the Q dependencies can help to distinguish between very different coupling topographies and rule out the scheme that cannot satisfactorily reproduce the experimental INS data. Furthermore, they calculated the INS intensity of the $M = \pm 10 \rightarrow \pm 9$ transition as a function of Q_x and Q_y for a single crystal of Mn_{12}. The calculated spectra present various features as a function of **Q** that provide additional information relevant for deducing the correct exchange scheme. Single crystal experimental data that can support the reported calculations are not yet available, but are within reach of today's cold neutron spectrometers that employ position sensitive detectors (IN5 at ILL and LET at ISIS).

5.1.1. *Beyond the giant spin approximation*

The Giant Spin Hamiltonian (GSH), with the inclusion of higher order zero-field splitting parameters to model the effects of S-mixing, generally represents a good approximation for modelling the magnetic properties of MNMs. However this model has been found to fail to describe the physics of a number of compounds where the isotropic exchange parameters have the same order of magnitude as the single ion anisotropy parameters (intermediate exchange limit). A striking example is given by the recently studied family of Mn_6 molecules, where the GSA has been found to breakdown in explaining the magnetic relaxation (Carretta *et al.*, 2008). There exist a large number of variants of Mn_6 which display a rich variety of spin ground states and anisotropy energy barriers (Inglis *et al.*, 2009; Milios *et al.*, 2008, 2007a). One of them, $[Mn_6O_2(Et\text{-}sao)_6(O_2CPh(Me)_2)_2(EtOH)_6]$ (spin ground state $S = 12$), presents to date the highest anisotropy barrier of 86.4 K for a single molecule magnet (Milios *et al.*, 2007b).

Susceptibility and magnetization experiments were performed to characterize the spin Hamiltonian of several Mn_6 variants. However, the GSH deduced from the analysis of the data was not able to satisfactorily reproduce the value of the measured anisotropy barrier. The discrepancies were tentatively attributed to the presence of low lying excited levels. In order to investigate this possibility, detailed spectroscopic studies on three selected members of the family were performed using INS and FDMR techniques (Carretta *et al.*, 2008; Pieper *et al.*, 2010). The INS spectra, together with the energy level diagrams calculated using the full spin Hamiltonian (left) and the GSH (right) for the Mn_6 $U_{eff} = 86.4$ K are reported in Fig. 9. The experiments have indeed revealed the presence of low-lying excited levels nested within the anisotropy split spin ground state. This energy level structure cannot be described at all using the GSH model (see Fig. 9b). Indeed, the GSH does not account for a number of spin states different from the

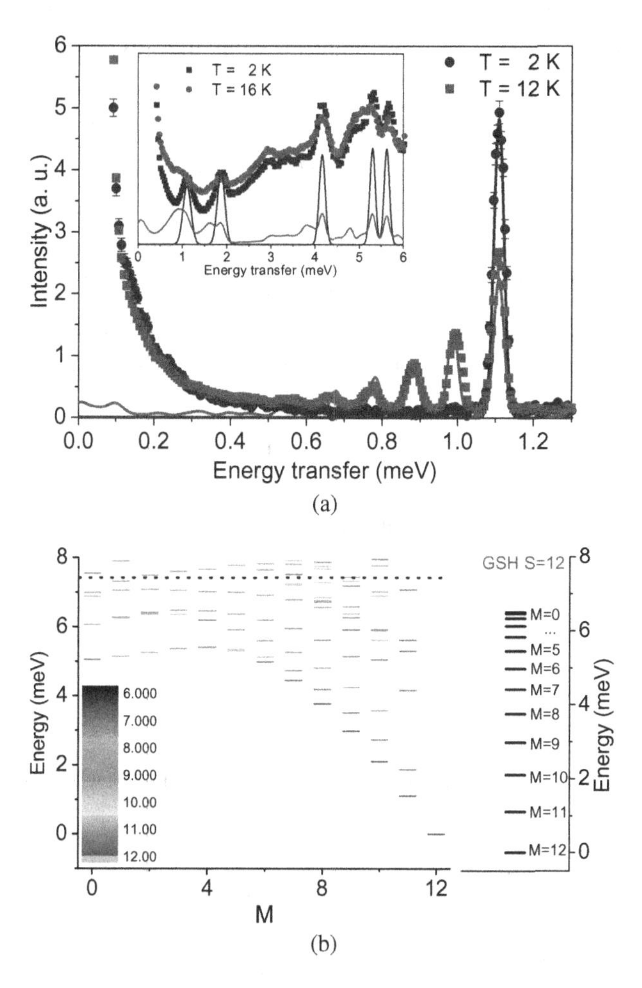

Fig. 9. (a) INS spectra collected on IN5 at different temperatures for Mn_6 $S = 12$ $U_{eff} = 86.4$ K. (b) Calculated energy level diagram using the full spin Hamiltonian (left) and the GSH (right). The color maps S_{eff}, where $\langle S^2 \rangle := S_{eff}(S_{eff} + 1)$. The black dashed line corresponds to the observed value of $U_{eff} = 86.4$ K. From Carretta *et al.* (2008).

ground state multiplet that lie within the split $S = 12$ energy level diagram. The excited levels play a crucial role in the spin dynamics providing additional paths in the relaxation process. As a consequence the anisotropy barrier cannot be simply calculated as the difference in energy between the lowest and highest states of the anisotropy split ground state. An accurate theoretical analysis of the spectroscopic experimental data carried out by Carretta *et al.* (2008) using the full microscopic spin Hamiltonian has succeeded to give a comprehensive description of the relaxation dynamics and to satisfactorily reproduce the experimental value for the effective energy barriers. The nesting of different spin states manifests its effect also in the quantum tunneling of the magnetization. Magnetization measurements show additional steps as a function of the magnetic field which are absent in the giant spin model and are attributed to quantum tunneling between spin states with different total spin S (Carretta *et al.*, 2008; Bahr *et al.*, 2008). Effects of the presence of low lying excited states and consequent strong S-mixing in the QTM have been reported also for Ni_4 (Wilson *et al.*, 2006) and Mn_{12} molecular wheel (Ramsey *et al.*, 2008).

These results emphasize the importance of using complementary techniques for investigating the magnetic properties of MNMs and INS is certainly very effective for shedding light on systems that present complex behavior.

References

Affronte, M., Carretta, S., Timco, G.A. and Winpenny, R.E.P. (2007). *Chem. Commun.* 1789.

Amoretti, G., Caciuffo, R., Combet, J., Murani, A. and Caneschi, A. (2000). *Phys. Rev. B* **62**, 3022.

Aubin, S., Sun, Z., Eppley, H., Rumberger, E., Guzei, I., Folting, K., Gantzel, P., Rheingold, A., Christou, G. and Hendrickson, D. (2001). *Inorg. Chem.* **40**, 2127.

Bahr, S., Milios, C.J., Jones, L.F., Brechin, E.K., Mosser, V. and Wernsdorfer, W. (2008). *Phys. Rev. B* **78**, 132401.

Barbara, B. and Chudnovsky, E.M. (1990). *Phys. Lett. A* **145**, 205.

Basler, R., Boskovic, C., Chaboussant, G., Güdel, H.U., Murrie, M., Ochsenbein, S.T. and Sieber, A. (2003). *Chem. Phys. Chem.* **4**, 910.

Bianchi, A., Carretta, S., Santini, P., Amoretti, G., Guidi, T., Qiu, Y., Copley, J.R.D., Timco, G., Muryn, C. and Winpenny, R.E.P. (2009). *Phys. Rev. B* **79**, 144422.

Bircher, R., Chaboussant, G., Sieber, A., Güdel, H.U. and Mutka, H. (2004). *Phys. Rev. B* **70**, 212413.

Borras-Almenar, J.J., Clemente-Juan, J.M., Coronado, E. and Tsukerblat, B.S. (1999). *Inorg. Chem.* **38**, 6081.

Caciuffo, R., Amoretti, G., Murani, A., Sessoli, R., Caneschi, A. and Gatteschi, D. (1998). *Phys. Rev. Lett.* **81**, 4744.

Caciuffo, R., Guidi, T., Amoretti, G., Carretta, S., Liviotti, E., Santini, P., Mondelli, C., Timco, G., Muryn, C.A. and Winpenny, R.E.P. (2005). *Phys. Rev. B* **71**, 174407.

Carretta, S., Guidi, T., Santini, P., Amoretti, G., Pieper, O., van Slageren, J., Hallak, F.E., Wernsdorfer, W., Mutka, H., Russina, M., Milios, C.J. and Brechin, E.K. (2008). *Phys. Rev. Lett.* **100**, 157203.

Carretta, S., Liviotti, E., Magnani, N., Santini, P. and Amoretti, G. (2004a). *Phys. Rev. Lett.* **92**, 207205.

Carretta, S., Santini, P., Amoretti, G., Affronte, M., ghirri, A., Sheikin, I., Piligkos, S., Timco, G. and Winpenny, R.E.P. (2005). *Phys. Rev. B* **72**, 060403(R).

Carretta, S., Santini, P., Amoretti, G., Guidi, T., Caciuffo, R., Candini, A., Cornia, A., Gatteschi, D., Plazanet, M. and Stride, J.A. (2004b). *Phys. Rev. B* **70**, 214403.

Carretta, S., Santini, P., Amoretti, G., Guidi, T., Copley, J.R.D., Qiu, Y., Caciuffo, R., Timco, G. and Winpenny, R.E.P. (2007). *Phys. Rev. Lett.* **98**, 167401.

Carretta, S., Santini, P., Amoretti, G., Guidi, T., Dyson, J., Caciuffo, R., Stride, J.A., Caneschi, A. and Copley, J. (2006). *Phys. Rev. B* **73**, 144425.

Carretta, S., Santini, P., Liviotti, E., Magnani, N., Guidi, T., Caciuffo, R. and Amoretti, G. (2003a). *Eur. Phys. J. B* **36**, 169.

Carretta, S., van Slageren, J., Guidi, T., Liviotti, E., Mondelli, C., Rovai, D., Cornia, A., Dearden, A.L., Carsughi, F., Affronte, M., Frost, C.D., Winpenny, R.E.P., Gatteschi, D., Amoretti, G. and Caciuffo, R. (2003b). *Phys. Rev. B* **67**, 094405.

Chaboussant, G., Sieber, A., Ochsenbein, S., Güdel, H.-U., Murrie, M., Honecker, A., Fukushima, N. and Normand, B. (2004). *Phys. Rev. B* **70**, 104422.

Chiolero, A. and Loss, D. (1998). *Phys. Rev. Lett.* **80**, 169.

Cornia, A., Sessoli, R., Sorace, L., Gatteschi, D., Barra, A.L. and Daiguebonne, C. (2002). *Phys. Rev. Lett.* **89**, 257201.

Dreiser, J., Waldmann, O., Dobe, C., Carver, G., Ochsenbein, S.T., Sieber, A., Güdel, H.U., van Duijn, J., Taylor, J. and Podlesnyak, A. (2010). *Phys. Rev. B* **81**, 024408.

EPAPS (2006). See EPAPS Document No. E-APPLAB-88-214603 for an animated movie. The document may be reached via the EPAPS homepage (http://www.aip.org/pubservs/epaps.html) or directly from the link http://netserver.aip.org/cgi-bin/epaps?ID=E-APPLAB-88-214603.

Furrer, A. and Güdel, H. (1979). *J. Magn. Mag. Mat.* **14**, 256.

Furrer, A. and Güdel, H.U. (1977). *Phys. Rev. Lett.* **39**, 657.

Furukawa, Y., Kiuchi, K., i. Kumagai, K., Ajiro, Y., Narumi, Y., Iwaki, M., Kindo, K., Bianchi, A., Carretta, S., Timco, G.A. and Winpenny, R.E.P. (2008). *Phys. Rev. B* **78**, 092402.

Garlea, V.O., Nagler, S.E., Zarestky, J.L., Stassis, C., Vaknin, D., Kögerler, P., McMorrow, D.F., Niedermayer, C., Tennant, D.A., Lake, B., Qiu, Y., Exler, M., Schnack, J. and Luban, M. (2006). *Phys. Rev. B* **73**, 024414.

Ghirri, A., Candini, A., Evangelisti, M., Affronte, M., Carretta, S., Santini, P., Amoretti, G., Davies, R.S.G., Timco, G. and Winpenny, R.E.P. (2007). *Phys. Rev. B* **76**, 214405.

Güdel, H. and Furrer, A. (1977). *Mol. Phys.* **33**, 1335.

Guidi, T., Carretta, S., Santini, P., Liviotti, E., Magnani, N., Mondelli, C., Waldmann, O., Thompson, L.K., Zhao, L., Frost, C.D., Amoretti, G. and Caciuffo, R. (2004). *Phys. Rev. B* **69**, 104432.

Haraldsen, J.T., Barbes, T. and Musfeldt, J.L. (2005). *Phys. Rev. B* **71**, 064403.

Hennion, M., Pardi, L., Mirebeau, I., Suard, E., Sessoli, R. and Caneschi, A. (1997). *Phys. Rev. B* **56**, 8819.

Hill, S., Edwards, R.S., Jones, S.I., Dalal, N.S. and North, J.M. (2003). *Phys. Rev. Lett.* **90**, 217204.

Hill, S., Maccagnano, S., Park, K., Achey, R.M., North, J.M. and Dalal, N.S. (2002). *Phys. Rev. B* **65**, 224410.

Honecker, A., Meier, F., Loss, D. and Normand, B. (2002). *Eur. Phys. J. B* **27**, 487.

Inglis, R., Jones, L.F., Milios, C.J., Datta, S., Collins, A., Parsons, S., Wernsdorfer, W., Hill, S., Perlepes, S.P., Piligkos, S. and Brechin, E.K. (2009). *Dalton Trans.* 3403.

Larsen, F.K., McInnes, E.J.L., Mkami, H.E., Overgaard, J., Piligkos, S., Rajaraman, J., Rentschler, E., Smith, A.A., Smith, G.M., Boote, V., Jennings, M., Timco, G.A. and Winpenny, R.E.P. (2003a). *Angew. Chem. Int. Ed.* **42**, 101.

Larsen, F.K., Overgaard, J., Parsons, S., Rentschler, E., Smith, A.A., Timco, G.A. and Winpenny, R.E.P. (2003b). *Angew. Chem. Int. Ed.* **42**, 5978.

Liviotti, E., Carretta, S. and Amoretti, G. (2002). *J. Chem. Phys.* **117**, 3361.

Lovesey, S.W. (1987). *Theory of Neutron Scatteirng from Condensed Matter, Vol. 1 and 2* (Clarendon Press, Oxford).

Meier, F. and Loss, D. (2001). *Phys. Rev. B* **64**, 224411.

Milios, C.J., Inglis, R., Vinslava, A., Bagai, R., Wernsdorfer, W., Parsons, S., Perlepes, S.P., Christou, G. and Brechin, E.K. (2007a). *J. Am. Chem. Soc.* **129**, 12505.

Milios, C.J., Piligkos, S. and Brechin, E.K. (2008). *Dalton Trans.* 1809–1817.

Milios, C.J., Vinslava, A., Wernsdorfer, W., Moggach, S., Parsons, S., Perlepes, S.P., Christou, G. and Brechin, E.K. (2007b). *J. Am. Chem. Soc.* **129**, 2754.

Mirebeau, I., Hennion, M., Casalta, H., Andres, H., Güdel, H., Irodova, A. and Caneschi, A. (1999). Neutron scattering mn12, *Phys. Rev. Lett.* **83**, 628–631.

Mukhin, A., Gorshunov, B., Dressel, M., Sangregorio, C. and Gatteschi, D. (2001). *Phys. Rev. B* **63**, 214411.

Ochsenbein, S.T., Tuna, F., Rancan, M., Davies, R.S.G., Muryn, C.A., Waldmann, O., Bircher, R., Sieber, A., Carver, G. and Mutka, H. (2008). *Chem. Eur. J.* **14**, 5144.

Ochsenbein, S.T., Waldmann, O., Sieber, A., Carver, G., Bircher, R., Güdel, H.U., Davies, R.S.G., Timco, G.A., Winpenny, R.E.P., Mutka, H. and Fernandez-Alonso, F. (2007). *Europhys. Lett.* **79**, 17003.

Pieper, O., Guidi, T., Carretta, S., van Slageren, J., Hallak, F.E., Lake, B., Santini, P., Amoretti, G., Mutka, H., Koza, M., Russina, M., Schnegg, A., Milios, C.J., Brechin, E.K., Julia, A. and Tejada, J. (2010). *J. Phys. Rev. B* **81**, 174420. *arXiv:1003.0537.*

Raghu, C., Rudra, I., Sen, D. and Ramasesha, S. (2001). *Phys. Rev. B* **64**, 064419.

Ramsey, C.M., Del Barco, E., Hill, S., Shah, S.J., Beedle, C.C. and Hendrickson, D.N. (2008). *Nature Phys.* **4**, 277–281.

Regnault, N., Jolicoeur, T., Sessoli, R., Gatteschi, D. and Verdaguer, M. (2002). *Phys. Rev. B* **66**, 054409.

Santini, P., Carretta, S., Amoretti, G., Guidi, T., Caciuffo, R., Caneschi, A., Rovai, D., Qiu, Y. and Copley, J.R.D. (2005). *Phys. Rev. B* **71**, 184405.

Schnack, J. and Luban, M. (2001). *Phys. Rev. B* **63**, 014418.

Sessoli, R., Tsai, H.L., Schache, A.R., Wang, S., Vincent, J.B., Folting, K., Gatteschi, D., Christou, G. and Hendrickson, D.N. (1993). *J. Am. Chem. Soc.* **115**, 1804.

Sieber, A., Bircher, R., Waldmann, O., Carver, G., Chaboussant, G., Mutka, H. and Güdel, H.U. (2005). *Angew. Chem., Int. Ed.* **44**, 4239.

Squires, G.L. (1978). *Introduction to the theory of thermal neutron scatteirng* (Cambridge University Press, Cambridge).

Taft, K.L., Delfs, C.D., Papefthymiou, G.C., Foner, S., Gatteschi, D. and Lippard, S.J. (1994). *J. Am. Chem. Soc.* **116**, 823.

Taft, K.L. and Lippard, S.J. (1990). *J. Am. Chem. Soc.* **112**, 9629.

Thomas, L., Lionti, F., Ballou, R., Gatteschi, D., Sessoli, R. and Barbara, B. (1996). *Nature* **383**, 145.

van Slageren, J., Sessoli, R., Gatteschi, D., Smith, A.A., Helliwell, M., Winpenny, R.E.P., Cornia, A., Barra, A.L., Jansen, A.G.M., Rentschler, E. and Timco, G.A. (2002). *Chem. Eur. J.* **8**, 277.

van Slageren, J., Vongtragool, S., Gorshunov, B., Mukhin, A. and Dressel, M. (2009). *Phys. Rev. B* **79**, 224406.

van Slageren, J., Vongtragool, S., Gorshunov, B., Mukhin, A., Karl, N., Krzystek, J., Telser, J., Muller, A., Sangregorio, C., Gatteschi, D. and Dressel, M. (2003). *PCCP* **5**, 3837–3843.

Waldmann, O. (2001). *Phys. Rev. B* **65**, 024424.

Waldmann, O. (2002). *Europhys. Lett.* **60**, 302308.

Waldmann, O. (2003). *Phys. Rev. B* **68**, 174406.

Waldmann, O. (2007). *Phys. Rev. B* **75**, 012415.

Waldmann, O., Bircher, R., Carver, G., Sieber, A., Güdel, H.U. and Mutka, H. (2007a). *Phys. Rev. B* **75**, 174438.

Waldmann, O., Carretta, S., Santini, P., Koch, R., Jansen, A.G.M., Amoretti, G., Caciuffo, R., Zhao, L. and Thompson, L.K. (2004). *Phys. Rev. Lett.* **92**, 096403.

Waldmann, O., Carver, G., Dobe, C., Biner, D., Sieber, A., Güdel, H.U., Mutka, H., Olivier, J. and Chakov, N.E. (2006a). *Appl. Phys. Lett.* **88**, 042507.

Waldmann, O., Carver, G., Dobe, C., Sieber, A., Güdel, H.U. and Mutka, H. (2007b). *J. Am. Chem. Soc.* **129**, 1526.

Waldmann, O., Dobe, C., Güdel, H. and Mutka, H. (2006b). *Phys. Rev. B* **74**, 054429.

Waldmann, O., Dobe, C., Mutka, H., Furrer, A. and Güdel, H. (2005). *Phys. Rev. Lett.* **95**, 057202.

Waldmann, O. and Güdel, H. (2005). *Phys. Rev. B* **72**, 094422.

Waldmann, O., Guidi, T., Carretta, S., Mondelli, C. and Dearden, A.L. (2003). *Phys. Rev. Lett.* **91**, 237202.

Waldmann, O., Schülein, J., Koch, R., Müller, P., Bernt, I., Saalfrank, R.W., Andres, H.P., Güdel, H.U. and Allenspach, P. (1999). *Inorg.Chem.* **38**, 5879.

Waldmann, O., Stamatatos, T.C., Christou, G., Güdel, H., Sheikin, I. and Mutka, H. (2009). *Phys. Rev. Lett.* **102**, 157202.

Wernsdorfer, W. and Sessoli, R. (1999). *Science* **284**, 133.

Wilson, A., Lawrence, J., Yang, E., Nakano, M., Hendrickson, D.N. and Hill, S. (2006). *Phys. Rev. B* **74**, 140403(R).

Zhong, Y., Sarachik, M.P., Friedman, J.R., Robinson, R.A., Kelley, T.M., Nakotte, H., Christianson, A.C., Trouw, F., Aubin, S.M. and Hendrickson, D.N. (1999). *J. Appl. Phys.* **85**, 5636.

Chapter 5

RECENT DEVELOPMENTS IN EPR SPECTROSCOPY OF MOLECULAR NANOMAGNETS

ERIC J. L. McINNES

School of Chemistry
The University of Manchester
Manchester M13 9PL, UK

This chapter is intended as an update to previous reviews on electron paramagnetic resonance (EPR) spectroscopy of exchange coupled transition ion cluster complexes (McInnes, 2006), a class of compounds now commonly known as molecular nanomagnets (Gatteschi *et al.*, 2006a). Much of the work in this field was, and is, driven by the study of a particular class of clusters: the single molecule magnets (SMMs) which display the remarkable property of a molecular magnetic memory effect (Gatteschi *et al.*, 2006a). This effect derives from a large ground state total spin quantum number S, split by a negative zero-field splitting (ZFS) such that the $M = \pm S$ substates are lowest in energy. At low enough temperature these states can be preferentially populated by application of an external magnetic field along the $\pm$z directions. There is an energy barrier to relaxation defined by the energy gap to $M = 0$ (for integer spin) or $\pm 1/2$ (for half-integer spin) states, which depends on the magnitudes of both S and the ZFS. Since these are both in principle EPR observables this technique has been important in the understanding and development of the fascinating low temperature physics of such materials. However, the interesting physics is not restricted to SMMs and many important quantum magnetic effects have been found in other classes of molecular nanomagnets, e.g. anti-ferromagnetically coupled rings and grids, high symmetry polyhedra, etc. (Gatteschi *et al.*, 2006a; McInnes and Winpenny, 2010).

More recent reviews of EPR spectroscopy of SMMs and related materials are available from some of the foremost groups in the field (Gatteschi *et al.*, 2006b; Feng *et al.*, 2008). The theory of EPR of exchange coupled species is comprehensively covered in Bencini and Gatteschi's textbook (1989), and an excellent early tutorial on the application of high-field/frequency techniques is available (Barra *et al.*, 1998). We have previously published a comprehensive review of EPR of SMMs and other high spin ground state complexes incorporating the literature up to mid-2005 (McInnes, 2006), and also a series of reviews on the EPR of exchange coupled compounds in general (Collison and McInnes, 2002, 2004, 2007; Boeer *et al.*, 2008). This current work is not intended to be comprehensive, but rather to cover what I regard as some of the most interesting developments in the last five years. These include:

(i) Modelling of EPR spectra beyond the giant spin approximation, thence investigation of mixing between total spin states and the origins of important higher order ZFS terms.
(ii) EPR studies of discrete clusters-of-clusters (linked molecular nanomagnets).
(iii) Pulsed EPR studies of molecular nanomagnets, thence direct investigation of spin dynamics and coherence.

1. Beyond the Giant Spin Approximation (GSA)

The most common approach that has been taken to the interpretation of EPR data of molecular nanomagnets is the so-called giant spin approximation (GSA). Spectra are treated as arising from an isolated spin S and modelled with spin Hamiltonian (1) or equivalent. The advantages of this approach are the conceptual simplicity of the model with a limited number of easily understandable parameters, and the restriction of calculations to a Hamiltonian matrix of dimension $2S+1$ with basis functions being the $|M>$ states.

$$H = \mu_B \mathbf{H} \cdot \mathbf{g} \cdot \hat{\mathrm{S}} + D\left[\hat{\mathrm{S}}_z^2 - S(S+1)/3\right] + E\left(\hat{\mathrm{S}}_+^2 + \hat{\mathrm{S}}_-^2\right)/2 + B_k^q \hat{O}_k^q \quad (1)$$

μ_B is the Bohr magneton, $\mathbf{H}$ is the applied magnetic field, $\mathbf{g}$ is the electronic g-matrix, and D and E are the axial and rhombic second order ZFS parameters. $\hat{\mathrm{O}}_{\mathrm{k}}^{\mathrm{q}}$ are Stevens operators describing higher order ZFS effects, allowed up to order $k = 2S$: a full list can be found in Abragam and Bleaney's textbook (1986). In general, these higher order ZFS terms are assumed to become less important as the order increases; most studies have not considered terms beyond 4th order. These are defined as:

$$\hat{\mathrm{O}}_4^0 = 35\hat{\mathrm{S}}_z^4 - [30S(S+1) - 25]\hat{S}_z^2 - 6S(S+1) + 3S^2(S+1)^2$$
$$\hat{\mathrm{O}}_4^2 = \{[7\hat{\mathrm{S}}_z^2 - S(S+1) - 5](\hat{\mathrm{S}}_+^2 + \hat{\mathrm{S}}_-^2) + (\hat{\mathrm{S}}_+^2 + \hat{\mathrm{S}}_-^2)[7\hat{\mathrm{S}}_z^2 - S(S+1) - 5]\}/4$$
$$\hat{\mathrm{O}}_4^4 = (\hat{\mathrm{S}}_+^4 + \hat{\mathrm{S}}_-^4)/2$$

Note that the B_4^0 and B_4^4 parameters can be non-zero in axial symmetry, the latter introducing transverse anisotropy, while B_4^2 can only be non-zero in rhombic symmetry.

It has long been established that these higher order ZFS terms cannot be neglected in simulation of EPR [and inelastic neutron scattering (INS)] data of SMMs. Gatteschi and co-workers pointed out that this has great significance because transverse $\hat{\mathrm{O}}_{\mathrm{k}}^{\mathrm{q}}$ terms lead to a mixing of M states with $\Delta M = \pm \mathrm{nq}$ where n is an integer, which provides efficient mechanisms for quantum tunnelling of magnetisation (QTM) (Barra *et al.*, 1997). Carretta *et al.* (2004) showed that such terms can arise from mixing between different total S-states: 'S-mixing'. This implies a breakdown of the GSA.

A more complete description of the electronic structure is given by a spin Hamiltonian of the form (2):

$$H = \sum_{j>i} -2J_{ij}\hat{\mathrm{s}}_{\mathrm{i}} \cdot \hat{\mathrm{s}}_{\mathrm{j}} + \sum_{j>i} \hat{\mathrm{s}}_{\mathrm{i}} \cdot \mathbf{d}_{ij} \cdot \hat{\mathrm{s}}_{\mathrm{j}} + \sum_{i} \mu_B \mathbf{H} \cdot \mathbf{g}_i \cdot \hat{\mathrm{s}}_{\mathrm{i}} + \sum_{i} \{d_i[\hat{\mathrm{s}}_z^2 - s(s+1)/3] + e_i(\hat{\mathrm{s}}_+^2 + \hat{\mathrm{s}}_-^2)/2\} \quad (2)$$

where $\mathbf{g}_i$, d_i, e_i describe the g and second-order ZFS parameters of the individual spin s_i, and J_{ij} and $\mathbf{d}_{ij}$ describe isotropic and anisotropic components of the exchange interaction between centres i and j. (In principle we could include higher order local ZFS terms and antisymmetric exchange terms in Hamiltonian (2).)

The GSA presupposes that the isotropic exchange is the utterly dominant term in (2), generating a set of non-interacting multiplets which can be grouped according to a total spin quantum number S. This is the strong exchange limit, and straightforward numerical relationships relate how the anisotropic terms in (2) project onto the total spin S parameters in the GSA Hamiltonian (1) (Bencini and Gatteschi 1989). In the weak exchange regime, where $J_{ij} << \mathbf{d}_i$, then the states are thoroughly scrambled and the total spin S has no meaning. However, even when J_{ij} is dominant in (2), such that states group according to a total S to a good approximation, the small mixing between the multiplets arising from the anisotropic terms ($\mathbf{d}_i$, $\mathbf{d}_{ij}$) cannot be neglected: this is the situation described by Carretta *et al.* (2004). Then using Hamiltonian (2) will provide a more physical description of the system.

However, a severe limitation of the applicability of (2) is the full matrix dimension of $\Pi(2s_i + 1)$, i.e. the dimension scales rapidly with the nuclearity of the cluster. Hence, calculations rapidly become prohibitively large. A further limitation is that there are several potential unknowns such as the local ZFS parameters which are difficult to determine for large nuclearity clusters. In the last few years several groups have attempted to address these problems in the context of EPR spectroscopy.

Accorsi *et al.* (2006) reported detailed studies on $[Fe_4\{R(CH_2O)_3\}_2(diketonate)_6]$ (R=Me, CH_2Br, Ph), $[Fe_4\{^tBu(CH_2O)_3\}(OEt)_3(diketonate)_6]$, and $[Fe_4(OMe)_6(diketonate)_6]$, part of a well known family of SMMs (Fig. 1). Well-resolved 230 GHz powder EPR spectra arising from the $S = 5$ ground states are modelled according to a GSA Hamiltonian equivalent to (1) including the axial $\hat{O}_4^0$ operator. The ground state D varies between 0.21 and $-0.45\,cm^{-1}$, with B_4^0 varying from -1.1 to $+2 \times 10^{-5}\,cm^{-1}$. $|D|$ is found to decrease as the dihedral angle between the mean $\{Fe_4\}$ plane and the $\{Fe_2O(R)_2\}$ planes (the helical pitch) increases.

To investigate this effect, the principal axes and values of the local $\mathbf{d}_{Fe}$ tensors were estimated from angular overlap model (AOM) calculations for the central ion (Fe1), and from EPR studies on $[Fe_2(OMe)_2(diketonate)_4]$ (Abbati *et al.*, 2007) for the peripheral ions (Fe2). The former are axial and easy-axis type (co-parallel with Z, Fig. 1), the latter rhombic and hard-axis type. These parameters were found to reasonably reproduce the ground state D of $[Fe_4\{Me(CH_2O)_3\}_2(dpm)_6]$, which has the largest $|D|$ experimentally,

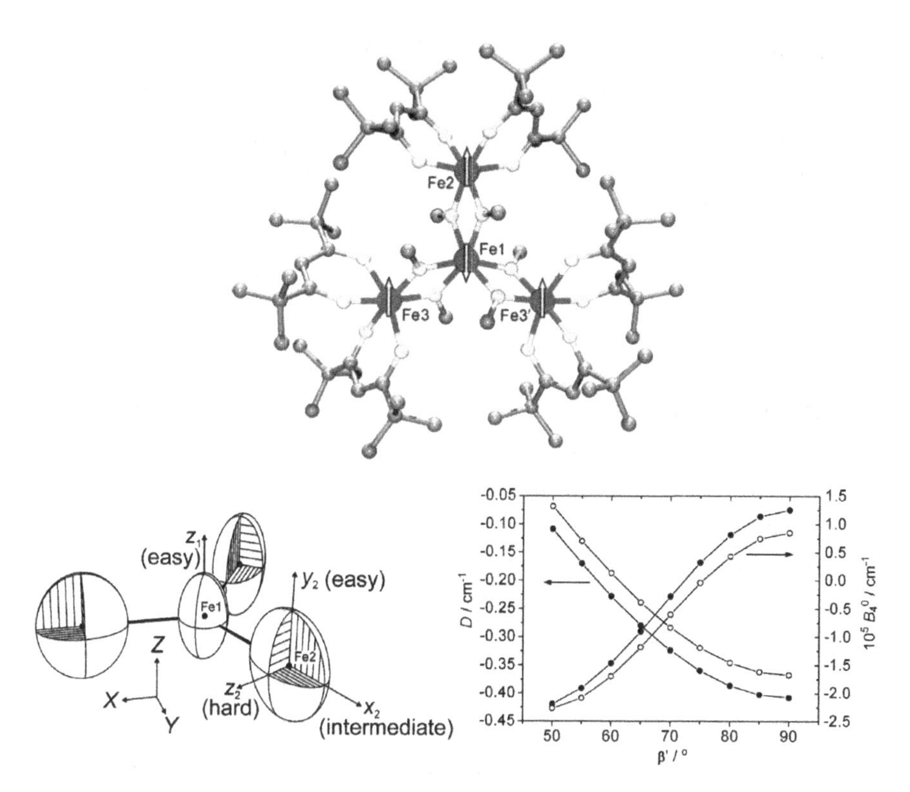

Fig. 1. Top: structure of $[Fe_4\{MeO)_6(dpm)_6]$ with ground state spin configuration. Bottom, left: Representation of orientation of local (xyz) and molecular (XYZ) ZFS tensors for $[Fe_4\{Me(CH_2O)_3\}_2(dpm)_6]$. Bottom, right: Variation of D and B_4^0 calculated by diagonalisation of zero-field Hamiltonian with varying angle (β') between z_2 and Z. The closed and open circles are for different values of d_1. Reproduced with permission from Accorsi *et al.*, 2006.

based on a simple strong exchange limit calculation assuming the hard and easy axes of the peripheral ions are perpendicular and parallel, respectively, to Z (Fig. 1, bottom left). They then diagonalise the 6^4 dimension microscopic Hamiltonian [equivalent to (2), neglecting $\mathbf{d}_{ij}$] for zero magnetic field, systematically decreasing the angle β' between z_1 (Z) and z_2, and fixing all other parameters. The lowest 11 eigenvectors correspond to the $S = 5$ ground state, and their eigenvalues were numerically fit to:

$$E(M) - E(0) = AM^2 + BM^4 + CM^6$$

where the coefficients A and B are related to the GSA ZFS parameters by $D = A + 875B_4^0$ and $B_4^0 = B/35$ (Fig. 1, bottom right). Decreasing

β' from 90° decreases $|D|$ as the projection of the local hard axis on Z is increased. More interestingly (in the present context), B_4^0 is calculated to be very sensitive to β' (i.e. the orientation of $\mathbf{d}_{Fe2}$), even changing sign at $\beta' \approx 72°$, qualitatively reproducing the experimental trend.

Wilson *et al.* (2006) probed similar effects in the heterocubane $[Ni_4(hmp)_4(^tBuCH_2CH_2OH)_4Cl_4]$ ("Ni_4"; hmpH = 2-hydroxymethyl pyridine), an $S = 4$ ground state SMM with $D = -0.589$, $B_4^0 = -1.2 \times 10^{-4}$ and $|B_4^4| = 4 \times 10^{-4}\,cm^{-1}$. These parameters come from fits of variable high-frequency EPR data to a GSA Hamiltonian and the latter parameter derived from the four-fold periodicity in 101 GHz single-crystal data in the hard plane. The molecule has S_4 (i.e. axial) symmetry. The great advantages of this system are the small matrix dimension (3^4) and that the single-ion $\mathbf{d}_{Ni}$ parameters (principal values and orientations) can be determined experimentally from the Ni-doped diamagnetic Zn analogue, giving $d_{Ni} = -5.30$ and $e_{Ni} = -1.20\,cm^{-1}$, with 15° between the local and molecular easy axes (Yang *et al.*, 2005). Feeding this information into the equivalent of Hamiltonian (2) for zero applied field ($\mathbf{d}_{NiNi}$ fixed from a dipolar model), the authors fit to the zero-field separations within the bottom 9-fold multiplet to obtain the isotropic exchange J. These separations were determined experimentally from extrapolation of multi-frequency single-crystal EPR data to zero-field. They find a best fit for $J = -5.9\,cm^{-1}$, i.e. of similar magnitude to d_{Ni}. In this intermediate exchange regime there is significant mixing of the total spin multiplets (giving rise to the apparent 4th-order behaviour of the $S = 5$ states in the GSA) and the authors note that this regime has allowed spectroscopic determination of J.

This regime was also observed by Prescimone *et al.* (2007) for the linear $Mn^{II}Mn^{III}Mn^{II}$ complex $[Mn_3(Hcht)_2(bipy)_4](ClO_4)_3$, which is an $S = 7$ ground state SMM, where powder 94 GHz EPR spectra can be modelled either by Hamiltonian (1) with 4th order effects, or via Hamiltonian (2) with $J \approx d_{Mn(III)}$. Calculation of the powder EPR spectrum using (2) is relatively straightforward given the matrix dimension of 180.

The origin of the various ZFS terms in the archetypal "Mn_{12}" family of SMMs has been debated for some time, particularly the non-zero rhombic terms (E, B_4^2 in the GSA) which are observed in the apparently tetragonally symmetric parent system $[Mn_{12}O_{12}(O_2CMe)_{16}(H_2O)_4] \cdot 2MeCO_2H \cdot 4H_2O$ (Gatteschi *et al.*, 2006a). In fact, it turns out that solvent disorder in the crystal lattice breaks the local four-fold symmetry (Cornia *et al.*, 2002). Two groups have reported single crystal EPR measurements of axially symmetric Mn_{12} derivatives where there is no apparent solvent disorder and the rhombic ZFS terms are absent (Chakov *et al.*, 2006; Barra *et al.*, 2007). This has allowed accurate determination of the remaining transverse anisotropy terms that are allowed in axial symmetry (B_4^4, etc.) through single crystal studies with the applied field in the hard plane. These papers give the parameters $D = -0.468$, $B_4^0 = -2.5 \times 10^{-5}$, $B_4^4 = \pm 3.0 \times 10^{-5}\,\text{cm}^{-1}$ for $[Mn_{12}O_{12}(O_2CCH_2Br)_{16}(H_2O)_4] \cdot 4CH_2Cl_2$ (Chakov *et al.*, 2006), and $D = -0.459$, $B_4^0 = -2.34 \times 10^{-5}$, $B_4^4 = +2.0 \times 10^{-5}\,\text{cm}^{-1}$ for $[Mn_{12}O_{12}(O_2CCH_2^tBu)_{16}(MeOH)_4] \cdot MeOH$ (Barra *et al.*, 2007). In the latter work, a remarkable feature is found in 115 GHz single-crystal EPR data in the hard plane (the crystallographic *ab* plane): although the highest field transition corresponding to $M = -10 \rightarrow -9$ in the GSA exhibits the expected four-fold periodicity, the $M = -9 \rightarrow -8$ transition is essentially invariant with angle (Fig. 2). Modelling this in the GSA necessitated introducing 6th-order terms allowed in tetragonal symmetry, giving $B_6^0 = -1.0 \times 10^{-8}$ and $B_6^4 = -1.0 \times 10^{-7}\,\text{cm}^{-1}$. They note that in order to reproduce the angular variation it is necessary to have opposite signs of the two transverse anisotropy terms B_4^4 and B_6^4.

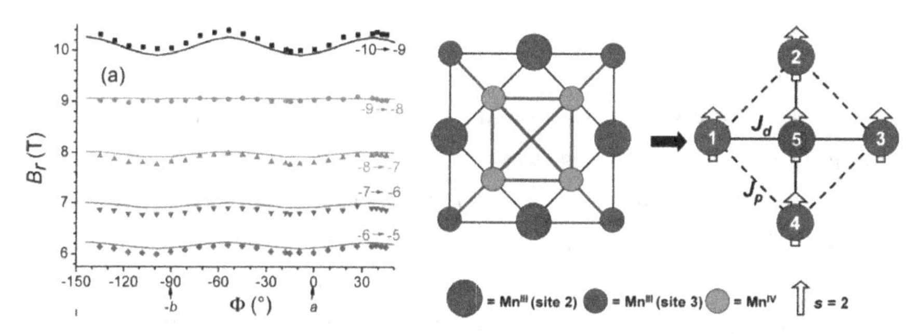

Fig. 2. Left: Angular dependence of resonance fields of $[Mn_{12}O_{12}(O_2CCH_2^tBu)_{16}(MeOH)_4] \cdot MeOH$, measured at 115 GHz and 5 K with the applied field in the hard plane. Solid lines are fits to the GSA Hamiltonian for $S = 10$. Right: Schematic of the reduced dimension spin model. Reproduced with permission from Barra *et al.*, 2007.

They then probed the origin of these terms using an exchange Hamiltonian. They adopt a reduced dimension model in order to overcome the 10^8 dimension of the full Hamiltonian matrix. The four Mn(III) ions that exhibit the greatest angle (36°) between their Jahn-Teller distortion axes and the tetragonal molecular axis are assumed to be the dominant contributors to the transverse anisotropy. The remaining four Mn(III) ions are taken as strongly coupled with the four Mn(IV) ions to give an intermediate spin of $s = 2$ (Fig. 2). This reduces the problem to a matrix of dimension 5^5. Fictitious J values are introduced such that the $S = 10$, $M = \pm 10$ ground state to $S = 9$, $M = \pm 9$ first excited state gap reproduces the value observed by INS; local $\mathbf{d}_{\mathbf{Mn(III)}}$ values and orientations are taken from AOM calculations; and the axial ZFS of the $s = 2$ intermediate spin is adjusted to reproduce the correct splitting in the $S = 10$ ground state. Diagonalising the zero-field matrix and identifying the 21 low-lying states that correspond to the $S = 10$ ground state in the GSA shows that the $\pm M$ pairs are degenerate for odd M, but non-degenerate for even M. They focus on the splitting of the $M = \pm 4$ pair, and calculate this as a function of the exchange J and the tilting angle of the $d_{Mn(III),z}$ axis: they find the latter parameter to have far greater effect for small changes. Moreover, the projection of the $d_{Mn(III),z}$ axes on the crystallographic *ab* plane match the orientations of the maxima in resonance fields in this plane. EPR spectra calculated on the basis of the simplified exchange Hamiltonian reproduce all the observed hard plane angular variations. The agreement with experiment is remarkable given the lack of any free parameters affecting the transverse anisotropy.

Datta *et al.*, (2007) report a complex in which S-mixing is so extensive that the $\Delta S = 0$ EPR selection rule breaks down, with the observation of inter-multiplet transitions. This is in the antiferromagnetically coupled [3 × 3]-grid-like complex $[Mn^{II}_9(2\text{-POAP-2H})_6](ClO_4)_6 \cdot 3.57MeCN \cdot H_2O$ (see the review by L. K. Thomson elsewhere in this volume), where antiferromagnetic coupling leads to an $S = 5/2$ ground state with an $S = 7/2$ first excited state. 50–60 GHz EPR single-crystal spectra with the field parallel to the easy axis of magnetization show the expected ground state intra-multiplet transitions and a much higher field transition. Resonance field-frequency plots show this latter transition shifts to lower fields for higher frequencies; extrapolation to zero-field gives a frequency of 233 GHz which

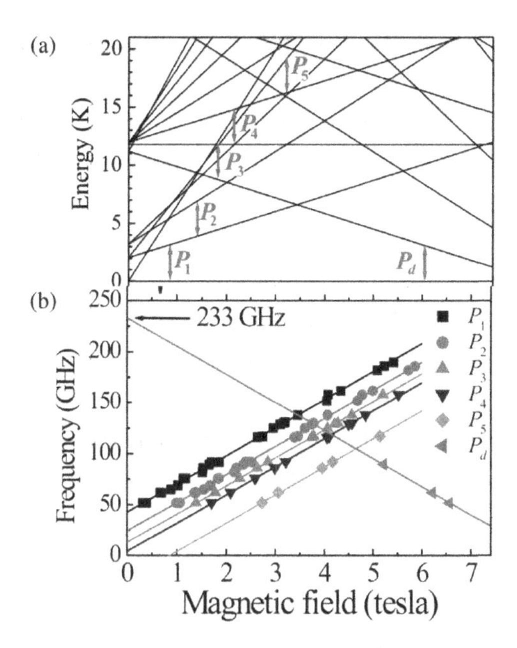

Fig. 3. Top: Calculated level scheme for $[Mn_9^{II}(2\text{-POAP-2H})_6](ClO_4)_6.3.57$ $MeCN.H_2O$, for field parallel to z, using the simplified exchange Hamiltonian described in the text. 62 GHz EPR transitions in red, "P_d" marks the intermultiplet transition. Bottom: Experimental resonance field-frequency plot showing inverse frequency dependence of P_d. Adapted with permission from Datta *et al.*, 2007.

is very close to the $S = 5/2$ to 7/2 energy gap determined by INS. The full exchange Hamiltonian has dimension 6^9, hence a simplified model is used where the outer Mn(II) ions are divided into two sublattices (the corner spins and the edge spins) which are then each treated as single $s = 10$ spins. These are coupled with each other and with the central Mn ion ($s = 5/2$), reducing the matrix dimension to 2646. A single J is assumed for coupling between the sublattice spins and with the central ion, and a single axial d for all spins, and these are determined from fitting to the experimentally determined (by extrapolation of field-frequency plots) zero-field separations for all observed transitions. The 62 GHz single-orientation EPR spectra are also calculated using this Hamiltonian, and match both the '$S = 5/2$' intra-multiplet transitions and the high-field transition (resonance field and intensity), and confirm the latter as $S = 5/2$, $M = -5/2 \rightarrow S = 7/2$, $M = -7/2$ in a GSA description. Analysis of the wavefunctions *cf.* strong exchange limit functions reveals that the ground state has 2.6% admixture of $S = 7/2$.

Piligkos *et al.* (2007 and 2009) have used full exchange coupled Hamiltonian methods to address a different problem: modelling EPR spectra where multiple total spin states are observed. This is unusual for high nuclearity clusters where well-resolved spectra are usually restricted to the ground state, but has been observed for a few antiferromagnetically coupled ring complexes (van Slageren *et al.*, 2002; Pilawa *et al.*, 2003; Piligkos *et al.*, 2007; Piligkos *et al.*, 2009). The challenge is that simulation of each of these states independently using the GSA generates a large number of independent parameters, where simulation of higher excited states may be based only on a few resonances. Furthermore, analysis of the results via strong exchange limit assumptions presupposes minimal S-mixing and is dependent on the coupling scheme chosen. It is then preferable to use an exchange Hamiltonian akin to (2), to generate all the relevant spin states from a limited number of parameters.

Piligkos *et al.* (2007 and 2009) broach this problem for a family of heterometallic $(NH_2Et_2)[Cr_7^{III}M^{II}F_8(O_2C^tBu)_{16}]$ ring complexes ('Cr_7M', Fig. 4). Antiferromagnetic coupling gives rise to discrete but non-zero ground state S depending on the choice of M^{II}. 9–94 GHz powder and single crystal EPR spectra are very well resolved with resolution of up to the 4th

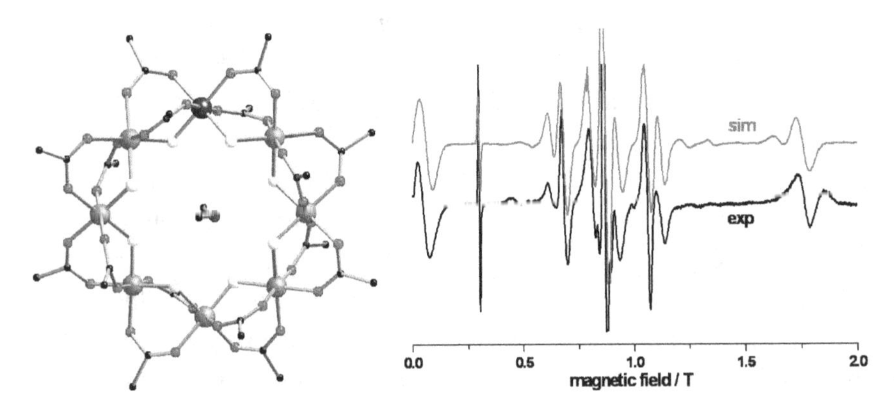

Fig. 4. Left: Structure of $(NH_2Et_2)[Cr_7MF_8(O_2C^tBu)_{16}]$ (^{t}Bu groups not shown). Right: 24 GHz EPR spectrum of a single-crystal of $(NH_2Et_2)[Cr_7CdF_8(O_2C^tBu)_{16}]$ at 5 K with field parallel to z (black), and calculated spectrum using full exchange Hamiltonian (2) (red).

or 5th excited states. To overcome the problem of the matrix dimensions, varying from 4^7 (= 16284) for Cr_7Cd to 6×4^7 (= 98304) for Cr_7Mn, they exploit the Davidson algorithm — an iterative subspace diagonalisation method that allows the exact calculation of a defined number of low-lying eigenstates of the full matrix. These can then be used in a unitary transformation of the full Hamiltonian to generate an effective operator accurate in the subspace. This allows relatively efficient calculation even of powder spectra where repeated diagonalisation is required over the applied field and orientation degrees of freedom.

The simplest system magnetically is Cr_7Cd. Using an isotropic exchange Hamiltonian [i.e. neglecting $\mathbf{d}_i$ and $\mathbf{d}_{ij}$ in (2)], the Hamiltonian matrix is diagonalised for a fixed J known accurately from INS studies. The lowest 12 total spin states (comprising 42 eigenvectors) span ca. 40 cm^{-1} in energy ($S = 3/2$ ground state, then $S = 1/2$, 5/2, 1/2, 3/2,... excited states): all population is in these states up to temperatures of ca. 10 K, and they are used to define the subspace for modelling the low temperature EPR. In order to limit the number of free variables a number of parameters were fixed: J_{CrCr} from INS modelling (-5.77 cm^{-1}), the $\mathbf{d}_{Cr}$ were assumed to be axial with z perpendicular to the Cr_7Cd plane, $\mathbf{d}_{CrCr}$ was also assumed to be axial but with z fixed along the Cr...Cr vectors. Hence, other than the g-values which were fixed at 1.96, the problem was reduced to only two free parameters: d_{Cr} and d_{CrCr}. The full matrix is diagonalised with initial estimates of these parameters and with zero applied field, using the Davidson algorithm, to obtain the lowest lying 42 eigenstates. These eigenstates are used to generate an effective operator for this subspace, which is then used to refine d_{Cr} and d_{CrCr} by numerical fitting to the multi-frequency EPR resonance fields. The best fit parameters are used in a new full matrix diagonalisation and the procedure repeated until the parameters converge. Excellent fits are obtained to both single orientation and powder multi-frequency spectra (e.g. Fig. 4). They note that the anisotropic component of the exchange interaction is significant since this is the dominant contribution to the ZFS in the $S = 5/2$ second excited state: this had previously been noted in detailed studies on dimetallics (ter Heerdt *et al.*, 2006). The Cr_7Ni ($S = 1/2$ total spin ground state) and Cr_7Mn ($S = 1$ total spin ground state) spectra are modelled using a similar methodology, fixing d_{Cr} and d_{CrCr} from Cr_7Cd and introducing the new free parameters d_M, D_{CrM} and J_{CrM}. Analysis of the zero-field eigenfunctions, cf. those from an isotropic version of Hamiltonian (2), shows that the ground state multiplets are well-described (>99%) by a total spin $S = 3/2$ (M = Cd), 1/2 (Ni) or 1 (Mn). However, increasing effects of S-mixing are observed as the level scheme is ascended; the authors go on to examine the effects of these in terms of modelling in the GSA approach.

2. Discrete Clusters-of-Clusters

Over and above the synthetic challenge of preparing ever more complex molecular systems, there is increasing interest in how to link molecular nanomagnets together in order to manipulate their quantum magnetic properties. An early example was from Wernsdorfer *et al.* (2002) who studied a hydrogen-bonded dimer of $S = 9/2$ ground state SMMs, $[Mn_4O_3Cl_4(O_2CEt)_3(py)_3]_2$ ('$(Mn_4)_2$'; py = pyridine; Fig. 5), and showed that a weak antiferromagnetic exchange between the two halves of the dimer quenches the QTM steps at zero applied magnetic field by shifting

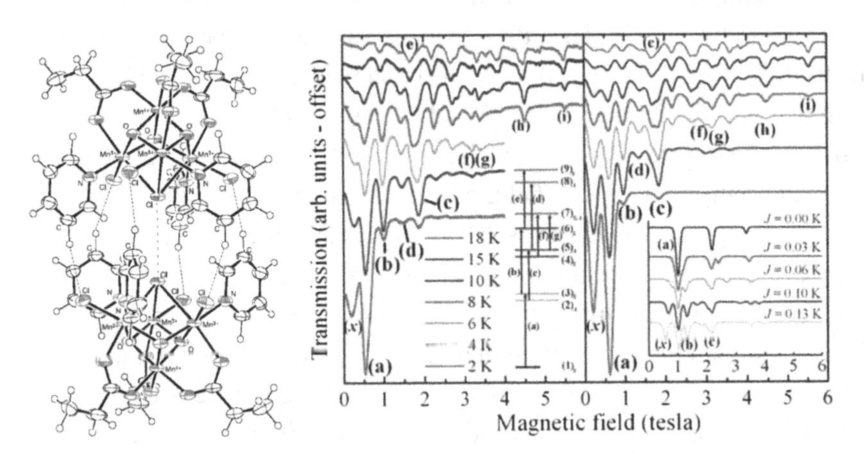

Fig. 5. Left: Structure of $[Mn_4O_3Cl_4(O_2CEt)_3(py)_3]_2$ dimer. Right: Experimental (left) and simulated (right) 145 GHz single crystal EPR spectra with magnetic field parallel to z. Inset, left: Level scheme for states with $M_1 + M_2 = -9$ (labelled **1**), -8 (**2**,**3**), -7 (**4**-**6**) and -6 (**7**-**9**); subscripts S and A denote symmetric and antisymmetric superposition states. Inset, right: Evolution of calculated spectrum with J (right). Reproduced with permission from Wernsdorfer *et al.*, 2002 and Hill and Wilson, 2007.

the tunnelling resonances to finite field values. They also noted that the resulting low-lying superposition states (between states of the two halves of the dimer) could potentially be exploited in quantum computing, building on ideas for SMM-based quantum information processing first proposed by Leuenberger and Loss (2001).

Hill *et al.* (2003) probed these states in $(Mn_4)_2$ directly by single crystal EPR at 145 GHz, providing spectroscopic evidence for, and quantification of, the exchange interaction between the two halves of the dimer.

Previous studies on the related, but isolated, 'Mn_4' complex $[Mn_4O_3(OSiMe)_3(O_2CEt)_3(dbm)_3]$ (dbmH = dibenzoylmethane) showed spectra that can be modelled simply with the fine structure expected for $S = 9/2$ within the GSA (Edwards *et al.* 2003). For $(Mn_4)_2$, with the field applied parallel to the S_6 axis of the dimer, additional structure is observed (Fig. 5). The spectra are modelled assuming an isotropic exchange between axially anisotropic $S_{1,2} = 9/2$ centres, with $|M_1,M_2>$ basis states, i.e. using an exchange Hamiltonian of the form (2) where the 'local spins' are the result of applying the GSA to the ground state of the Mn_4 "monomers" (Hill *et al.*, 2003; Hill and Wilson 2007). The extra structure is very sensitive to the magnitude of the intra-dimer exchange, and modelling gives $D_{Mn4} = -0.521$, $B_4^0(Mn4) = -3 \times 10^{-5}\,\text{cm}^{-1}$ with $J_{Mn4Mn4} = 0.042\,\text{cm}^{-1}$ [the latter parameter has been converted to be consistent with Hamiltonian (2)]. The authors point out that the although J is modelled as isotropic, the J_z component merely shifts the energies of the $|M_1,M_2>$ states but does not mix them (hence EPR resonance fields are shifted but transitions are not split); it is the transverse component of the exchange ($J_x = J_y$) that mixes states ($|M_1,M_2{\pm}1>$ with $|M_1{\pm}1,M_2>$) and hence is the origin of the additional splitting.

Winpenny and co-workers report studies on two families of controllably and covalently linked clusters (Timco *et al.*, 2008 and 2009) this time inspired by an alternative quantum computing proposal from Meier *et al.* (2003) based on $S = 1/2$ ground state clusters. The Cr_7Ni species mentioned in Sec. 1 meet this criterion and EPR spectra at 5 K are consistent with all population in the $S = 1/2$ ground state. Such a species can in principle act as a Qubit, but construction of logic gates would require the controlled electronic coupling of the molecular Qubits.

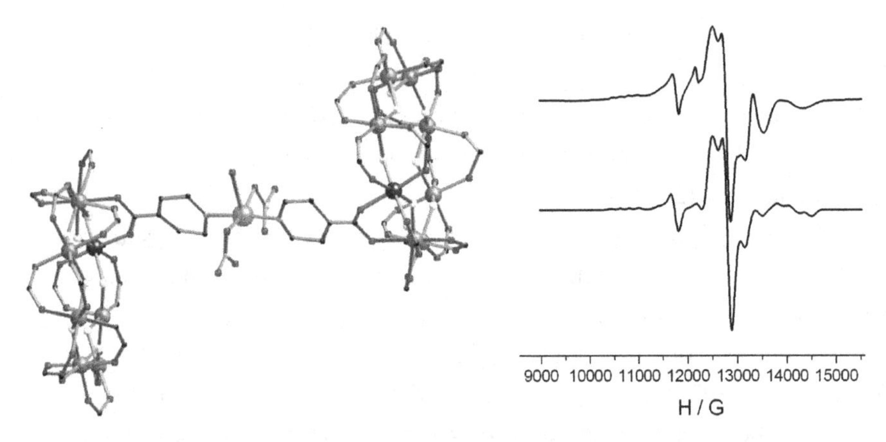

Fig. 6. Left: Structure of $[\{Cr_7NiF_8(O_2C^tBu)_{15}(O_2CC_5H_4N)\}_2\{Cu(NO_3)_2(H_2O)\}]^{2-}$ dianion ('Cr_7Ni-Cu-Cr_7Ni'; ^tBu not shown). Right: 34 GHz powder EPR spectra at 5 K of Cr_7Ni-Cu-Cr_7Ni (top) and simulation using $S_{Cr7Ni} = 1/2$ (bottom) approximation.

Timco *et al.*, (2009) exploit the chemical stability of the $[Cr_7NiF_8(O_2C^tBu)_{16}]^-$ anion to substitute a single carboxylate for a pyridyl-functionalised carboxylate on one of the Cr...Ni edges; two of these are then coordinated to a linking metal complex. One example is $[\{(NH_2Pr_2)[Cr_7NiF_8(O_2C^tBu)_{15}(O_2CC_5H_4N)]\}_2\{Cu(NO_3)_2(H_2O)\}]$ where two Cr_7Ni rings are linked via a single Cu(II) ion (Fig. 6). EPR immediately reveals that there is an exchange interaction between the rings and Cu. These spectra were modelled in two ways. The first way treated the rings as isolated $S = 1/2$ centres (i.e. the GSA), and couples them with the central $s = 1/2$ of Cu(II) — the problem has been reduced to that of a linear trimer of "s" $= 1/2$ in a three-spin version of Hamiltonian (2). This gives two doublets and a quartet as "total" spin states. These have very different g-values because of the very different g-values of the monomers [$g_{x,y} = 1.83$ and $g_z = 1.79$ for Cr_7Ni and $g_{x,y} = 2.07$ and $g_z = 2.27$ for Cu(II)] and all three states are resolved separately in the 34 GHz powder EPR (Fig. 6) allowing determination of the magnitude and sign of $J_{ring\text{-}Cu}$. Because $\mathbf{g}_i$ values and orientations (and Cu hyperfine which is resolved experimentally) can be fixed from studies on the isolated components and structural considerations, there are only two free variables, giving $J_{ring\text{-}Cu} = +0.15\ \text{cm}^{-1}$ (ferromagnetic) with $d_{ring\text{-}Cu} = -0.03\ \text{cm}^{-1}$. The anisotropic exchange is necessary to introduce the experimentally observed ZFS of the $S = 3/2$ ground state in this very simple model (its principal axis is assumed to be oriented along the Cu-ring axes). The second model uses the full Hamiltonian (2) for 14 Cr(III), 2 Ni(II) and a Cu(II) ion. In order to overcome the enormous matrix dimension (ca. 5×10^9) they first diagonalise the Hamiltonian for a single Cr_7Ni ring, fixing all parameters from previous studies on the separate components, and express the basis of the full Hamiltonian as $|\phi_{ring1}, \phi_{ring2}, \phi_{Cu}>$. For the next step they truncate the product basis to include only the four lowest lying multiplets of the rings, i.e. only a very limited number of the $\phi_{ring1,2}$ states are incorporated in the next diagonalisation to obtain the lowest eigenstates of the full Hamiltonian. This is justified because the $J_{Cr/Ni-Cu}$ exchange is much weaker than couplings within the ring. $J_{Cr/Ni-Cu} = +0.35\,\text{cm}^{-1}$ is fixed from modelling a Schottky anomaly in specific heat data, leaving no free variables. Remarkably, the calculated spectrum is very similar to that from the three spin model. In this model there is no need to introduce any anisotropic exchange as the ground state ZFS arises as a consequence of the excited states of the rings. Note that the J-values in the three-spin and 17-spin model are not equivalent; a perturbation treatment predicts that $J_{Cu-Cr/Ni}/J_{ring\text{-}Cu} \approx 2$, as observed. The authors go on to discuss potential quantum computation protocols based on this system and others based on dimetallic linkers.

In their second paper (Timco *et al.*, 2008), the same group present a much more direct route to linking Cr_7Ni rings, where again the coupling between them is proved by EPR. This is based on a modified 'monomer', $[Cr_7NiF_3(Etglu)(O_2C^tBu)_{15}(H_2O)]$ {$EtgluH_5$ = N-ethyl-D-glucamine; "Cr_7Ni(glu)"}, where there is a terminal ligand (water) on the Ni(II) ions which can be substituted with diimines (4,4'-bipyridine, *trans*-1,2-dipyridylethene) to give '$\{Cr_7Ni(glu)\}_2$' dimers (Fig. 7). The "monomer" still has an $S = 1/2$ ground state, this is the only state observed in EPR spectra at 5 K ($g_{x,y} = 1.84$ and $g_z = 1.78$; Fig. 7). The dimer should then have lowest lying $S = 0$ and 1 states and indeed a triplet is observed in 34 GHz powder EPR (Fig. 7). Since EPR is blind to the singlet state, the spectra were simulated simply on the basis of Hamiltonian (1) for $S = 1$. This gives $D = +0.013\,\text{cm}^{-1}$ for the 4,4'-bipy linked dimer. The form of the triplet spectrum is slightly unusual because, in this field regime for

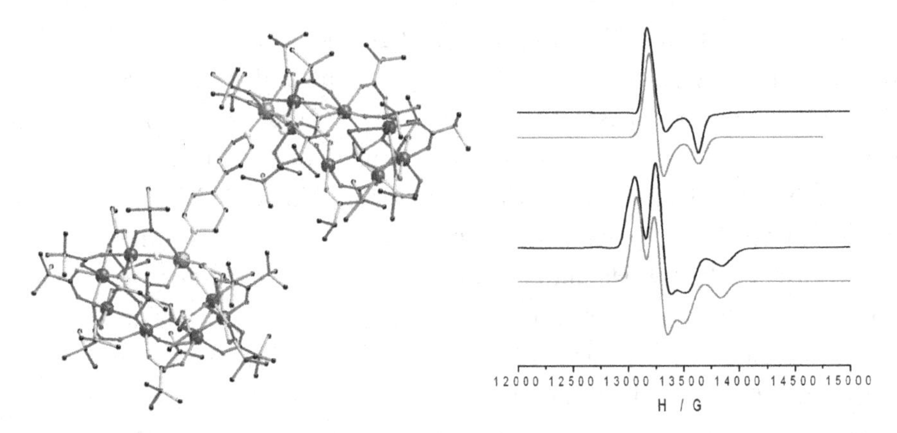

Fig. 7. Left: structure of [{Cr_7NiF_3(Etglu)(O_2C^tBu)$_{15}$}$_2$(4,4-bipy)] (^tBu groups not shown). Right: 34 GHz powder EPR spectra (black) at 5 K of [Cr_7NiF_3(Etglu) (O_2C^tBu)$_{15}$(Ph-C_6H_5N)] (top) and [{Cr_7NiF_3(Etglu)(O_2C^tBu)$_{15}$}$_2$(4,4-bipy)] (bottom). Simulations (red) using GSA for $S = 1/2$ and 1, respectively.

34 GHz EPR, the *g*-anisotropy is greater than the ZFS. The positive sign of *D* implies that the ZFS is dominated by exchange rather than dipolar interactions. A smaller $|D|$ is found for the longer *trans*-1,2-dipyridylethene linker ($+0.009\,cm^{-1}$) in keeping with the longer exchange pathway.

3. Pulsed EPR

The proposals for exploitation of SMMs (Leuenberger and Loss 2001), or $S = 1/2$ ground state molecular nanomagnets (Meier *et al.*, 2003), as Qubits in quantum computing (for a recent review see Stepanenko *et al.*, 2008) stimulated investigation into the spin dynamics of such materials. The feasibility of any QIP scheme is critically dependent on the intrinsic relaxation times being much longer than the times required for coherent spin manipulation in QIP. Before we discuss these investigations it is useful to recall some definitions: a spin precesses about a static applied field $\mathbf{B}_0$ and hence has a fixed component of its magnetic moment along the field axis (z) and a rotating component in the xy plane. For an ensemble of spins, the spin-lattice relaxation time constant T_1 characterises the restoration of the z component of the magnetisation (i.e. establishment of equilibrium populations) after a perturbation. At thermal equilibrium there is no x or y component of magnetisation since there is no phase coherence between the individual spin moments. If coherence is forced, e.g. by an appropriate microwave pulse, the time constant T_2 characterises the decay of the xy magnetisation due to loss of phase coherence. Some authors prefer to call this the phase memory time T_m, with T_2 being explicitly the spin-spin relaxation time, since there are other important mechanisms for decoherence.

Several groups attempted to probe the magnetisation dynamics by coupling "standard" magnetisation techniques (SQUID, micro-SQUID or Hall bar arrays) whilst simultaneously exposing the sample to pulsed microwave irradiation. These methods were applied to the high spin ground state SMMs $[Fe_8O_2(OH)_{12}(tacn)_6]$ ('Fe_8'; tacn = 1,4,7-triazacyclononane; Petukhov *et al.*, 2005; Bal *et al.*, 2005 and 2008; Petukhov *et al.*, 2007; Bahr *et al.*, 2007) and Ni_4 (see above; del Barco *et al.*, 2004), and to the $S = 1/2$ ground state clusters $K_6[V^{IV}_{15}As_6O_{42}(H_2O)]$ ('V_{15}'; Wernsdorfer *et al.*, 2004) and Cr_7Ni (see above; Wernsdorfer *et al.*, 2005). Dips are observed in the low temperature magnetisation vs. applied field curves due to resonant microwave absorption as the $M \rightarrow M+1$ gaps match $h\nu$. These are simply EPR transitions, albeit measured in a rather unconventional way. When coupled with the extraordinary sensitivity of, for example, micro-SQUID methods (Wernsdorfer *et al.*, 2004) this has the advantage of allowing measurement on micron-sized crystals and at mK temperatures. A lower bound for decoherence times T_2 is given by absorption linewidths, as with standard EPR spectra, and in principle the spin-lattice relaxation T_1 can be monitored via magnetisation recovery after the microwave pulse. However, in reality linewidths are dominated by inhomogeneous broadening, and the intrinsic T_1 contribution to magnetisation recovery is masked by phonon bottleneck phenomena, associated with the much slower phonon relaxation from the lattice to the heat bath (i.e. from crystal to cryostat) (Petukhov *et al.*, 2007; Bal *et al.*, 2008). One route round this is presented by Bahr *et al.*, (2007) who use a two-pulse pump-probe experiment on the $S = 10$ ground state SMM Fe_8. The first pulse pumps the $M = -10 \rightarrow -9$ transition then after a set delay the second pulse pumps $M = -9 \rightarrow -8$. The pump and probe pulses have frequencies of 118 and 107 GHz, respectively, tuned to these energy gaps for an applied field of 0.2 T. The second pulse only has an effect on the magnetisation if there is still population in $M = -9$ after the delay; by varying this delay the characteristic lifetime of this state can be determined.

However, even the early papers using these magnetisation-detected EPR methods recognised that attempts to study coherent spin dynamics would require shorter pulse lengths and spin echo methods (Wernsdorfer *et al.*, 2004; Bal *et al.*, 2005). Surprisingly it was not until 2007 that the first report of 'conventional' pulsed EPR to measure the intrinsic T_1, T_2 of molecular nanomagnets was published (Ardavan *et al.*, 2007). This is surprising because pulsed EPR had long been used to probe metalloenzymes with exchange coupled cluster active sites (e.g. iron-sulfur clusters, the Mn_4 cluster in photosystem II) (for recent reviews see: Collison and McInnes, 2002; 2004; 2007; Boeer *et al.*, 2008). T_1 and T_2 (or T_M) can be measured directly by standard EPR spin echo methods and here we briefly describe two common experiments. These are more easily visualised in the rotating coordinate frame [observing the system sitting on the resonant circularly polarised component of the microwave field $\mathbf{B}_1$, defined as x, rotating at the microwave frequency ν about z] (Schweiger and Jeschke, 2001; for an excellent short review see Schweiger, 1991). In this frame, the effect of a microwave pulse of duration t_p and field amplitude B_1 is to rotate the magnetisation vector from z (defined by $\mathbf{B}_0$) about the x-axis by an angle $\theta = g\mu_B B_1 t_p$, or $\omega_1 t_p$ where $\omega_1 = g\mu_B B_1$ is the microwave field amplitude in angular frequency units. Hence, the tip angle can be set by choice of B_1 and t_p. Measurement of relaxation time constants can be achieved by, for example:

(i) T_2 by a two-pulse Hahn-echo sequence: $\pi/2-\tau-\pi-\tau-$ echo. The first ($\pi/2$) pulse flips the magnetisation vector from z to y; the spins dephase in the xy plane during the delay time τ; then the second (π) pulse refocuses the spins along –y after an identical time τ thus generating the spin echo. The decay of the echo amplitude as a function of τ gives T_2, where the effects of interactions that give rise to inhomogeneous broadening have been removed by the refocusing pulse.

(ii) T_1 by an echo-detected magnetisation recovery after inversion sequence: $\pi - t - \pi/2 - \tau - \pi - \tau-$ echo. The first (π) pulse inverts the magnetisation to –z; the system relaxes during a variable time t; the remaining magnetisation along –z is flipped to the xy plane by the $\pi/2$ pulse and detected as above with a short and fixed τ. Hence, the echo intensity as a function of t gives T_1.

Ardavan *et al.* (2007) applied these methods at X-band to Cr_7Ni and Cr_7Mn (see above) which have $S = 1/2$ and 1 ground states, respectively, in dilute toluene solution. For Cr_7Ni they found $T_2 = 379$ ns at 4.5 K (Fig. 8a), orders of magnitude longer than had been estimated previously from magnetisation-detected EPR measurements (Wernsdorfer *et al.*, 2005). They also observed modulation of the echo decay with a characteristic (Larmor) frequency of ^1H nuclei (Fig. 8a). This is the electron spin echo envelope modulation (ESEEM) effect due to interaction of the electron spin with nearby nuclear spins, and suggests that these hyperfine interactions are a major contributor to decoherence. (Surprisingly there is no evidence of

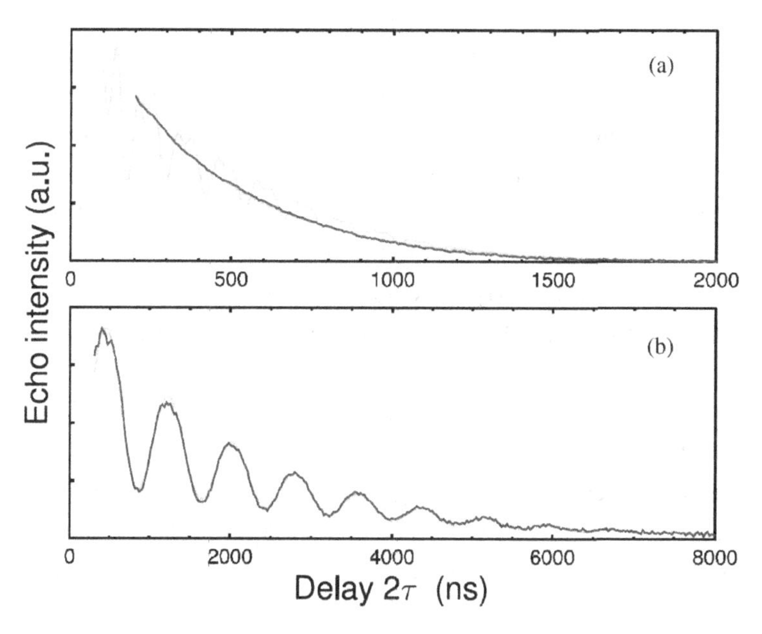

Fig. 8. (a) Decay of Hahn echo intensity with τ for $(NH_2Me_2)[Cr_7NiF_8(O_2C^tBu)_{16}]$ (dilute toluene solution, X-band, 4.5 K) using short, broadband pulses (cyan) and long, selective pulses (dark blue; suppressing the ESEEM effect); (b) Decay of Hahn echo intensity with τ for the per-deuterated version $[ND_2(CD_3)_2][Cr_7NiF_8\{O_2CC(CD_3)_3\}_{16}]$ with long, selective pulses (dark blue) and fit (green). Reproduced with permission from Ardavan *et al.*, 2007.

coupling to the ^{19}F nuclei.) This is confirmed by measurements on the perdeuterated analogue $[ND_2(CD_3)_2][Cr_7NiF_8\{O_2CC(CD_3)_3\}_{16}]$ ("d-Cr_7Ni") showing a lower ESEEM frequency characteristic of ^{2}D (Fig. 8b), and $T_2 = 2.21\ \mu s$ is *ca.* six times longer than for the isotopically natural abundance Cr_7Ni at the same temperature. This factor is the ratio of the ^{1}H to ^{2}D nuclear gyromagnetic ratios. Extending these measurements to lower temperatures, T_2 reaches as long as 0.55 and 3.8 μs for Cr_7Ni and d-Cr_7Ni, respectively, at 1.8 K. T_1 reaches *ca.* 1 ms at these temperatures. The authors note that T_2 is orders of magnitude longer than required for spin manipulation, hence this important criterion for QIP is met for these molecules.

Further surprises are thrown up from measurements on Cr_7Mn. T_1 and T_2 are similar to those for Cr_7Ni; and while T_1 varies by a factor of two depending on which part of the EPR spectrum of the $S = 1$ ground state is excited, T_2 does not vary significantly. One implication is that the ZFS of the triplet is not a significant factor in decoherence. A further demonstration of the ability to coherently manipulate the electronic spin comes from the observation of Rabi oscillations via transient nutation experiments. The first report on molecular nanomagnets was from Bertaina *et al.* (2008)[1] quickly followed by others (Mitrikas *et al.*, 2008; Schlegel *et al.*, 2008). Nutation is the complicated path that the spin takes (in the fixed coordinate frame) when precessing under the influence of both the static field $\mathbf{B}_0$ (precession about z) and the microwave field $\mathbf{B}_1$ (precession about x). The result is a cycling from +z to −z in a spiral motion (nutation) with a characteristic frequency which, at resonance ($h\nu = g\mu_B B_0$), is $\omega_1 = g\mu_B B_1$ for $S = 1/2$ (things are more complicated off-resonance, see Schweiger and Jeschke (2001) for full detail). The resulting oscillations between maximum and minimum values of the z component of magnetisation, also known as Rabi oscillations, are transient since they are damped by relaxation, but they can be observed by spin echo methods (Schweiger and Jeschke, 2001). In a

[1] In their paper, Bertaina *et al.* (2008) comment that in Ardavan *et al.*,'s (2007) pulsed EPR studies of Cr_7Ni "the important Rabi quantum oscillations were not observed, probably because the electronic and nuclear degrees of freedom were too strongly linked to each other." This is not the case: Rabi oscillations are observed for Cr_7Ni, they were simply not reported by Ardavan *et al.* in this paper (2007).

simple (pulse $-\tau-\pi-\tau-$ echo) sequence the spin is rotated through angle $\omega_1 t_p$ by the first (nutation) pulse of variable duration t_p then refocused by the π pulse and echo detected. This effectively measures the y component of the magnetisation after the nutation pulse. Alternatively, in the (pulse $-T-\pi/2-\tau-\pi-\tau-$ echo) sequence, a two-pulse echo detection can be used after a delay $T >> t_p$, T_2 in which the coherence decays completely. This measures the remaining magnetisation along z after the nutation pulse. Either way, modulation of the echo intensity as function of t_p gives the nutation or Rabi frequency (which of course depends on the choice of B_1). The observation of Rabi oscillations is a demonstration of coherent manipulation of the spin. Since the orientations $\pm$z correspond to the pure M states involved in the EPR transition (treating any pair of states as a fictitious $S = 1/2$), rotation to any other position corresponds to the creation of superposition states $|\Psi> = c_1|M> + c_2|M+1>$. The coefficients then have a cyclic dependence on the pulse duration t_p. This ability to create arbitrary superpositions is a key requisite for QIP.

Bertaina *et al.* (2008) report Rabi oscillations in studies of the anionic V_{15} cluster (see above) diluted by use of a cationic surfactant $[NMe_2(C_{18}H_{37})_2]^+$ which also imparts solubility in chloroform. V_{15} has a lowest-lying pair of $S = 1/2$ states (mixed by antisymmetric exchange interactions) with an $S = 3/2$ first excited state at 2.5 cm^{-1} higher in energy; all these states contribute to the unresolved low temperature X-band EPR spectrum. Transient nutation experiments revealed two different sets of Rabi frequencies, corresponding to transitions: (i) within the bottom four states (a very low frequency, complicated by the four possible EPR transitions depending on the mixing between the two $S = 1/2$ states), and (ii) within the essentially isotropic $S = 3/2$ state. Clear oscillations are seen for the latter (Fig. 9). A decoherence time of $T_2 = 340$ ns is measured at 4 K for this group of transitions, probably dominated by the ^{51}V hyperfine interactions.

Mitrikas *et al.* (2008) report studies on a much simpler compound, the trimetallic $[Fe^{III}_3O(O_2CPh)_5(salox)(EtOH)(H_2O)]$ ($saloxH_2$ = salicylaldehyde oxime). Antiferromagnetic coupling in the isosceles triangle gives an isolated, near-isotropic $S = 1/2$ ground state. This state has long T_1 (693 to 0.55 μs between 5.5 and 11 K) and T_2 (2.6 μs below 7 K) and Rabi oscillations are again observed via transient nutation experiments.

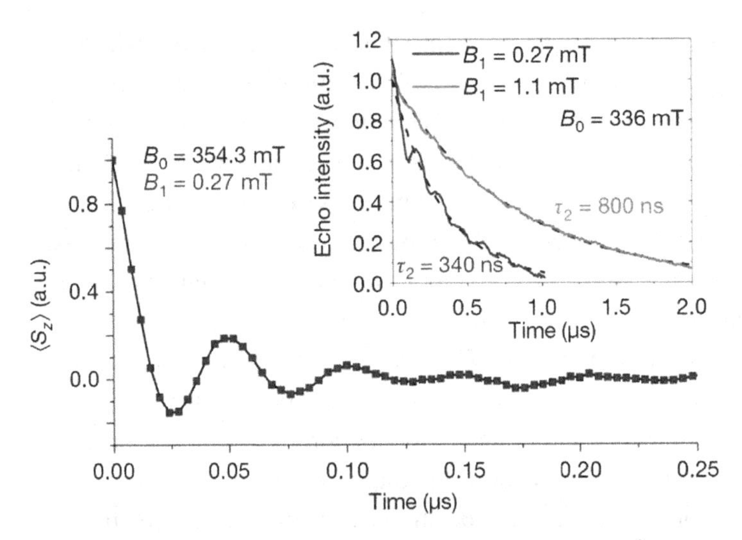

Fig. 9. Echo amplitude (presented as the expectation value of $\hat{S}_z$; detected by two-pulse echo) as a function of nutation pulse length for transitions within the $S = 3/2$ first excited state of $[NMe_2(C_{18}H_{37})_2]_6[V_{15}As_6O_{42}(H_2O)]$ in $CHCl_3$ at 4 K (X-band), showing Rabi oscillations. Inset: Hahn echo decays to obtain T_2 for transitions within the $S = 1/2$ (red) and $S = 3/2$ states (blue). Reproduced with permission from Bertaina *et al.*, 2008.

Interestingly, these oscillations are of much shorter duration (<120 ns) than T_2; they propose that B_1-field inhomogeneities and limited excitation bandwidths are the limiting factors. They also note a ^{1}H ESEEM effect in the transient nutation signal. An interesting aside of this work was that T_1 studies allowed them to discriminate between two different magnetic models for the isosceles Fe_3 triangle, with $|J| >$ or $< |J'|$, where J' is the unique interaction. Fitting variable temperature T_1 data to an Orbach relaxation model (a two-phonon process via an excited state) gives the energy gap to the excited state of 57 cm^{-1} which only fits their $|J| > |J'|$ model.

Schlegel *et al.* (2008) reported the first pulsed EPR studies, including Rabi oscillations, on an SMM, the $S = 5$ ground state $[Fe_4(acac)_6(Br\text{-}mp)_2]$ {'Fe_4'; acacH = acetylacetone; Br-mpH_3 = $BrCH_2$-$C(CH_2OH)_3$}. A further novelty is that the experiments are performed in zero static field. The energy gap between the $M = \pm 5$ and ± 4 sublevels in zero-field is 3.08 cm^{-1}, or 92.4 GHz, and therefore this transition comes into resonance at zero-field in a W-band EPR experiment. Pulsed W-band studies give $T_1 = 1.06\,\mu$s and

$T_2 = 307$ ns at 4.3 K, much longer than previously estimated for SMMs. They found ^{1}H and ^{2}D ESEEM modulations when measuring in natural abundance and perdeutero-toluene, respectively, and measurement in CS_2 did not reveal ESEEM modulations. Hence, they are observing hyperfine effects with the solvent rather than with ligands. In another twist, they do not observe any significant enhancement of T_2 in d-toluene but they do in CS_2 ($T_2 = 527$ ns), showing the importance of eliminating all nuclear spins as far as possible. In transient nutation experiments in CS_2, Rabi oscillations are observed in zero applied field — this origin of the echo modulation is confirmed by the linear frequency dependence on B_1.

In all the pulsed experiments above the (electron) spin-spin relaxation mechanism is suppressed as far as possible by measuring in dilute media (frozen solutions). This is relaxation due to dipolar coupling with fluctuating neighbouring electron spins, and is dominated by so-called flip-flop process where one spin 'flips' from $M \rightarrow M+1$ while another 'flops' from $M+1 \rightarrow M$, thus conserving energy (Schweiger and Jeschke 2001). These processes contribute to T_2 because phase coherence of the transverse moment (with respect to other spins) is lost in the spin flip process. Dilution simply moves the spins as far apart as possible to minimise these effects. Alternatively, they can be eliminated by fully polarising all spins in the ensemble (the 'spin bath' in physics literature). Takahashi *et al.* (2009) exploit this to measure pulsed EPR on a *neat* single-crystal of Fe_8, by operating at high enough field and low enough temperature such that the magnetisation is saturated. The static field is applied along the easy axis. At 240 GHz the $M = -10 \rightarrow -9$ transition comes into resonance at 4.6 T; at this field and at 1.27 K more than 99% of spins are in the $M = -10$ sublevel. Hahn echo measurements give $T_2 = 714$ ns rising rapidly with temperature; a simple flip-flop model reproduces this behaviour. Once the flip-flop mechanism is quenched the coherence times are limited by hyperfine interactions.

References

Abbati, G., Brunel, L-C., Casalta, H., Cornia, A., Fabretti, A.C., Gatteschi, D., Hassan, A.H., Jansen, A.G.M., Maniero, A.L., Pardi, L., Paulsen, C. and Segre, U. (2007). Single-ion versus dipolar origin of the magnetic anisotropy in iron(III)-oxo clusters: a case study, *Chem. Eur. J.* **7** 1796–1807.

Abragam, A. and Bleaney, B. (1986). *Electron Paramagnetic Resonance of Transition Ions*, Dover; New York.

Accorsi, S., Barra, A-L., Caneschi, A., Chastanet, G., Cornia, A., Fabretti, A.C., Gatteschi, D., Mortalò, C., Oilvieri, E., Parenti, F., Rosa, P., Sessoli, R., Sorace, L., Wernsdorfer, W. and Zobbi, L. (2006). Tuning anisotropy barriers in a family of tetrairon(III) single-molecule magnets with an $S = 5$ ground state, *J. Amer. Chem. Soc.* **128**, 4742–4755.

Ardavan, A., Rival, O., Morton, J.J.L., Blundell, S.J., Tyryshkin, A.M., Timco G.A. and Winpenny, R.E.P. (2007). Will spin-relaxation times in molecular magnets permit quantum information processing?, *Phys. Rev. Lett.* **98**, art. 057201.

Bahr, S., Petukhov, K., Mosser, V. and Wernsdorfer, W. (2007). Pump-probe measurements on the single-molecule magnet Fe_8: measurement of excited level lifetimes, *Phys. Rev. Lett.* **99**, art. 147205.

Bal, M., Friedman, J.R., Suzuki, Y., Rumberger, E.M., Hendrickson, D.N., Avraham, N., Myasoedov, Y., Shtrikman, H. and Zeldov, E. (2005). Non-equilibrium magnetization dynamics in the Fe_8 single-molecule magnet induced by high-intensity radiation, *Europhys. Lett.* **71**, 110–116.

Bal, M., Friedman, J.R., Chen, W., Tuominen, M.T., Beedle, C.C., Rumberger E.M. and Hendrickson, D.N. (2008). Radiation- and phonon- bottleneck-induced tunnelling in the Fe_8 single-molecule magnet, *Europhys. Lett.* **82**, art. 17005.

Barra, A-L., Gatteschi, D. and Sessoli, R. (1997). High-frequency EPR spectra of a molecular nanomagnet: understanding of the quantum tunneling of magnetization, *Phys. Rev. B* **56**, 8192–8298.

Barra, A-L., Brunel, L-C., Gatteschi, D., Pardi, L.A. and Sessoli, R. (1998). High-frequency EPR spectroscopy of large metal ion clusters: from zero field splitting to quantum tunneling of magnetization, *Acc. Chem. Res.* **31**, 460–466.

Barra, A-L., Caneschi, A., Cornia, A., Gatteschi, D., Gorini, L., Heininger, L-P., Sessoli, R. and Sorace, L. (2007). The origin of transverse anisotropy in axially symmetric single molecule magnets, *J. Amer. Chem. Soc.* **129**, 10754–10762.

Bencini, A. and Gatteschi, D. (1989). *EPR of Exchange Coupled Systems*, Springer-Verlag, Berlin.

Bertaina, S., Gambarelli, S., Mitra, T., Tsukerblat, B., Müller, A. and Barbara, B. (2008). Quantum oscillations in a molecular magnet, *Nature* **453**, 203–206.

Boeer, A.B., Collison D. and McInnes, E.J.L. (2008). EPR of exchange coupled oligomers, *Royal Society of Chemistry Specialist Periodical Reports, Electron Spin Resonance* **21**, 131–161.

Carretta, S., Liviotti, E., Magnani, N., Santini, P. and Amoretti, G. (2004). S mixing and quantum tunneling of the magnetization in molecular nanomagnets, *Phys. Rev. Lett.* **92**, art. 207205.

Chakov, N.E., Lee, S-C., Harter, A.G., Kuhns, P.L., Reyes, A.P., Hill, S.O., Dalal, N.S., Wernsdorfer, W., Abboud, K.A. and Christou, G. (2006). The properties of the $[Mn_{12}O_{12}(O_2CR)_{16}(H_2O)_4]$ single-molecule magnets in truly axial symmetry: $[Mn_{12}O_{12}(O_2CCH_2Br)_{16}(H_2O)_4]\cdot 4CH_2Cl_2$, *J. Amer. Chem. Soc.* **128**, 6975–5989.

Collison, D. and McInnes, E.J.L. (2002). EPR of exchange coupled oligomers, *Royal Society of Chemistry Specialist Periodical Reports, Electron Spin Resonance* **18**, 161–182.

Collison, D. and McInnes, E.J.L. (2004). EPR of exchange coupled oligomers, *ibid.* **19**, 374–397.

Collison, D. and McInnes, E.J.L. (2007). EPR of exchange coupled oligomers, *ibid.* **20**, 157–191.

Cornia, A., Sessoli, R., Sorace, L., Gatteschi, D., Barra, A-L. and Daiguebonne, C. (2002). Origin of second-order transverse magnetic anisotropy in Mn_{12}-acetate, *Phys. Rev. Lett.* **89**, art. 257201.

Datta, S., Waldmann, O., Kent, A.D., Milway, V.A., Thomson, L.K. and Hill, S. (2007). Direct observation of mixing of spin multiplets in an antiferromagnetic molecular nanomagnet by electron paramagnetic resonance, *Phys. Rev. B* **76**, art. 052407.

del Barco, E., Kent, A.D., Yang, E.C. and Hendrickson, D.N. (2004). Quantum superposition of high spin states in the single molecule magnets Ni_4, *Phys. Rev. Lett.* **93**, art. 157202.

Edwards, R.S., Hill, S., Bhaduri, S., Aliaga-Alcade, N., Bolin, E., Maccagnano, S., Christou G. and Hendrickson, D.N. (2003). A comparative high frequency EPR study of monomeric and dimeric Mn_4 single-molecule magnets, *Polyhedron*, **22**, 1911–1916.

Feng, P.L., Beedle, C.C., Koo, C., Lawrence, J., Hill, S. and Hendrickson, D.N. (2008). Origin of magnetization tunneling in single-molecule magnets as determined by single-crystal high-frequency EPR, *Inorg. Chim. Acta.* **361**, 3465–3480.

Gatteschi, D., Sessoli, R. and Villain, R. (2006a). *Molecular Nanomagnets*, Oxford University Press, Oxford.

Gatteschi, D., Barra, A-L., Caneschi, A., Cornia, A., Sessoli, R. and Sorace, L. (2006b). EPR of molecular nanomagnets, *Coord. Chem. Rev.* **250**, 1514–1529.

ter Heerdt, P., Stefan, M., Goovaerts, E., Caneschi, A. and Cornia, A. (2006). Single-ion and molecular contributions to the zero-field splitting in an iron(III)-oxo dimer studies by single crystal W-band EPR, *J. Magn. Reson.* **179**, 29–37.

Hill, S., Edwards, R.S., Aliaga-Alcade, N. and Christou, G. (2003). Quantum coherence in an exchange-coupled dimer of single-molecule magnets, *Science*, **302**, 1015–1018.

Hill, S. and Wilson, A. (2007). Calculation of the EPR spectrum for an entangled dimer of $S = 9/2$ Mn_4 single-molecule magnets, *J. Low Temp. Phys.* **142**, 271–276.

Leuenberger, M.N. and Loss, D. (2001). Quantum computing with molecular magnets, *Nature*, **410**, 789–793.

Meier, F., Levy, J. and Loss, D. (2003). Quantum computing with spin cluster qubits, *Phys. Rev. Lett.* **90**, art. 047901.

McInnes, E.J.L. (2006). Spectroscopy of single molecule magnets, *Structure&Bonding*, **122**, 69–102.

McInnes, E.J.L. and Winpenny, R.E.P. (2010). Molecular Nanomagnets, in "Contempory Inorganic Materials" book series, Volume 3: "Molecular Materials" Volume Three: Molecular Materials" Eds. Bruce, D.W., Walton, R.I. and O'Hare, D., Wiley.

Mitrikas, G., Sanakis, Y., Raptopoulou, C.P., Kordas G. and Papavasiliou, G. (2008). Electron spin-lattice and spin-spin relaxation study of a trinuclear iron(III) complex and its relevance in quantum computing, *Phys. Chem. Chem. Phys.* **10**, 743–748.

Petukhov, K., Wernsdorfer, W., Barra, A-L. and Mosser, V. (2005). Resonant photon absorption in Fe_8 single-molecule magnets detected via magnetization measurements, *Phys. Rev. B* **72**, art. 052401.

Petukhov, K., Bahr, S., Wernsdorfer, W., Barra, A-L. and Mosser, V. (2007). Magnetization dynamics in the single-molecule magnet Fe_8 under pulsed microwave irradiation, *Phys. Rev. B* **75**, art. 064408.

Pilawa, B., Keilhauer, I., Fischer, G., Knorr, S., Rahmer, J., Grupp, A. (2003). Magnetic properties of the cyclic spincluster Fe6:bicine, *Eur. Phys. J. B* **33**, 321–330.

Piligkos, S., Bill, E., Collison, D., McInnes, E.J.L., Timco, G.A., Weihe, H., Winpenny, R.E.P. and Neese, F. (2007). Importance of the anisotropic exchange interaction for the magnetic anisotropy of polymetallic systems, *J. Amer. Chem. Soc.* **129**, 760–761.

Piligkos, S., Weihe, H., Bill, E., Neese, F., Mkami, H. El, Smith, G.M., Collison, D., Rajaraman, G., Timco, G.A., Winpenny R.E.P. and McInnes E.J.L. (2009). EPR spectroscopy of a family of $Cr_7^{III}M^{II}$ (M = Cd, Zn, Mn, Ni) 'wheels': studies of isostructural compounds with different spin ground states, *Chem. Eur. J.* **15**, 3152–3167.

Prescimone, A., Wolowska, J., Rajaraman, G., Parsons, S., Wernsdorfer, W., Murugesu, M., Christou, G., Piligkos, S., McInnes, E.J.L. and Brechin, E.K. (2007). Studies of a linear single-molecule magnet, *Dalton Trans.* 5282–5289.

Schlegel, C., van Slageren, J., Manoli, M., Brechin, E.K. and Dressel, M. (2008). Direct observation of quantum coherence in single-molecule magnets, *Phys. Rev. Lett.* **101**, art. 147203.

Schweiger, A. and Jeschke, G. (2001). *Principles of Pulse Electron Paramagnetic Resonance*, OUP, New York.

Schweiger, A. (1991). Pulsed electron spin resonance spectroscopy: basic principles, techniques and examples of applications, *Angew. Chem. Int. Ed.* **30**, 265–292.

van Slageren, J., Sessoli, R., Gatteschi, D., Smith, A.A., Helliwell, M., Winpenny, R.E.P., Cornia, A., Barra, A-L., Jansen, A.G.M., Rentschler, E. and Timco, G.A. (2002). Magnetic anisotropy of the antiferromagnetic ring $[Cr_8F_8Piv_{16}]$, *Chem. Eur. J.* **8**, 277–285.

Stepanenko, D., Trif, M. and Loss, D. (2008). Quantum computing with molecular magnets, *Inorg. Chim. Acta.* **361**, 3740–3745.

Takahashi, S., van Tol, J., Beedle, C.C., Hendrickson, D.N., Brunel, L-C. and Sherwin, M.S. (2009). Coherent manipulation and decoherence of $S = 10$ single-molecule magnets, *Phys. Rev. Lett.* **102**, art. 087603.

Timco, G.A., McInnes, E.J.L., Pritchard, R.G., Tuna, F. and Winpenny, R.E.P. (2008). Sugared donuts made from chromium stick together easily, *Angew. Chem. Int. Ed.* **47**, 9681–9684.

Timco, G.A., Carretta, S., Troiani, F., Tuna, F., Pritchard, R.J., McInnes, E.J.L., Ghirri, A., Candini, A., Santini, P., Amoretti, G., Affronte, M. and Winpenny, R.E.P. (2009). Engineering coupling between spin qubits by coordination chemistry, *Nature Nanotech.* **4**, 173–178.

Wernsdorfer, W., Aliaga-Alcade, N., Hendrickson, D.N. and Christou, G. (2002). Exchange-biased quantum tunnelling in a supramolecular dimer of single-molecule magnets, *Nature*, **416**, 406–409.

Wernsdorfer, W., Müller, A., Mailly, D. and Barbara, B. (2004). Resonant photon absorption in the low spin molecule V_{15}, *Europhys. Lett.* **66**, 861–867.

Wernsdorfer, W., Mailly, D., Timco, G.A. and Winpenny, R.E.P. (2005). Resonant photon absorption and hole burning in Cr_7Ni antiferromagnetic rings, *Phys. Rev. B* **72**, art. 060409.

Wilson, A., Lawrence, J., Yeng, E-C., Nakano, M., Hendrickson, D.N. and Hill, S. (2006). Magnetization tunneling in high-symmetry single-molecule magnets: limitations of the giant spin approximation, *Phys. Rev. B* **74**, art. 140403.

Yang, E-C., Kirman, C., Lawrence, J., Zakharov, L.N., Rheingold, A.L., Hill, S. and Hendrickson, D.N. (2005). Single-molecule magnets: high-field electron paramagnetic resonance evaluation of the single-ion zero-field interaction in a $Zn_3^{II}Ni^{II}$ complex, *Inorg. Chem.* **44**, 3827–3836.

Chapter 6

SIMULATING COMPUTATIONALLY COMPLEX MAGNETIC MOLECULES

LARRY ENGELHARDT

Department of Physics and Astronomy, Francis Marion University, Florence, South Carolina, 29501, USA

CHRISTIAN SCHRÖDER

Department of Engineering Sciences and Mathematics, University of Applied Sciences Bielefeld, D-33602 Bielefeld, Germany

1. Introduction

1.1. *Scope and purpose*

The process of making quantitative determinations of the interactions that are present within a magnetic molecule has been, and remains, a non-trivial task. In principle, if the detailed chemical structure of a molecule were known precisely, one could calculate the many-electron wavefunction of the system, and from this all physical properties could be determined. However, *ab initio* methods to carry out such calculations have not yet progressed to the point for this to be a feasible option for most magnetic molecules (de Graaf and Sousa, 2006). Instead, we take a model-based approach to the analysis of magnetic molecules. The purpose of this chapter will be to describe and demonstrate how (relatively simple) theoretical models have been constructed and successfully used for describing a variety of specific magnetic molecule systems using quantum Monte Carlo and classical spin dynamics simulations.

Throughout this chapter, our starting point for the theoretical models will always be the Heisenberg model, which is introduced in Sec. 1.2 and is applied to several specific magnetic molecules in Secs. 2 and 3. This model of interacting magnetic moments is attractive due to its relative simplicity. It is not complicated in form, but it can be very computationally *complex*. Even using present-day computers, the process of extracting results from a simple-looking model Hamiltonian can be extremely computationally demanding. To address this challenge, we apply quantum Monte Carlo and classical spin dynamics techniques, which are described in Secs. 2 and 3, respectively. We emphasize the fact that there is not a single 'best' theoretical method. Instead, the details of the model that is being used determine which method is the most appropriate, and the most useful, method for each specific situation. Finally, we have already alluded to the fact that the process of developing a theoretical model to describe a specific magnetic molecule system is heavily reliant on experimental measurements. There are several standard measurements that are typically used, that each provides somewhat different insights into the physical system. These include the weak-field magnetic susceptibility versus temperature, $\chi(T)$, the low-temperature magnetization versus field, $M(H)$, and the low-temperature differential susceptibility versus field $dM/dH(H)$, which are analyzed in detail in Secs. 2 and 3.

Analytic solutions are typically not possible in the study of magnetic molecules, so matrix diagonalization is the standard tool for numerically determining the energy eigenvalues (and associated eigenstates) for a model. However, the matrix diagonalization method is subject to a serious constraint: Roughly stated, its use is only possible for relatively small systems (as described in Sec. 1.3). When matrix diagonalization is not possible, the quantum Monte Carlo (QMC) method is proving to be a valuable tool for studying such models. This method is discussed in Sec. 2.1, and applications of the method to specific magnetic molecule systems are presented in Secs. 2.2, 2.3, and 2.4. It can be used for arbitrarily large (complex) systems, but it does have two significant limitations. First, the complete energy spectrum cannot be calculated using the QMC method, so it is not able to provide predictions for certain measurements. Second, for frustrated systems, the 'negative sign problem' can limit the usefulness of the QMC method, as described in Sec. 2.4.

Finally, if none of the quantum methods mentioned above are applicable, it is possible to obtain approximate results by replacing the quantum model with an analogous *classical* model. The big advantage of classical spin dynamics methods is the fact that *technically* they do not have any restrictions whatsoever. With classical methods one can handle arbitrarily large systems and complex interaction scenarios (including anisotropies, higher order interaction terms, frustration, dipole-dipole interaction, etc.). Moreover, classical methods enable one to simulate dynamical equilibrium and non-equilibrium (e.g. relaxation) properties to get access to important experimental techniques like field-cooled (fc) and zero-field-cooled (zfc), ac or thermoremanent decay (TRM) measurements. We will "derive" the classical Heisenberg model from the quantum model in Sec. 3.1 and illustrate the differences between both descriptions. Based on the classical Heisenberg model we present the classical Monte Carlo method in Sec. 3.2, followed by the introduction of the (quantum and classical) spin equations of motion in Sec. 3.2.1, and the presentation of a classical Monte Carlo algorithm for determining equilibrium dynamical properties. In Sec. 3.3, we give an overview about heat bath simulational methods and discuss specifically the so-called stochastic Landau-Lifshitz approach which is a powerful tool for the simulation of equilibrium and non-equilibrium dynamical properties. Based on the aforementioned methods we present results for various molecular magnetic systems in Sec. 3.4.

Collectively, this set of methods is providing a significant arsenal for the theoretical understanding of numerous magnetic molecules.

1.2. *Introduction to the Heisenberg Hamiltonian*

Throughout this chapter calculations are performed using the Heisenberg model and its classical analog (introduced in Sec. 3.1). In its simplest form, this model is described by the Hamiltonian[1]

$$\underset{\sim}{\mathcal{H}} = -J \sum_{\langle j,k \rangle} \vec{\underset{\sim}{s}}_j \cdot \vec{\underset{\sim}{s}}_k + g\mu_B \vec{H} \cdot \sum_{j=1}^{N} \vec{\underset{\sim}{s}}_j, \tag{1}$$

where the summation $\langle j, k \rangle$ is over all distinct pairs of interacting spins, the spin operators $\vec{\underset{\sim}{s}}_j$ are given in units of $\hbar$, g is the spectroscopic splitting factor, μ_B is the Bohr magneton, and J represents the exchange constant. The magnitude of J determines the strength of the interaction between a pair of interacting spins, and its sign determines whether the interaction is antiferromagnetic ($J < 0$) or ferromagnetic ($J > 0$).

In Eq. (1), we note that writing J outside of the first summation assumes that all bonds in the molecule are equivalent, and writing g outside the second summation assumes that all magnetic moments in the molecule are equivalent. Examples of two specific magnetic molecules for which these assumptions can be justified are shown in Figs. 1(a) and 1(b). The $\{Fe_{12}\}$ ring represented in Fig. 1(a) possesses a particularly simple geometry, consisting of 12 Fe^{III} ions, each interacting equally with its two neighboring Fe^{III} ions. An example of a geometry with lower symmetry is shown in Fig. 1(c), which represents the $\{Ni_{12}\}$ magnetic molecule. This structure requires four distinct bonds (labeled J_1, J_2, J_3, J_4), and it can be represented by the more general Hamiltonian

$$\underset{\sim}{\mathcal{H}} = -\sum_{\langle j,k \rangle} J_{j,k} \vec{\underset{\sim}{s}}_j \cdot \vec{\underset{\sim}{s}}_k + g\mu_B \vec{H} \cdot \sum_{j=1}^{N} \vec{\underset{\sim}{s}}_j, \tag{2}$$

where $J_{j,k}$ represents the exchange constant that couples spin j and spin k.

Several magnetic molecules (e.g. $\{Mo_{72}Fe_{30}\}$) have been synthesized in the form of an icosidodecahedron — shown in Fig. 1(b) — with a paramagnetic ion appearing at each of the 30 vertices. This structure is more complex than the ring. However, for high enough temperatures it still possesses very high symmetry; all vertices and edges in Fig. 1(b) are equivalent, so Eq. (1) is still appropriate. For low temperatures it has been found that $\{Mo_{72}Fe_{30}\}$ exhibits exchange disorder and consequently it requires the more general description using Eq. (2) (see Sec. 3.4 for details).

---

[1] Note that the tilde beneath the characters $\mathcal{H}$ and $\vec{s}$ are used to distinguish the character as representing a quantum operator. This notation is used throughout this chapter.

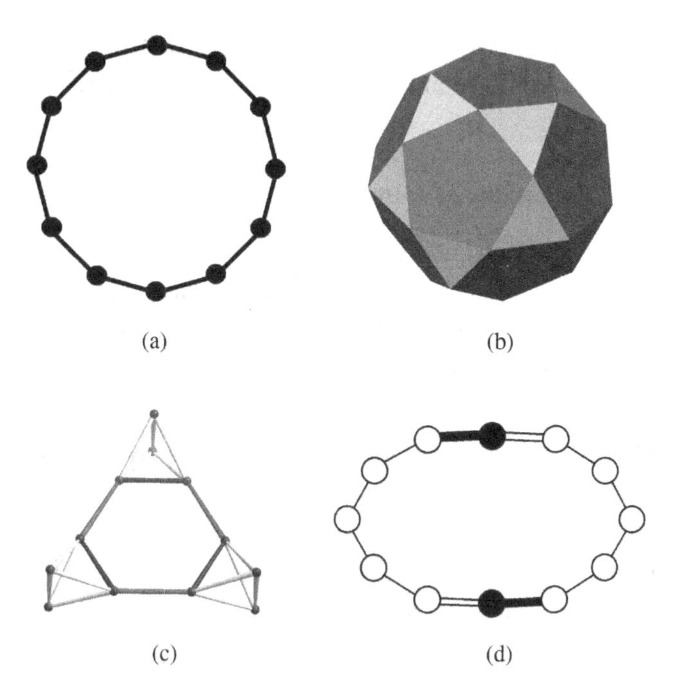

Fig. 1. Geometric representations of the model Hamiltonian for the magnetic molecules (a) $\{Fe_{12}\}$ (b) $\{Mo_{72}Fe_{30}\}$ (c) $\{Ni_{12}\}$ and (d) $\{Cr_{10}Cu_2\}$.

Finally, Fig. 1(c) shows the structure of the heterometallic $\{Cr_{10}Cu_2\}$ ring that consists of two different paramagnetic ions and requires three distinct exchange constants. This system (as well as the other $\{Cr_xCu_2\}$ rings described in Sec. 2.2) is represented with the Hamiltonian

$$\underset{\sim}{\mathcal{H}} = -\sum_{\langle j,k\rangle} J_{j,k}\vec{\underset{\sim}{s}}_j \cdot \vec{\underset{\sim}{s}}_k + \mu_B \vec{H} \cdot \sum_{j=1}^{N} g_j \vec{\underset{\sim}{s}}_j, \tag{3}$$

where the value of g_j is different for the Cr sites than the Cu sites. These four examples, along with others, are considered in detail in Secs. 2 and 3.

Using the Heisenberg Hamiltonian as a starting point, there are various properties that one can obtain. For sufficiently small systems, it is possible to calculate the eigenstates of $\underset{\sim}{\mathcal{H}}$, as discussed in Sec. 1.3. These eigenstates have the important property that the total spin, S, and its z-component, M_S, are good quantum numbers, where M_S takes on values ranging in integer steps from $-S$ to S. For more complex systems, quantum Monte Carlo calculations can be used to calculate thermodynamic properties directly, and these quantum numbers are particularly relevant to those data, as described in Sec. 2. Finally, dynamical equilibrium and non-equilibrium properties are accessible by classical spin dynamics methods as discussed in Sec. 3.

1.3. *Usefulness and limitations of matrices*

Despite its simple form, the Heisenberg Hamiltonian can rarely be solved analytically, so instead calculations must be carried out numerically. The most straightforward numerical method is that of matrix diagonalization, sometimes referred to as 'exact diagonalization' (ED). In its simplest form, the process of ED proceeds as follows: (1) Determine every possible way in which the spins in the system can be combined together. These are the basis states to be used for the ED. (2) Create a matrix, having matrix elements of the form $\langle\psi_i|\underset{\sim}{\mathcal{H}}|\psi_j\rangle$, where $\langle\psi_i|$ and $|\psi_j\rangle$ are two of the basis states, and $\underset{\sim}{\mathcal{H}}$ is the Hamiltonian of the system. (3) Diagonalize this matrix in order to obtain the eigenvalues (the energies) and the eigenvectors (the eigenstate wavefunctions).

The process briefly outlined above is desirable for two significant reasons. First, it is not limited to specific geometries, which is the case for analytic solutions. Second, once the energy eigenvalues and the eigenstate wavefunctions have been determined, it is possible to calculate *all* properties associated with the model, including dynamical properties. However, there is also a serious limitation to this method: The matrices can get very large! For a model that involves N magnetic moments, each having spin s, the number of basis states is given by $D = (2s + 1)^N$. For example, $\{Fe_{12}\}$ and $\{Mo_{72}Fe_{30}\}$ [introduced in Figs. 1(a) and 1(b)] each have spins $s = 5/2$ (Fe^{III} ions), giving $D = 6^{12} > 2 \times 10^9$ and $D = 6^{30} > 2 \times 10^{23}$, respectively. More generally, if a system consists of different values of s, D is given by the product $D = \prod(2s_i + 1)^{N_i}$, where s_i represents the spin of the ith species, and N_i represents the number of spins associated with this species. For example, $\{Cr_{10}Cu_2\}$ [see in Fig. 1(d)] has 10 Cr^{III} ions with $s = 3/2$ and two Cu^{II} ions with $s = 1/2$, giving $D = 4^{10} \times 2^2 \approx 4.1 \times 10^6$.

Typical modern computers cannot handle matrices that are larger than $\approx 10^4 \times 10^4$. This would restrict the usefulness of matrix diagonalization to extremely small systems, but there are steps that can be taken in order to use this method for somewhat larger models. Generally, the entire $D \times D$ matrix does not need to be created, as large portions of the matrix would contain zeros. Instead, smaller sub-matrices are created and diagonalized. The size of these smaller matrices depends on both the symmetries of the model system and the choice of how the system is to be represented (the basis states). However, the current (Schnalle and Schnack, 2009) rule of thumb is that ED can be used for models if $D < 10^6$ or 10^7. In addition to ED, there are other approximate methods of matrix diagonalization, e.g. the Davidson and Lanczos methods, that can be used to determine restricted subsets of energy eigenvalues and eigenstates, rather than taking all states into account. These methods can be applied to larger systems, up to $D \approx 10^9$.

Matrix diagonalization is very useful, but the very nature of the method necessitates a limiting value of D. This is because all of the basis states must be determined and stored in the computer's memory. The development of faster computers with more memory can provide incremental increases to this maximum value of D, as can the development of improved algorithms to take advantage of symmetries. However, the treatment of models with very large values of D (several orders of magnitude larger than the current limiting values) would require a fundamentally different computer architecture, and this will hence not be possible in the foreseeable future. For large systems, the quantum Monte Carlo method provides a viable alternative, which is described in Sec. 2.

2. Quantum Monte Carlo Simulations

2.1. *Avoiding the 'roadblock' of large matrices*

As we have described in Sec. 1.3, matrix diagonalization is a powerful tool for performing calculations for models that can be used to describe magnetic molecules. In fact, for models that involve a relatively small value of the Hilbert space dimension, D, matrix diagonalization is the ideal method to use — provided an analytical solution is not possible. However, it is very common for magnetic molecule systems to be synthesized for which D is so large that matrix diagonalization is either impossible or impractical. In these situations, the quantum Monte Carlo (QMC) method is proving to be an extremely useful tool.

The power of the QMC method is a result of how the necessary computational resources (both computation time and computer memory) scale with the system size, N. For matrix diagonalization, this scaling is exponential since $D = (2s + 1)^N$. For the QMC method, this scaling depends *linearly* on N, which is a vast improvement over exponential scaling. For example, a typical QMC calculation [calculating $\chi(T)$ for $T = 2 - 300$ K] for a model that consists of $N = 10$ spins, each having $s = 5/2$, might take five minutes of computation time. Due to the linear scaling, the analogous calculation for $N = 20$ spins, $s = 5/2$, would require 10 minutes of computation; and the calculation for $N = 50$ spins, $s = 5/2$, would require 50 minutes. Using matrix diagonalization, calculations for $N = 10$ spins with $s = 5/2$ are pushing the limits of what is currently possible; and calculations for $N = 20$ spins with $s = 5/2$ are far beyond the realm of possibility.

It is easy to understand why computations scale linearly with N — rather than exponentially — for the QMC method. The reason is that the method does not require large matrices (representing the system Hamiltonian) to be created, and matrices are also not diagonalized. Instead, the thermodynamic quantities of interest (e.g. magnetization, magnetic susceptibility, specific heat) are determined by a process of statistical sampling. This sampling process is controlled by the system's partition function, bypassing any calculation of the energy spectrum.[2] The main advantage of the QMC method — linear scaling with system size — therefore comes at a price: The QMC method does not directly provide information about the energy spectrum. However, by calculating the low-temperature magnetization versus field (or dM/dH), one can identify the fields at which ground-state level crossings occur. These fields can then be used to infer the energies of the lowest "rotational band". (See Sec. 2.2.) The energy of a state that lies above this rotational band cannot be determined using the QMC method, and eigenvectors also cannot be determined. Because the QMC method involves statistical sampling, the calculated data necessarily have some statistical uncertainty associated with them. The magnitudes of these uncertainties are calculated

[2] To carry out the statistical sampling, the partition function is expressed as a series expansion (Handscomb, 1964). The algorithm that is used to sample the terms of the expansion was introduced in (Syljuåsen and Sandvik, 2002) and is described in greater detail in (Engelhardt, 2006).

during the course of each QMC simulation by observing the variance in the sampled data, and very small uncertainties can typically be achieved within a reasonable amount of computation time. If greater precision is needed, this simply requires additional sampling, which in turn involves additional computation time. The scaling of these uncertainties is inversely proportional to the square-root of the computation time, so extreme precision would require very long computation times. However, uncertainties that are a fraction of a percent can be achieved in a matter of minutes for a typical calculation of $\chi(T)$.

As described above, the system size and the desired level of precision play important roles in determining the computation time that will be required for a given QMC calculation. Additionally, the temperature at which QMC data are calculated has a significant effect on computation time. The QMC process involves a mathematical expansion of the partition function that causes an increase in computation time as the temperature is lowered. From a practical perspective, this has the consequence that a calculation of $\chi(T)$ for T in the range from 2 K to 300 K (the typical range of a SQUID measurement) requires a few minutes of computation time, while a calculation of $M(H)$ for $T < 1$ K can take an hour or longer. The precise computation times involve both the details of the model system and the details of the computer being used.

Given the relative quickness of QMC calculations, along with the limitation that the QMC method does not provide the complete energy spectrum, the use of QMC and matrix diagonalization as complementary techniques is an attractive option for certain systems. Namely, QMC can be used to carry out a detailed search of the parameter space by comparing measured $\chi(T)$ data to calculations for a model Hamiltonian, and looking for the best fit. Since each calculation takes only a few minutes of computation, thousands of sets of model parameters can be explored in a reasonable amount of time, on the order of a few days of computation. As a result of this fitting procedure, we obtain a set of model parameters that can then be used for calculating other quantities. However, not all quantities are accessible using QMC; whereas if the complete energy spectrum were known, it would be possible — at least in principle — to calculate any quantity that is desired. Hence, if the Hilbert space dimension is not too large, it is possible to carry out a single, very time consuming matrix diagonalization. (An example of this complementary use of QMC and matrix diagonalization is described in Sec. 2.4 for the $\{Ni_{12}\}$ magnetic molecule.)

Finally, the QMC method has an important limitation. Namely, QMC calculations cannot be carried out below a certain temperature in some systems, due to the so-called 'negative sign problem' (NSP) (Troyer and Wiese, 2005). For the present purpose (i.e. to assess whether or not the QMC method will be useful for the analysis of a given magnetic molecule system) there are two important questions that need to be answered regarding the NSP: (1) How does one determine whether or not the NSP occurs for a given magnetic molecule? (2) If the NSP does occur, to what extent does this limit the usefulness of the QMC method for that system? These questions are addressed in detail in Sec. 2.4.

2.2. *Energy spectrum for symmetric rings*

The purpose of this section is two-fold: (1) to demonstrate the power of the QMC method for simple — but computationally demanding — geometries, and (2) to demonstrate how a certain subset of a magnetic molecule's energy spectrum can be explored using QMC calculations. These issues are introduced here in the context of the $\{Fe_{12}\}$ ring (Caneschi *et al.*, 1999; Inagaki *et al.*, 2003; Engelhardt and Luban, 2006), but these results are also of relevance to the systems that are studied in Sec. 2.3.

The $\{Fe_{12}\}$ ring introduced in (Caneschi *et al.*, 1999) can be described by the particularly simple Hamiltonian,

$$\underset{\sim}{\mathcal{H}} = -J \sum_{i=1}^{12} \vec{\underset{\sim}{s}}_i \cdot \vec{\underset{\sim}{s}}_{i+1} + g\mu_B \vec{H} \cdot \sum_{i=1}^{12} \vec{\underset{\sim}{s}}_i, \tag{4}$$

where $\vec{\underset{\sim}{s}}_{13} \equiv \vec{\underset{\sim}{s}}_1$. Letting each site have spin $s = 5/2$ (so as to represent Fe^{III} ions), the QMC method can be used to calculate $M(H)$ and $dM/dH(H)$; and these data are shown in Fig. 2 for the fixed temperature, $k_BT/|J| = 0.1$. The $M(H)$ data of Fig. 2(a) appear roughly linear until saturating to $30\, g\mu_B$, whereas individual peaks are clearly visible in dM/dH, shown in Fig. 2(b). For these calculations, specific values of the parameters J and g have not been specified, so the value of H at which the peaks occur will depend on these values. In particular, measurements of the first four $\{Fe_{12}\}$

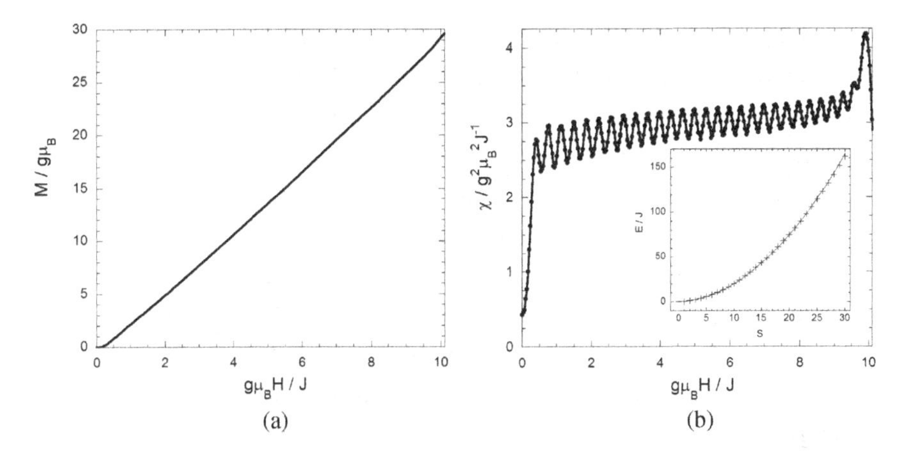

Fig. 2. Quantum Monte Carlo calculations for a ring of 12 spins $s = 5/2$ for a fixed temperature, $k_B T/|J| = 0.1$. (a) $M(H)$, and (b) dM/dH (main figure) and the lowest rotational band of the energy spectrum (inset).

level-crossing fields were reported in (Inagaki *et al.*, 2003), and the values of these level-crossing fields are in excellent agreement with the values of $g\mu_B H/J$ at which the first four peaks occur in Fig. 2(b), provided we let $J/k_B = -40.7\,\text{K}$ (assuming $g = 2.02$, as used in (Inagaki *et al.*, 2003)). We note that this QMC estimate (Engelhardt and Luban, 2006) of J results directly from Eq. (4) without making any approximations, and for this reason it should be more accurate than other estimates (Caneschi *et al.*, 1999; Inagaki *et al.*, 2003) that involved various approximations due to the large matrix dimensionality ($D \approx 2.2 \times 10^9$) of Eq. (4).

In addition to using dM/dH data to infer the value of J from experimental data, these calculations also supply more general information: From the level-crossing fields, H_n, we can infer the energies of the lowest "rotational band" of the energy spectrum (i.e. the smallest energy eigenvalue for each value of the total spin, S). For an antiferromagnetic ring of N spins s, these energies are given by

$$E(S) = E_G + g\mu_B \sum_{n=1}^{S} H_n, \quad (1 \leq S \leq Ns), \tag{5}$$

where E_G is the energy of the $S = 0$ ground state, which we define to be $E_G \equiv 0$. These energies are shown in the inset of Fig. 2(b) for $N = 12$ and $s = 5/2$ (of relevance for $\{Fe_{12}\}$), where the solid curve represents the best fit using the parabola $E(S) = cS^2$, for which $c = 4.219$. We note that this rotational band is roughly — but not exactly — parabolic. In particular, the curvature of $E(S)$ decreases (becoming more linear) with increasing S, and this is associated with fact that the peaks in dM/dH become more closely spaced with increasing H. This effect has been explored in detail in (Engelhardt and Luban, 2006) for different ring sizes (up to $N = 80$) and different intrinsic spins (up to $s = 7/2$), which provides a striking example of the power of the QMC method: For $s = 7/2$ and $N = 80$ the matrix dimensionality is $8^{80} \approx 1.7 \times 10^{72}$. Yet, using the QMC method, we have been able to calculate dM/dH at low temperatures and subsequently obtain all of the energies of the lowest rotational band.

2.3. *Applications to heterometallic rings*

In this section, we demonstrate the usefulness of the QMC method for studying magnetic molecule systems that involve multiple exchange constants. To introduce this process, we consider the family of $\{Cr_nCu_2\}$ rings (with $n = 8, 10, 11, 12$) that have been introduced in (Engelhardt *et al.*, 2008). Each of these rings involve two "horseshoes" of Cr^{III} sites ($s = 3/2$) that are coupled together via two Cu^{II} sites ($s = 1/2$), and a simplified representation of the structure for the system with $n = 10$ (i.e. $\{Cr_{10}Cu_2\}$) is shown in Fig. 1(d).[3] Within the Cr horseshoes, each of the Cr-Cr bonds are taken to be identical due to the crystallographic symmetry, and these interactions are represented with the exchange constant J_1. The structures also have 180° rotational symmetry, so the Cr-Cu bonds on opposite sides of the molecules are identical. However — due to asymmetry of the ligands — these bonds are not identical to the Cu-Cr bonds, so two additional exchange constants, J_2 and J_3, are needed to represent these Cr-Cu and Cu-Cr interactions.

With three exchange constants (J_1, J_2, J_3), the challenge of determining the nature of the magnetic interactions within the $\{Cr_nCu_2\}$ rings becomes that of searching a three-dimensional parameter space and assessing the

[3] $\{Cr_{10}Cu_2\}$ was the first member of this family of rings to be synthesized, and it was introduced in (Shanmugam *et al.*, 2006).

Table 1. Exchange constants that provide the best fit to $\chi(T)$ for $\{Cr_nCu_2\}$ rings.

System	J_1 (meV)	J_2 (meV)	J_3 (meV)
$\{Cr_8Cu_2\}$	-1.30 ± 0.04	-4.8 ± 0.9	0.9 ± 0.4
$\{Cr_{10}Cu_2\}$	-1.28 ± 0.04	-4.6 ± 0.5	1.0 ± 0.2
$\{Cr_{11}Cu_2\}$	-1.29 ± 0.04	-5.6 ± 0.7	0.4 ± 0.4
$\{Cr_{12}Cu_2\}$	-1.25 ± 0.04	-3.7 ± 0.9	0.21 ± 0.17

agreement between experiment and theory at different points in this space. To properly survey such a parameter space requires calculations for *many* choices of (J_1, J_2, J_3), and — thanks to its quickness — the QMC method is ideally suited for this task. For each system, the following process was used: First, $\chi(T)$ was measured in the range 2–300 K. Next, QMC calculations of $\chi(T)$ were carried out for thousands of points in the (J_1, J_2, J_3) parameter space. Then the square of the deviation between these measured and calculated data were calculated. Using this process, the best fit was obtained from the J values listed in Table 1, and the corresponding $\chi(T)$ data are shown in Fig. 3. In addition to the excellent agreement between experiment and theory, we also note that for $\{Cr_{12}Cu_2\}$ $D \approx 6.7 \times 10^7$. Hence matrix diagonalization is not an option even for a *single* set of (J_1, J_2, J_3) values, let alone thousands of values.

In addition to finding the values of the exchange constants that provide the *best* fit, this process has also allowed us to assess the extent to which the goodness of fit (both the *uncertainty* and the *uniqueness*) depends on each parameter. To illustrate this, Fig. 4 shows the deviation between experiment and theory,[4] plotted as a function of both (a) J_1 and (b) J_2. We first note that the deviation has a much narrower minimum when plotted as a function of J_1 than it does versus J_2. This is reflected in the uncertainties that are listed in Table 1. Secondly, we note that — in addition to the global minimum in the deviation — there is a clear *local* minimum, corresponding to a weaker interaction (by a factor of 2) for the J_2 bond. At this local minimum, the

[4]The 'deviation' is the sum (over all temperatures) of the squares of the deviations between the measured $T\chi(T)$ data and the QMC calculations.

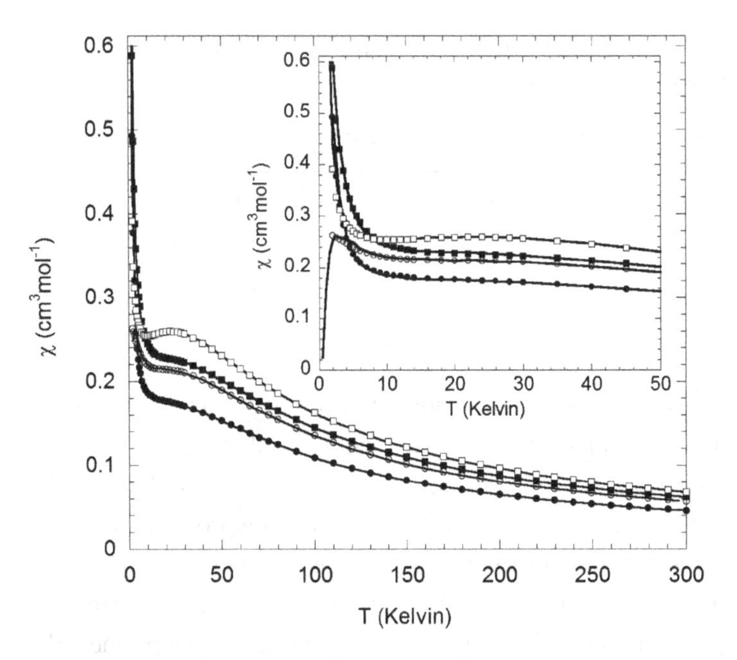

Fig. 3. $\chi(T)$ for $\{Cr_nCu_2\}$ rings, including experimental data for $n = 8$ (•), $n = 10$ (○), $n = 11$ (□), and $n = 12$ (■). The solid curves represent the QMC data that provide the best fit for each structure, as described in the text.

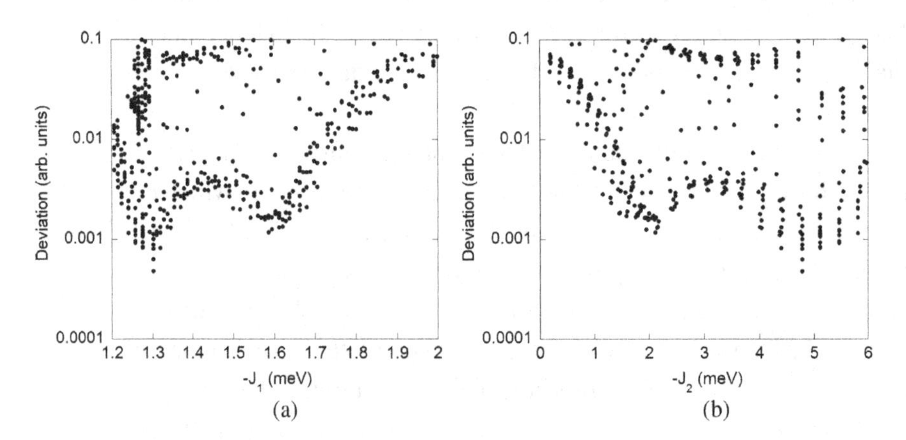

Fig. 4. Deviation between experiment and theory for $\{Cr_8Cu_2\}$, with the results for the three-dimensional parameter space projected onto (a) the J_1 axis and (b) the J_2 axis.

agreement between experiment and theory is reasonably good — although visibly inferior to the results shown in Fig. 3. Hence, without a thorough search of this parameter space, it would be easy to mis-identify this local minimum as "the best fit". The question of which minimum corresponds to the *true* set of exchange constants that are present in the molecule remains an open question, as both sets of parameters produce a nearly identical energy spectrum, yielding nearly identical predictions for other potential measurements.

In addition to its usefulness for fitting $\chi(T)$ data, the QMC method can also be used to generate predictions for other experiments. To demonstrate one such use, we now consider the $\{Cr_{12}Cu_2\}$ molecule that was introduced and analyzed in (Martin *et al.*, 2009). This molecule is a variation of the $\{Cr_{12}Cu_2\}$ ring that was previously introduced in (Engelhardt *et al.*, 2008). It has a geometry that is more planar than the first version, resulting in subtle differences in the measured $\chi(T)$ data as compared to Fig. 3.

As with the other $\{Cr_nCu_2\}$ rings, the QMC method was used to fit $\chi(T)$, and again an excellent fit was obtained. This time the best fit was obtained for ($J_1 = -1.3$ meV, $J_2 = -4.9$ meV, $J_3 = 1.7$ meV), showing that the ferromagnetic (J_3) interaction is much stronger for this version of $\{Cr_{12}Cu_2\}$. Using these exchange constants, we have subsequently calculated the low-temperature differential susceptibility versus field, $dM/dH(H)$. These data show two peaks for $H < 15$ T, associated with two ground state level crossings, which occur at $H_1 = 8.15$ T and $H_2 = 11.9$ T. Using Eq. (5), these values of H_1 and H_2 can be used to predict the excitation energies of the lowest $S = 1$ and $S = 2$ levels. They are 0.94 meV and 2.32 meV respectively, both measured relative to the $S = 0$ ground state, and the predicted field dependence of these energy levels is shown in Fig. 5(a).

To test the QMC predictions of Fig. 5(a), tunnel diode oscillator (TDO) measurements were carried out for $H < 15$ T, and these data are shown in Fig. 5(b).[5] This TDO spectrum includes seven peaks, and the values of H at which the peaks occur are in excellent agreement with the QMC predictions.

[5]The use of the TDO method for studying magnetic systems is described in (Vannette *et al.*, 2008), and a detailed discussion of why peaks occur for excited-state level crossings is given in (Engelhardt *et al.*, 2009).

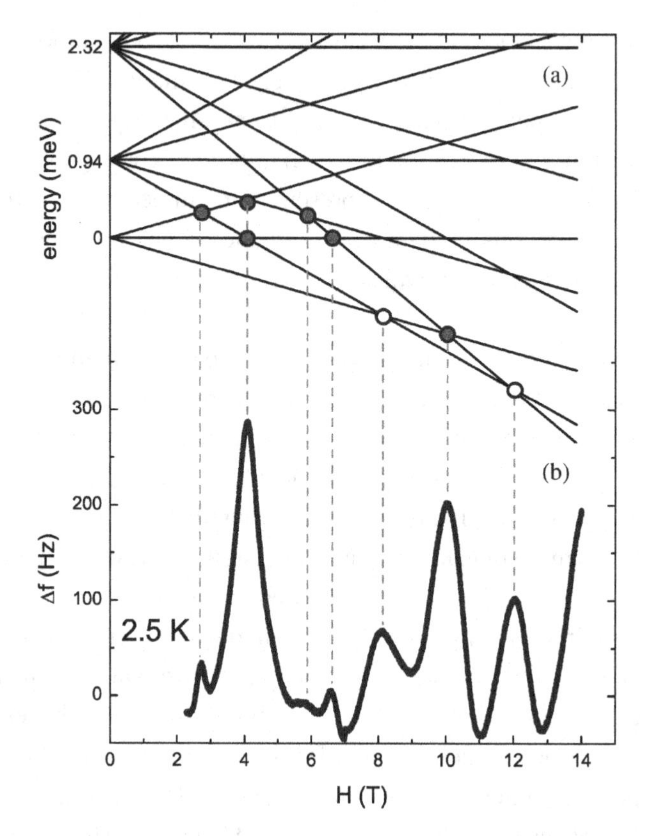

Fig. 5. (a) Zeeman splitting of predicted energy levels (b) TDO measurements versus magnetic field for 2.5 K (directly below the energy levels, with dashed lines to emphasize the agreement between peaks and crossings) (Martin *et al.*, 2009).

Five of these peaks are associated with excited state level crossings, and these peaks should only be observable if the relevant states have a significant thermal occupation. Hence, as the temperature is lowered below $T \approx 1$ K, these peaks should vanish. These measurements are shown in Fig. 6; and indeed, at very low temperatures, the only peaks that remain are those associated with the ground state level crossings.

Finally, it should be stressed that the QMC predictions shown in Fig. 5(a) are based on a model whose only experimental input was $\chi(T)$. Yet, this model is able to reproduce the detailed features of a totally different (TDO) experiment with astounding accuracy. These results demonstrate (1) the accuracy of the Heisenberg model for describing this system,

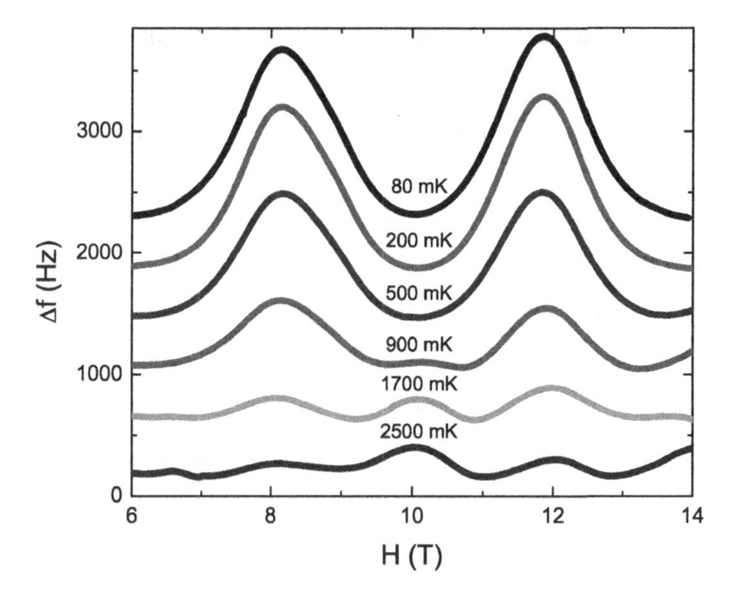

Fig. 6. TDO measurements as a function of H for different temperatures, from 80 mK to 2.5 K (Martin *et al.*, 2009).

(2) the usefulness of the TDO method as a spectroscopic tool, and (3) the power of the QMC method for this computationally complex magnetic molecule.

2.4. *Applications to frustrated magnetic molecules*

Despite the usefulness of the quantum Monte Carlo (QMC) method for simulating complex magnetic molecules, there are some types of molecules for which the usefulness of the QMC is significantly limited. This is due to the "negative sign problem" (NSP), which does not allow the QMC to be applied below a certain minimum temperature, T_m, for frustrated geometries. However, for some magnetic molecules (including one that was analyzed in Sec. 2.3), T_m is so low that the NSP does not actually present a 'problem' in any practical sense. For other magnetic molecules, T_m is much higher, such that QMC can provide only limited information. The purpose of the present section is to answer the following two questions (using several illustrative examples), which are of practical importance for the analysis of magnetic molecules:[6]

(1) How does one determine whether or not the NSP occurs for a given magnetic molecule?
(2) If the NSP does occur, to what extent does this limit the usefulness of the QMC method for that particular magnetic molecule?

The answer to the first question is quite straightforward. The condition that determines whether or not the NSP will occur is identical to the condition that determines whether or not a system will be frustrated:

Theorem 1. *The NSP will be present (as will frustration) for a particular system if and only if that system includes a polygon (or multiple polygons) composed of an odd number of anti-ferromagnetic (AFM) bonds.*

When only AFM bonds are present, it is easy to determine whether or not the NSP is present by simply looking at the geometry. If odd-sided polygons (e.g. triangles or pentagons) are present, so is the NSP. When both AFM and FM bonds are present, the situation is only slightly more complicated. For example, each of the $\{Cr_nCu_2\}$ rings described in Sec. 2.3 consist of a single polygon, made up of n AFM bonds and 2 FM bonds. Hence, according to Theorem 1, the NSP does occur for the $\{Cr_{11}Cu_2\}$ ring since it contains 11 AFM bonds. However, the minimum temperature is so low for this system ($T_m \approx 0.5$ K), that it does not present any problem for the analysis of $\chi(T)$ described in Sec. 2.3.

A second example, the icosidodecahedron [shown in Fig. 1(b)], will help us to further address question (2). This structure is composed of triangles and pentagons, so — assuming that all bonds are AFM — the NSP will occur. Magnetic molecules have indeed been synthesized in this structure with entirely AFM interactions, including $\{Cr_{30}\}$ (Todea *et al.*, 2007) (Cr^{III} ions, $s = 3/2$), and two versions of $\{V_{30}\}$ (each with V^{IV} ions, $s = 1/2$) that will be denoted $\{V_{30}\}_a$ (introduced in (Müller *et al.*, 2005)) and $\{V_{30}\}_b$

[6]For a detailed description of *why* the NSP occurs, see Ch. 8 of (Engelhardt, 2006). Very briefly, there is almost complete cancelation between the positive and negative terms that are encountered during the Monte Carlo sampling, leading to a very small signal-to-noise ratio.

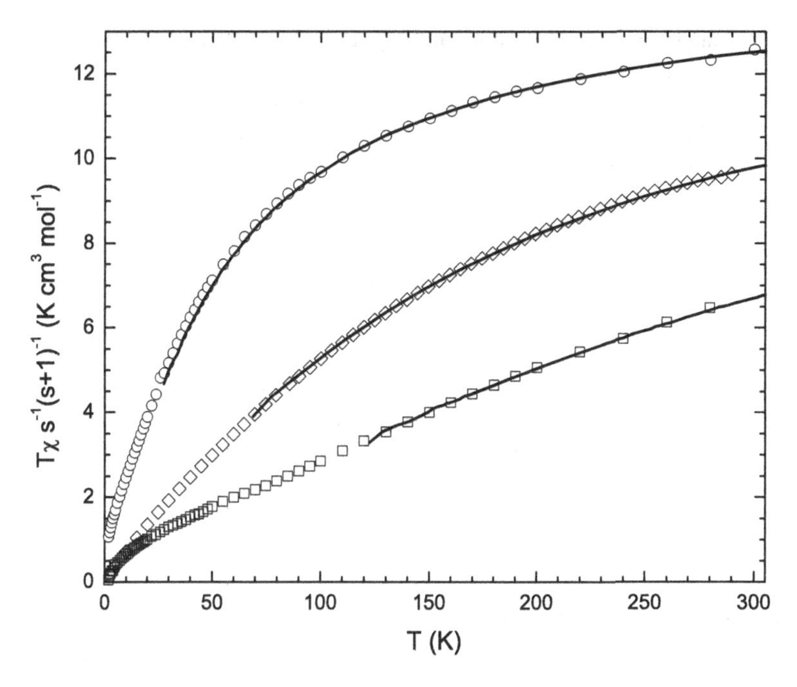

Fig. 7. $T\chi(T)/[s(s+1)]$ for $\{Cr_{30}\}$ (○), $\{V_{30}\}_a$ (□), and $\{V_{30}\}_b$ (◇), along with quantum Monte Carlo calculations (solid curves).

(introduced in (Todea *et al.*, 2009)). $T\chi(T)$ data for each of these systems are shown in Fig. 7. Since the Curie constant is proportional to $s(s+1)$, these data have all been normalized by plotting $T\chi/[s(s+1)]$ such that all data sets should coincide in the high temperature ($T \to \infty$) limit. By plotting $T\chi$ in this way, it is clear from the high temperature data that $\{Cr_{30}\}$ has the smallest J value and $\{V_{30}\}_a$ has the largest J value. To obtain precise J values for each system, the QMC method was used, assuming nearest-neighbor exchange with a single exchange constant. The J values that give the best fit are listed in the first column of Table 2, and the corresponding QMC data are shown as solid curves in Fig. 7.

With the help of Table 2, we are now prepared to give a detailed answer to question (2) that was posed at the beginning of this section. The values of T_m — and subsequent usefulness of the QMC method — depends on three properties of the magnetic molecule in question:

(i) the magnitude of the exchange constant, J,

Table 2. Exchange constant, J, minimum QMC temperature, T_m, and relevant ratios for specific magnetic molecules.

System	J/k_B (K)	T_m (K)	$k_BT_m/\|J\|$	s	$k_BT_m/[\|J\|s(s+1)]$
$\{V_{30}\}_a$	−245	120	0.5	1/2	0.7
$\{V_{30}\}_b$	−115	60	0.5	1/2	0.7
$\{Cr_{30}\}$	−8.7	30	3.4	3/2	0.9
$\{Cr_{11}Cu_2\}$[a]	−15	0.5	0.03	3/2	0.009
$\{Ni_{12}\}$[b]	−13.5	1.0	0.07	1	0.04

[a]For these $\{Cr_{11}Cu_2\}$ values, we have assumed $J = J_1$ and $s = 3/2$, since nine of the 13 bonds are described by J_1 and 11 of the 13 spins have $s = 3/2$.
[b]For $\{Ni_{12}\}$ (discussed below) we have assumed $J = (J_1 + J_2)/2$, since the NSP is caused by a hexagon formed by three J_1 bonds and three J_2 bonds.

(ii) the intrinsic spin, s, of the paramagnetic ions,
(iii) the geometry — in particular, the *size* of the relevant polygons.

For item (i), T_m is directly proportional to J. The two $\{V_{30}\}$ systems provide a clear example of this, since both J and T_m are twice as large for $\{V_{30}\}_a$ as compared to $\{V_{30}\}_b$. For $\{Cr_{30}\}$, T_m is smaller due to its small J value. However, compared the $\{V_{30}\}$ systems, $k_BT_m/|J|$ is much larger for $\{Cr_{30}\}$. This is due to its larger value of s. Specifically, for item (ii), T_m is approximately proportional to $Js(s+1)$. (For the icosidodecahedron, $k_BT_m/[|J|s(s+1)] \approx 1$ for very large s. For different geometries, this quantity approaches a different geometry-dependent constant value for large s (Engelhardt, 2006).) For item (iii), the NSP is more severe (i.e. $k_BT_m/[|J|s(s+1)]$ is larger) when small polygons are present. Thus, since triangles are the smallest possible polygon, the icosidodecahedron — composed of 20 triangles — represents a "worst case" of the NSP. For comparison, results for the $\{Cr_{11}Cu_2\}$ ring are also included in Table 2, showing that the NSP is much less severe for this large (13 sided) polygon. In fact, $k_BT_m/[|J|s(s+1)]$ is smaller by two orders of magnitude compared to the icosidodecahedron.

The final row of Table 2 contains data for the $\{Ni_{12}\}$ magnetic molecule that was introduced in Cooper *et al.* (2007), the structure of which is shown

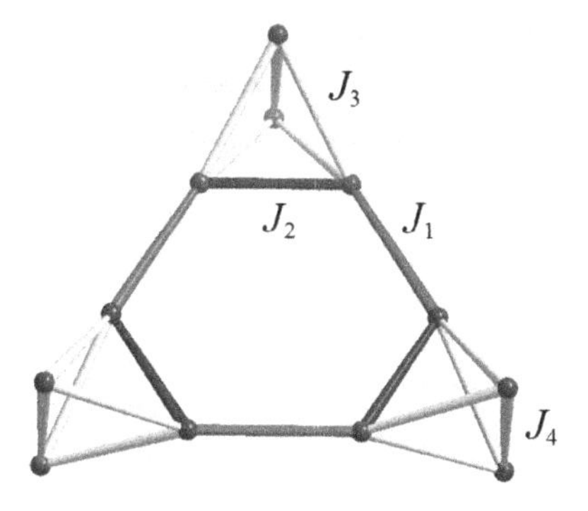

Fig. 8. Structure of the $\{Ni_{12}\}$ magnetic molecule (Cooper *et al.*, 2007).

in Fig. 8, where each site represents a V^{IV} ion ($s = 1$). This structure clearly includes several (12) triangles, but our analysis suggests that all of the bonds that form the triangles (J_2, J_3, J_4) are ferromagnetic. The only AFM bond is J_1, so the only polygon that causes the NSP is the central hexagon (composed of bonds J_1 and J_2). For this intermediately sized polygon, the value of $k_B T_m/[|J|s(s+1)]$ in Table 2 is also intermediate.

Figure 9 shows $T\chi(T)$ for $\{Ni_{12}\}$, where the solid curve represents the QMC data that provide the best fit ($J_1/k_B = -17.5 \pm 0.5\,\text{K}$, $J_2/k_B = 9.5 \pm 1.5\,\text{K}$, $J_3/k_B = 1.9 \pm 0.1\,\text{K}$, and $J_4/k_B = 22.0 \pm 1.0\,\text{K}$, with $g = 2.21$). Using such a large number of exchange constants as fitting parameters might seem inappropriate given the amount of information contained in the experimental data of Fig. 9, so we initially attempted to construct a simpler model with only two or three distinct J values. However, we found that it was not possible to obtain a reasonable fit with fewer than four distinct bonds. Furthermore, the values of (J_1, J_2, J_3, J_4) that are listed above represent a *unique* region of the four-dimensional parameter space in which a good fit exists (i.e. other local minima do not exist).

As described above, $\{Ni_{12}\}$ represents an example for which — despite the existence of frustration and the NSP — the QMC method has been used to construct a very detailed model by fitting $\chi(T)$ throughout the entire measured temperature range ($T = 2 - 300\,\text{K}$). Since $T_m \approx 1\,\text{K}$, the NSP did not hinder the calculations of $\chi(T)$, but it does prevent us from using the QMC method to calculate $M(H)$ at a sufficiently low temperature to clearly identify level-crossing fields. To accomplish this, matrix diagonalization can

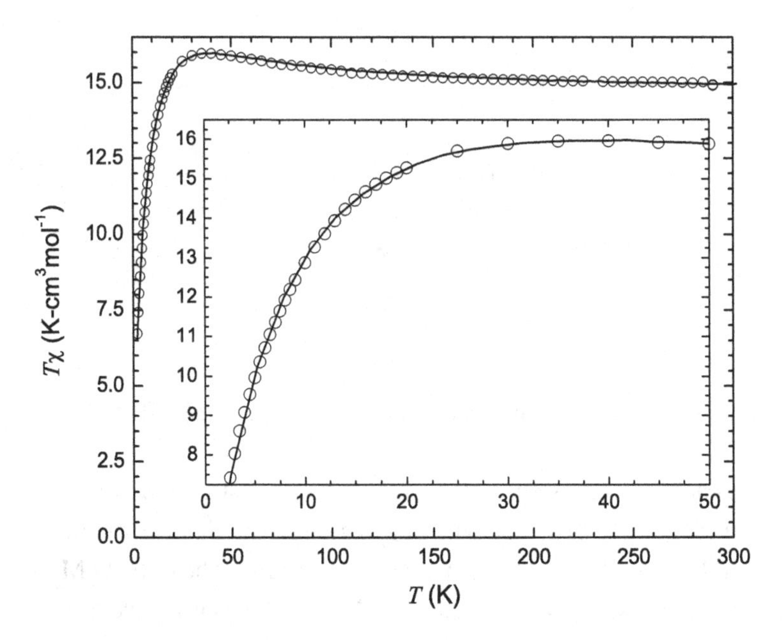

Fig. 9. $T\chi(T)$ for $\{Ni_{12}\}$, including both experiment (○) and theory (solid curve), with low temperature data shown in the inset (Cooper *et al.*, 2007).

be used to complement the QMC method. For 12 spins $s = 1$, the matrix dimensionality of $\{Ni_{12}\}$ is $D = 3^{12} = 531,441$, which is small enough to be solved by numerical diagonalization.[7] However, this was sufficiently time consuming that it would have been impractical to attempt to use this method — rather than QMC — to fit $\chi(T)$. Hence, we have first used the QMC method to fit $\chi(T)$, and then a single matrix diagonalization was carried out to produce the full set of energy eigenvalues using the exchange constants listed above. The low-energy portion of this spectrum is shown in Table 3.

3. Classical Spin Dynamics Simulations

'Everything should be made as simple as possible, but not simpler'. —This statement made by Albert Einstein about a century ago is very relevant to the

[7] We have used the MAGPACK software described in Borrás-Almenar *et al.* (1999) and Borrás-Almenar *et al.* (2001).

Table 3. Low-energy excitations for the $\{Ni_{12}\}$ model Hamiltonian.

S	Excitation energy (meV)	Multiplicity
0	0	1
1	0.0272	2
1	0.0327	1
2	0.0740	2
2	0.1030	2
2	0.1063	1
3	0.1414	1
3	0.1553	1
3	0.1697	2
4	0.2434	2
4	0.2606	1
5	0.3659	2
6	0.5162	1

balancing act that one has to face when dealing with classical spin dynamics methods within the magnetic molecule community. There is certainly no doubt about the tremendous success of the classical spin dynamics approach for the prediction of physical properties of (infinite) bulk magnetic systems over the past 30 years. However, when it comes to magnetic molecules it seems to be questionable to what extent an approach based on the 'classical version' of the Heisenberg model would be accurate enough to describe these zero-dimensional systems or whether one over-simplifies the problem by totally ignoring quantum effects. *Reality* however does not leave us a choice! Always limited by computational power, numerical simulations based on interacting classical spins appear just too tempting compared to exact quantum calculations which are still very often beyond today's most advanced computational capabilities, even for relatively small and simple systems as pointed out in the previous sections. Surprisingly, the classical Heisenberg model turns out to provide accurate quantitative results for static properties, such as magnetic susceptibility, down to thermal energies (k_BT) of the order of the exchange coupling or even below. Moreover, classical spin dynamics methods can easily handle large structures composed of thousands of spins and, most important, geometrical frustration can be handled as well.

This section gives an overview of the state-of-the-art classical spin dynamics approaches, namely classical Monte Carlo methods and heat bath coupling methods. Both approaches serve as excellent tools to supplement exact and approximate quantum methods. In fact, classical numerical simulations allow us to explore magnetic molecules very efficiently which has led to the discovery of a variety of new and surprising physical phenomena as will be described in Sec. 3.4.

3.1. *The classical Heisenberg Hamiltonian*

For the description of classical spin systems we need to find the proper classical counterparts of the quantum spin operators, the quantum Heisenberg model, and the quantum spin dynamics. Unfortunately, there is no way to "derive" the classical description from the exact quantum model. However, using proper definitions one can find a *correspondence* between the quantum and classical models. The starting point is the quantum spin operator — or short — spin $\vec{\underset{\sim}{s}}_j$ at site j which obeys the following commutation relation

$$[\underset{\sim}{s}_{j,\alpha}, \underset{\sim}{s}_{k,\beta}] = i\hbar\delta_{jk}\varepsilon_{\alpha\beta\gamma}\underset{\sim}{s}_{j,\gamma}, \quad \alpha, \beta, \gamma = x, y, z. \tag{6}$$

For an arbitrary spin quantum number s the z-component of such a spin $\underset{\sim}{s}_{j,z}$ is quantized and its $(2s+1)$ eigenvalues are discrete values in the interval

$$m_{j,z} = -s, -s+1, \ldots, s-1, s. \tag{7}$$

One can visualize this quantum property by a vector with constant length pointing in *discrete* directions as shown in the right panel of Fig. 10. For a quantum spin number of $s = 3/2$ one gets 4 different values of $m_{j,z}$ according to Eq. (7). The classical analogue of a quantum spin would be that of a simple three dimensional vector, however pointing *continuously* in any direction as shown in the left panel of Fig. 10. Using such a picture it is obvious that the correspondence between a quantum spin and a classical spin becomes closer with increasing quantum spin number s. On the one hand, this leads to a finer "stepping" of the quantum vector between $-s$ and s since the number of eigenvalues increases according to Eq. (7) as $(2s+1)$; on the other hand the maximum and minimum eigenvalue $m_{j,z}$ of $\underset{\sim}{s}_{j,z}$ becomes closer to its classical counterpart, and with $s \to \infty$ the spin pictures in the

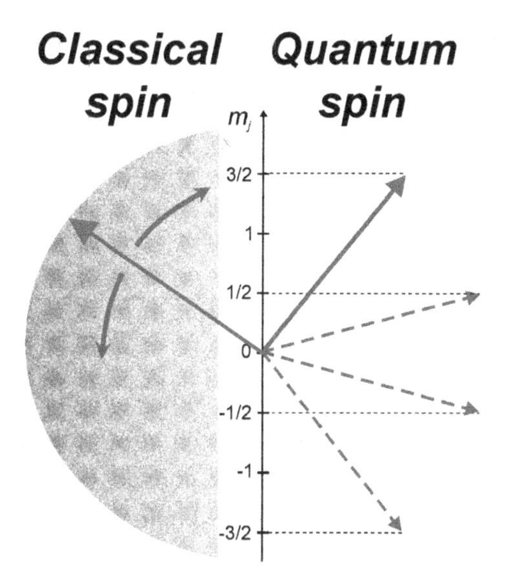

Fig. 10. Visualization of a quantum spin and its classical analogue. The z-component of a quantum spin can only have *discrete* values as shown in the right panel whereas its classical analogue can *continuously* point into any direction.

right and left panel would become identical. This correspondence between the quantum and classical spin can be expressed mathematically by introducing normalized spin operators

$$\vec{\underset{\sim}{e}}_j = \frac{\vec{\underset{\sim}{s}}_j}{\hbar\sqrt{s(s+1)}} \tag{8}$$

which depend on s (Fisher, 1964; Mentrup *et al.*, 1999). Given this, the commutation relation for these normalized spin operators becomes

$$[\underset{\sim}{e}_{j,\alpha}, \underset{\sim}{e}_{k,\beta}] = \frac{i}{\sqrt{s(s+1)}}\hbar\delta_{jk}\varepsilon_{\alpha\beta\gamma}\underset{\sim}{e}_{j,\gamma}, \quad \alpha, \beta, \gamma = x, y, z. \tag{9}$$

Using Eq. (9) the eigenvalues $m_{e_{jz}}$ of the z-component $\underset{\sim}{e}_{jz}$ become

$$m_{e_{jz}} = \frac{m_{jz}}{\sqrt{s(s+1)}} \tag{10}$$

which leads to an interval of the eigenvalues of $\underset{\sim}{e}_{jz}$ in the range

$$\left[\frac{-s}{\sqrt{s(s+1)}}, \frac{s}{\sqrt{s(s+1)}}\right]. \tag{11}$$

Finally, we replace the spin operators in the quantum Heisenberg model Hamiltonian

$$\underset{\sim}{H} = -\sum_{\langle j,k \rangle} J_{j,k} \underset{\sim}{\vec{s}}_j \cdot \underset{\sim}{\vec{s}}_k + g\mu_B \vec{B} \cdot \sum_{j=1}^{N} \underset{\sim}{\vec{s}}_j \qquad (12)$$

by the normalized spin operators Eq. (8) and obtain

$$\underset{\sim}{H} = -\sum_{\langle j,k \rangle} J_{j,k} s(s+1) \underset{\sim}{\vec{e}}_j \cdot \underset{\sim}{\vec{e}}_k + g\mu_B \sqrt{s(s+1)} \vec{B} \cdot \sum_{j=1}^{N} \underset{\sim}{\vec{e}}_j. \qquad (13)$$

Equation (11) demonstrates that for arbitrary values of s the spectrum of eigenvalues of the normalized spin operators is confined to the interval $[-1,1]$ and becomes *dense* for $s \to \infty$ and thus coincides with the continuous range of the z-component e_{jz} of a classical unit vector. Simultaneously by looking at Eq. (9) we find that for $s \to \infty$ the spin operators $\underset{\sim}{e}_{jx}$ and $\underset{\sim}{e}_{jy}$ commute corresponding to *independent* Cartesian components of a classical spin vector.

Hence it is *suggestive* (but not more!) to replace the spin operators in the quantum Heisenberg Hamiltonian Eq. (12) by classical spin vectors thus *defining* a classical Heisenberg Hamiltonian of the form

$$\mathcal{H}_C = -\sum_{\langle j,k \rangle} J^C_{j,k} \vec{e}_j \cdot \vec{e}_k + g\mu^C_B \vec{B} \cdot \sum_{j=1}^{N} \vec{e}_j, \qquad (14)$$

with *classical* exchange constants

$$J^C_{j,k} = J_{j,k} \cdot s(s+1), \qquad (15)$$

and the *classical* Bohr magneton

$$\mu^C_B = \mu_B \sqrt{s(s+1)}. \qquad (16)$$

If there is just a *single* exchange constant relevant in the system under study Eqs. (14) and (15) can be simplified to

$$\mathcal{H}_C = -J_C \sum_{\langle j,k \rangle} \vec{e}_j \cdot \vec{e}_k + g\mu^C_B \vec{B} \cdot \sum_{j=1}^{N} \vec{e}_j, \qquad (17)$$

with

$$J_C = J \cdot s(s+1). \qquad (18)$$

It is important to note that although the correspondence between the classical and quantum Heisenberg model only holds in the limit $s \to \infty$, in principle the classical Hamiltonian Eq. (14) can be applied to any system with arbitrary spin quantum number s. However, one expects significant differences between the exact quantum results and a classical calculation using Eq. (14) for small spin quantum numbers s and the question arises when it is allowed to apply classical methods. Unfortunately, there is no simple answer to that question and there exist merely some "rules of thumb" for special systems and parameter spaces (Engelhardt *et al.*, 2006; Mentrup *et al.*, 1999, 2000)

To illustrate the difference between quantum and classical results we show in Fig. 11 a comparison of a classical and quantum (using $s = 4$) calculation for the antiferromagnetically coupled spin triangle. For temperatures of $k_BT/|J| \gtrsim 0.6$ and above both calculations show almost identical results for the differential susceptibility dM/dB versus the external magnetic field B. On lowering the temperature the results differ significantly; the fact that the spins are quantized and level crossings occur lead to "wiggles" in the low temperature results for the quantum system (corresponding to a staircase like magnetization curve M vs. B).

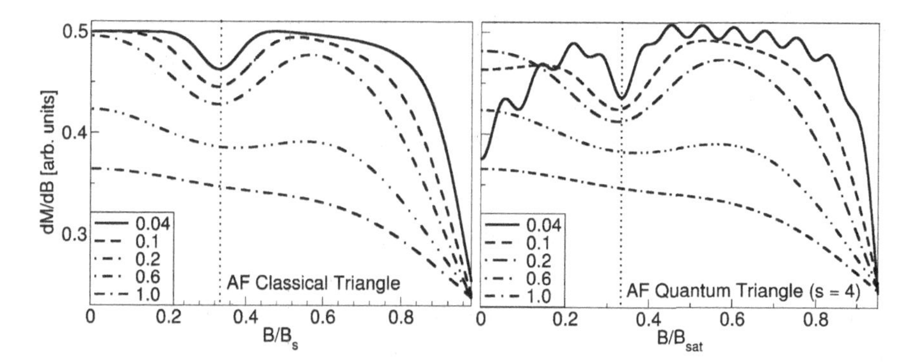

Fig. 11. Comparison of the differential susceptibility dM/dB versus the external magnetic field B of the antiferromagnetic classical and quantum triangle (using $s = 4$). The temperatures are given $k_BT/|J|$. For temperatures of $k_BT/|J| \gtrsim 0.6$ and above both calculations show almost identical results. On lowering the temperature the results differ significantly; the fact that the spins are quantized and level crossings occur lead to "wiggles" in the low temperature results for the quantum system.

3.2. *Classical Monte Carlo simulations*

As described in the previous sections one of the goals of the theoretical methods based on the Heisenberg model is to calculate experimentally accessible thermodynamical quantities. By far the most efficient way to do this on the basis of a classical Heisenberg model is to utilize the Metropolis algorithm (Metropolis *et al.*, 1953; Landau and Binder, 2009). Simply speaking, the Metropolis algorithm produces spin configurations $[S]_i$ according to a Boltzmann distribution which is exactly what we need in order to determine thermodynamical properties in the canonical ensemble, i.e. with the temperature T as the constant controlling parameter. For a system of N spins that interact via the classical Heisenberg interaction according to Eq. (14) the algorithm works as follows:

(1) Produce an initial, e.g. random, spin configuration for the N spins.
(1) Select a spin at site j (randomly or sequentially) and calculate the *local* energy E_{old} with the k interacting spins according to the Heisenberg Hamiltonian Eq. (14).
(3) Change the direction of the spin at site j.
(4) Calculate the local energy of the new configuration E_{new}.
(5) If $\Delta E = E_{new} - E_{old} < 0$ then accept the new configuration.
(6) If $\Delta E > 0$ then pick an evenly distributed random number r from the interval $[0, 1]$. If $r < \exp(-\Delta E/k_B T)$ then accept the change, otherwise reject it and keep the old configuration.
(7) Save the energy of the last configuration as E_{old}.
(8) Iterate 3 to 7.

Inspired by the use of random numbers the Metropolis algorithm belongs to the class of *Monte Carlo Methods*. A single spin flip and subsequent evaluations is commonly called a *Monte Carlo step*. A trial change of *all* N spins within the system is called a *Monte Carlo cycle*. The important question is, how can this algorithm be used to calculate thermal averages of a desired quantity, e.g. the magnetization of the system? The answer is by *sampling* data during the simulation. The thermal average of any quantity $A(T)$ is then calculated according to

$$A(T) \approx \frac{1}{M} \sum_{i=1}^{M} A([S]_i) \qquad (19)$$

where $A([S]_i)$ is the value of the desired quantity calculated by using the ith generated spin configuration of a total of M sampled spin configurations during the simulation. The "$\approx$" sign accounts for the fact that these calculations are statistical in origin, i.e. are based on ensemble averaging. For example, the calculation of the z-component of the magnetization per spin is determined by evaluating

$$M_z([S]_i) = \frac{1}{N}\sum_{j=1}^{N} e_{jz} \tag{20}$$

after each Monte Carlo cycle. The thermal average for a temperature T is then calculated by

$$M_z(T) \approx \frac{1}{M}\sum_{i=1}^{M} M_z([S]_i) \tag{21}$$

Since in the beginning of such a simulation the spin configuration does not reflect a typical state according to the Boltzmann distribution one discards the first few thousand steps. As soon as the system is "thermalized" one can start sampling data to generate thermally averaged properties. While there are statistical errors involved in this approach, the errors can be made arbitrarily small by performing many Monte Carlo cycles. The question of how many Monte Carlo cycles are necessary in order to calculate the desired physical quantity for a given temperature with a certain accuracy can not be answered in general. However, one can control the error by using standard statistical techniques and therefore check the convergence on the fly. In general, about $10^6 - 10^8$ Monte Carlo cycles generate acceptable data, but this strongly depends on the temperature, size of the molecule and the interactions. It is important to note, that there are no systematic errors introduced, and the statistical errors are accurately estimated during the course of a calculation.

This algorithm can be implemented very easily and efficiently on parallel computers, so that even for large systems ($N > 1000$) it just takes minutes to produce accurate thermal equilibrium values.

However, there is an alternative way to perform large scale Monte Carlo simulations. As described above, the computational problem of calculating thermodynamic properties using the Monte Carlo Metropolis algorithm requires the generation and evaluation of a certain number spin configurations of the *same* system. By averaging over *all* these configurations one obtains the desired thermodynamic properties. Therefore, a straightforward strategy is to use a *distributed* computer network where a server computer distributes copies of the system onto several client computers. Each client computer calculates a fraction of the total number of configurations and evaluates the associated *partial* averages of the desired thermodynamic properties. These partial averages are then collected from every client computer into a database on the server computer and the *global* average of all partial averages yields the thermodynamic properties with the same accuracy as one long Monte Carlo simulation on a single computer.[8]

Since each client performs its partial task independently of each other, that is no communication between clients is necessary, one can implement such a strategy by using the so-called "public resource" or 'volunteer' computing technology which originates from the famous SETI project (Anderson *et al.*, 2002).

This technology utilizes private computer resources provided voluntarily by people from all over the world. The only requirement is that a connection of the volunteer's computers with the Internet exists so that data can be exchanged with the project server. There are currently about 70 projects in the world that make use of this technology based on the so-called 'Berkeley Open Infrastructure for Network Computing' (BOINC).[9] Among these projects there is one called 'Spinhenge@home' which performs classical Monte Carlo simulations for magnetic molecules using the combined computational power of more than 170 000 registered computers.[10]

However, since the Metropolis algorithm works locally it has a serious drawback when applied to very large systems that undergo phase transitions. Since the information of a local change needs a certain time to propagate through the system, cooperative phenomena like phase transitions cause so-called slowing down effects, i.e. the number of samples that are needed

[8]However, a key requirement is that the random number series used for the client calculations are completely uncorrelated!

[9]See http://boinc.berkeley.edu

[10]See http://spin.fh-bielefeld.de

to produce thermally averaged properties is increasing dramatically in the vicinity of the transition temperature. For such situations modified Monte Carlo algorithms have been invented (Ferrenberg and Swendsen, 1988; Wang and Landau, 2001); however for characterizing single, uncoupled molecular systems that do not show cooperative phenomena the simple Metropolis algorithm is absolutely sufficient.

By modifying the Heisenberg Hamiltonian of Eq. (14), i.e. including anisotropies or higher order exchange terms, a wide variety of physical scenarios can be modeled. Furthermore, from Eq. (19) it is apparent that the temperature dependence for *any* physical quantity for a classical spin system can be calculated. However, in most cases we need to calculate experimentally observable properties like the magnetic susceptibility $\chi(T, B)$ or magnetization $M(T, B)$. We refer to these quantities as *static* thermal properties since there is no intrinsic time-dependence whatsoever.

On the other hand there are important experimentally observable *dynamical* quantities, e.g. the NMR relaxation time $1/T_1$ and the neutron scattering cross section σ. Both quantities require the calculation of the equilibrium time correlation function classically defined as

$$C_{ij}^{\alpha\beta}(t; T) = \langle e_{i\alpha}(0) \cdot e_{j\beta}(t) \rangle. \tag{22}$$

This quantity links the α-component of a spin at lattice site i at time 0 with the β-component of a spin at lattice site j at time t. Here $\langle \ldots \rangle$ denotes the canonical ensemble average. In order to evaluate Eq. (22) one needs to know how classical spins behave dynamically, i.e. which equations of motion govern the spin dynamics in time.

3.2.1. *The spin equations of motion*

Quantum mechanically the equations of motion for a spin operator $\underset{\sim}{\vec{s}}_j$ can be expressed within the Heisenberg picture by

$$\frac{d\underset{\sim}{\vec{s}}_j}{dt} = \frac{i}{\hbar}[\underset{\sim}{\mathcal{H}}, \underset{\sim}{\vec{s}}_j]. \tag{23}$$

Using the commutation relation Eq. (6) and the quantum Heisenberg operator Eq. (12) we can rewrite Eq. (23)

$$\frac{d\underset{\sim}{\vec{s}}_j}{dt} = \frac{\partial \underset{\sim}{\mathcal{H}}}{\partial \underset{\sim}{\vec{s}}_j} \times \underset{\sim}{\vec{s}}_j. \tag{24}$$

If we now replace the spin operators by the normalized spin operators Eq. (8) we get

$$\frac{d\vec{\underset{\sim}{e}}_j}{dt} = \frac{1}{\hbar\sqrt{s(s+1)}}\frac{\partial \underset{\sim}{\mathcal{H}}}{\partial \vec{\underset{\sim}{e}}_j} \times \vec{\underset{\sim}{e}}_j. \tag{25}$$

One can absorb the prefactor into the Heisenberg Hamiltonian and write

$$\frac{d\vec{\underset{\sim}{e}}_j}{dt} = \frac{\partial \underset{\sim}{\mathcal{H}}_s}{\partial \vec{\underset{\sim}{e}}_j} \times \vec{\underset{\sim}{e}}_j \tag{26}$$

with

$$\underset{\sim}{\mathcal{H}}_s = -\Omega_s \sum_{\langle j,k \rangle} \vec{\underset{\sim}{e}}_j \cdot \vec{\underset{\sim}{e}}_k \tag{27}$$

and

$$\Omega_s = \frac{J\sqrt{s(s+1)}}{\hbar}. \tag{28}$$

For the sake of simplicity we have assumed that there is just a single exchange constant within the system and external magnetic fields are absent. Ω_s can be expressed in units of Hertz and defines the time scale of the dynamics. Analogous to the procedure in Sec. 3.1 one can again *define* a classical counterpart to the quantum equations of motion by replacing the normalized spin operators with classical spin vectors and yields

$$\frac{d\vec{e}_j}{dt} = \frac{\partial \mathcal{H}_C}{\partial \vec{e}_j} \times \vec{e}_j \tag{29}$$

with the classical Heisenberg Hamiltonian

$$\mathcal{H}_C = -\Omega_C \sum_{\langle j,k \rangle} \vec{e}_j \cdot \vec{e}_k \tag{30}$$

and

$$\Omega_C = \frac{J_C}{\hbar\sqrt{s(s+1)}} \tag{31}$$

with J_C as defined by Eq. (18), again for the case of a single exchange constant only.

The interpretation of the classical spin equations of motion Eq. (29) is straightforward and describes a precession of the spin vector $\vec{e}_j$ about the *local* field $\partial \mathcal{H}_C / \partial \vec{e}_j$ generated by the nearest neighbors of $\vec{e}_j$ at the site j as shown in Fig. 12.

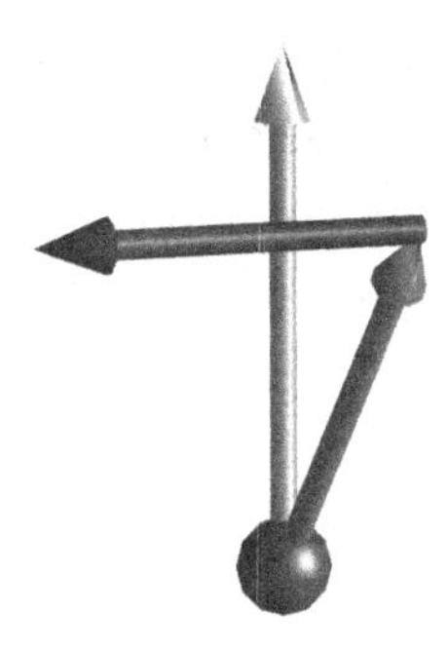

Fig. 12. Precession of a classical spin vector according to the classical spin equations of motion Eq. (29). The grey vector is the precessing spin $\vec{e}_j$ about the *local* field $\partial\mathcal{H}_C/\partial\vec{e}_j$ (shown as light grey vector) generated by the nearest neighbors of $\vec{e}_j$ at the site j. The dark grey vector displays the "velocity" $d\vec{e}_j/dt$.

Using Eqs. 29–31, one can now calculate the direction of a spin vector as a function of *time* $\vec{e}_j(t)$. However, in order to calculate the temperature and time dependent equilibrium correlation functions [Eq. (22)] we need to perform the ensemble average over a time-dependent quantity. This is done via the so-called 'Gibbs approach' (Luban and Luscombe, 1999; Schröder *et al.*, 2004), i.e. the thermal average is prescribed by the canonical ensemble over the initial configurations. In principal the algorithm works as follows:

(1) 'Thermalize' the spin system by performing a number of Monte Carlo cycles according to a temperature T.
(2) Store the current spin configuration.
(3) Integrate Eq. (29) using the current spin configuration as initial conditions over a certain time interval $[0; t]$ and store a number of N values of $\vec{e}_j(\tau)$ at discrete times $\tau \in [0; t]$.
(4) Calculate and store $e_{i\alpha}(0) \cdot e_{j\beta}(\tau)$ for all N.
(5) Perform a Monte Carlo cycle.
(6) Iterate 2 to 5 for M Monte Carlo cycles.
(7) Calculate the average over the M cycles of $e_{i\alpha}(0) \cdot e_{j\beta}(\tau)$ for all N values of $\tau \in [0; t]$.

Note, that compared to the standard Monte Carlo algorithm as described in Sec. 3.2 the algorithm here requires much more computational power! In addition to the effort of producing ensemble averages one has to do the integration of the spin equations of motion Eq. (29) over a time interval $[0; t]$. Depending on the value of t and the complexity of the Hamiltonian such integrations can become very time consuming. However, there exist a number of algorithms for an efficient time integration in the literature (Tsai *et al.*, 2005).

3.3. *Heat bath simulational methods*

As described in Sec. 3.2, thermal magnetic properties can be calculated by performing state space sampling of spin configurations produced by the Metropolis algorithm. For the determination of *static* equilibrium magnetic properties like the magnetization, susceptibility, specific heat etc. This approach works very efficiently. Furthermore, by using the 'Gibbs approach' as described in the previous section even *dynamical* equilibrium magnetic properties like the spin correlation functions [Eq. (22)] can be calculated.

However, the underlying mechanism for generating the ensemble averages depends on state space sampling of *randomly* generated spin configurations — or in other words — there is no relation between the *randomly* generated spin configurations and the spin equations of motion Eq. (29)! This limits the applicability of Monte Carlo simulations since the correct description of relaxation effects that are vitally important for the description of *non-equilibrium* properties are not accessible using such methods. Simulations of experimentally accessible properties that reveal information about the "energy-landscape" of the system's state space like the time decay of the thermoremanent dc magnetization (TRM), the field-cooled (fc), zero-field cooled (zfc) magnetization and ac susceptibility require the correct description of the underlying spin dynamics *coupled to a heat bath*. There are a number of approaches available in the literature that can be used to simulate such a heat bath as well as its coupling to the spin system.

Here we will focus on a Langevin-approach where the heat bath contact is modeled by the competition of frictional and fluctuational fields. The full theory behind this approach and the numerical approximation of the resulting equations is not trivial and beyond the scope of this book. In the following we will just sketch the main ideas and the reader is invited to study the references given below.

The starting point is the spin equation of motion Eq. (29). As shown in Sec. 3.2.1 this equation describes the precession of the spin about a local field. One can show that these dynamics not only conserves the energy but also the length of the individual spin vector, i.e. it represents a spin system that is completely isolated from its (thermal) environment. In order to build in the heat bath contact one has to modify the spin equations of motion according to

$$\frac{d\vec{e}_j}{dt} = \frac{\partial \mathcal{H}_C}{\partial \vec{e}_j} \times \vec{e}_j + \text{heat bath contact.} \tag{32}$$

We will explain how to build in the heat bath contact in two steps. First, we add a frictional or damping term and get

$$\frac{d\vec{e}_j}{dt} = \frac{\partial \mathcal{H}_C}{\partial \vec{e}_j} \times \vec{e}_j + \lambda \left(\frac{\partial \mathcal{H}_C}{\partial \vec{e}_j} \times \vec{e}_j \right) \times \vec{e}_j. \tag{33}$$

The term containing λ is called "Landau-Lifshitz damping" and provides a "force" causing the spin $\vec{e}_j$ to move towards the field direction as shown in Fig. 13. The damping factor λ is a phenomenological factor that accounts for the true physical mechanisms that are responsible for the damping.

By construction this additional damping term still conserves the spin vector length but now allows the system to loose its energy which leads to the following interesting application: It is now possible to find — numerically — the classical groundstate for a spin system as a result of the relaxation from a random initial state.[11]

The contact of a system with a heat bath is characterized by the exchange of energy with it. However, the Landau-Lifshitz damping only allows the system to loose energy. In a second step we build in an additional term which also allows the system to gain energy from the heat bath. This is done by adding fluctuating fields $\vec{f}_j$ to the equations of motion Eq. (33) yielding

$$\frac{d\vec{e}_j}{dt} = \left(\frac{\partial \mathcal{H}_C}{\partial \vec{e}_j} + \vec{f}_j \right) \times \vec{e}_j + \lambda \left(\frac{\partial \mathcal{H}_C}{\partial \vec{e}_j} \times \vec{e}_j \right) \times \vec{e}_j. \tag{34}$$

[11] By taking a closer look at Eq. (33) one realizes that for the simple Heisenberg model Eq. (17) with $B = 0$ a parallel or antiparallel initial configuration of all spins is metastable since the right hand side of Eq. (33) would become zero! Therefore it is necessary to start with for example a random initial configuration.

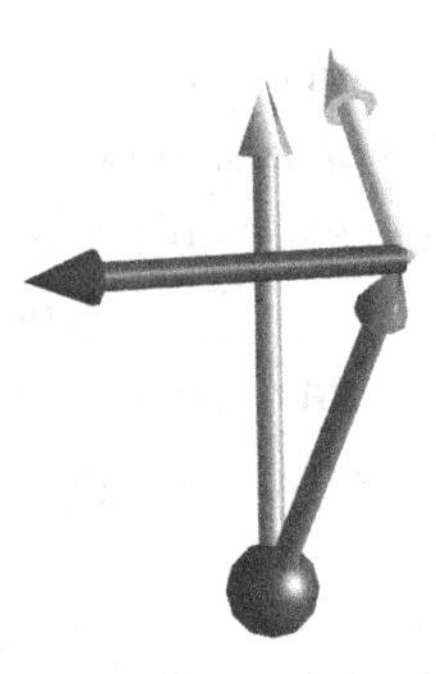

Fig. 13. Damped precession of a classical spin vector. The grey vector is the precessing spin $\vec{e}_j$ about the *local* field $\partial\mathcal{H}_C/\partial\vec{e}_j$ (shown as light grey vector) generated by the nearest neighbors of $\vec{e}_j$ at the site j. The dark grey vector displays the "velocity" $d\vec{e}_j/dt$. The white vector shows the effect of the damping term in Eq. (33), i.e. it causes the spin to move toward the local field.

This equation is called the *stochastic Landau-Lifshitz equation*. Equation (34) can be rewritten and classified as a stochastic differential equation with multiplicative white noise. The fluctuating fields $\vec{f}_j$ obey the following relation

$$\left\langle f_i^\alpha(t) f_j^\beta(t')\right\rangle = \epsilon^2 \delta_{ij}\delta_{\alpha\beta}\delta(t-t'), \tag{35}$$

where ϵ is the amplitude of the fluctuations and defined by

$$\epsilon^2 = 2\lambda k_B T. \tag{36}$$

With this last step the heat bath temperature T enters the equations of motion. Equation (35) is the mathematical description of the fact that the fluctuations occur *instantaneously*, i.e. their duration and the intervals between fluctuations are much smaller than the time scales of the precessional spin dynamics. By construction Eqs. (34)–(36) ensure the evolution to the stationary Gibbs distribution, i.e. by sampling the *time dependent trajectories* of the spin system as calculated by using Eq. (34) one can determine thermal averages in the very same way as is done using the Monte Carlo approach.

However, in contrast to the Monte Carlo algorithm which just produces spin configurations randomly, the stochastic Landau-Lifshitz equation generates a "chronologically ordered" sequence of spin configurations based on the underlying equations of motion.

Fig. 14. Solving the stochastic Landau-Lifshitz equation for a classical spin vector. The grey vector is the precessing spin $\vec{e}_j$ about the *local* field $\partial\mathcal{H}_C/\partial\vec{e}_j$ (shown as light grey vector) generated by the nearest neighbors of $\vec{e}_j$ at the site j. The blue vector displays the "velocity" $d\vec{e}_j/dt$. The white vector shows the effect of the damping term in Eq. (33), i.e. it causes the spin to move toward the local field. Fluctuating fields (lightest green vector) control the energy exchange with the heat bath.

As noted in the beginning of this section the theory of stochastic differential equations as well as the implementation of numerical schemes for solving them is beyond the scope of this book.[12] For additional background see references (Schröder, 1999; Antropov *et al.*, 1997; Milstein and Tretyakov, 2004; Kloeden and Platen, 1992; García-Palacios and Lázaro, 1998).

3.4. *Revealing novel physics in magnetic molecules with classical methods*

As stated in the beginning of this section the application of classical spin dynamics methods, as described in the previous sections, to systems which are composed of just a few interacting magnetic ions appears questionable. However, as pointed out in Sec. 1.3 the use of exact diagonalization methods is restricted to small systems described by rather simple interaction scenarios. Quantum Monte Carlo methods are very powerful to handle large systems, but 'suffer' inherently from the negative sign problem (see Sec. 2.4) when dealing with strongly frustrated systems.

[12]A tutorial with animations and examples is provided at http://ti.fh-bielefeld.de/~cschroed/spindyntutor.htm

Consequently, classical methods fit very well into the niche of simulating large and strongly frustrated magnetic systems where the application of quantum methods is restricted or even impossible.

In the following we will show that numerical simulations based on classical spin dynamics serve as an excellent tool to supplement exact and approximate quantum methods. Moreover, classical numerical simulations allow us to explore magnetic molecules very efficiently which has lead to the discovery of a variety of new and surprising physical phenomena.

3.4.1. *Competing spin phases and exchange disorder in the Keplerate type molecules $\{Mo_{72}Fe_{30}\}$ and $\{Mo_{72}Cr_{30}\}$*

The pair of Keplerate structural type magnetic molecules abbreviated as $\{Mo_{72}Fe_{30}\}$ and $\{Mo_{72}Cr_{30}\}$, each hosting a highly symmetric array of 30 exchange-coupled magnetic ions ("spin centers"), serve as highly attractive targets for the investigation of frustrated magnetic systems. In these molecules (Müller *et al.*, 1999; Todea *et al.*, 2007) the magnetic ions Fe^{III} (spin $s = 5/2$) and Cr^{III} (spin $s = 3/2$) occupy the 30 symmetric sites of an icosidodecahedron, a closed spherical structure consisting of 20 corner-sharing triangles arranged around 12 pentagons (diamagnetic polyoxomolybdate fragments) (see Fig. 15).

In fact these molecules represent a *zero-dimensional* analogue of the planar Kagomé lattice that is composed of corner-sharing triangles arranged around hexagons. A useful theoretical framework that has been employed (Axenovich and Luban, 2001; Müller *et al.*, 2001; Schnack *et al.*, 2001) in studying these magnetic molecules is based on an isotropic Heisenberg model, where each magnetic ion is coupled via intra-molecular isotropic antiferromagnetic exchange to its four nearest neighbor magnetic ions, and all of the 60 intra-molecular exchange interactions are of equal strength (henceforth, 'single-J model').

The numerical values of the nearest-neighbor exchange constant J (in units of Boltzmann's constant k_B), the spectroscopic splitting factor g, and the saturation field B_{sat} are (1.57 K, 1.974, 17.7 T) (Müller *et al.*, 2001) and (8.7 K 1.96, 60.0 T) (Todea *et al.*, 2007) for $\{Mo_{72}Fe_{30}\}$ and $\{Mo_{72}Cr_{30}\}$, respectively. The corresponding values of the exchange

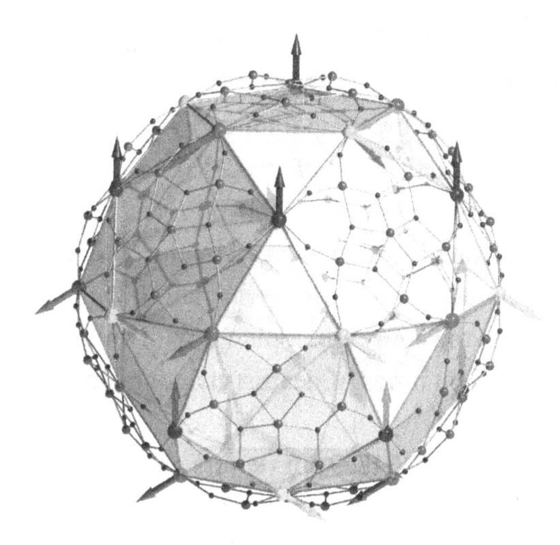

Fig. 15. Ball and stick representation of $\{Mo_{72}Fe_{30}\}$. For clarity, acetate and $\{Mo_2\}$-type ligands within the molecule are not shown. The Fe^{III} ions of the cluster span an icosidodecahedron, i.e. a spherical structure composed of corner-sharing triangles. The classical groundstate is characterized by three sets of Fe^{III} centers (shown in red, blue, green), each having parallel spin vectors. All 30 spin vectors are coplanar and nearestneighbors differ in angular orientation by 120°.

constants for the classical Heisenberg model are given by $J_C = Js(s + 1)$, namely 13.74 K and 32.63 K for the two molecules.

The quantum Heisenberg model of the two magnetic molecules is intractable using either analytical or matrix diagonalization methods. Nevertheless, the nearest-neighbor exchange constant for each molecule has been established by comparing experimental data above 30 K for the temperature-dependent zero-field susceptibility with data obtained by simulational methods using the quantum (Todea *et al.*, 2007) and classical (Müller *et al.*, 2001) Monte Carlo methods.

For temperatures below about 30 K, where the quantum Monte Carlo method proves to be ineffective for the two magnetic molecules due to frustration effects, the classical Heisenberg model is at present the only practical platform for establishing the dependence of the magnetization $M(B, T)$ on external magnetic field B and temperature T.

A rigorous analytical result (Axenovich and Luban, 2001) for the classical, nearest-neighbor, single-J Heisenberg model states that, in the zero temperature limit, M is linear in B until M saturates (saturation fields

$B_{sat} = 17.7\,\mathrm{T}$ and $60.0\,\mathrm{T}$ for $\{Mo_{72}Fe_{30}\}$ and $\{Mo_{72}Cr_{30}\}$, respectively). The practical relevance of the classical Heisenberg model in describing these magnetic molecules even at low temperatures was strikingly demonstrated in an earlier experiment (Müller *et al.*, 2001) on $\{Mo_{72}Fe_{30}\}$ at 0.4 K, showing an overall linear dependence of M on B and its saturation at approximately 17.7 T. The ground state envisaged by the classical single-J model is characterized by high-symmetry spin frustration. In particular, for $B = 0\,\mathrm{T}$, in the ground state the spins are coplanar with an angular separation of 120° between the orientations of nearest-neighbor spins as shown in the right panel of Fig. 15.

On increasing the external field B the spin vectors gradually tilt towards the field vector like a folding umbrella, until full alignment is achieved when $B = B_{sat}$, while their projections in the plane perpendicular to the field vector retain the 120° pattern for nearest-neighbor spins (Axenovich and Luban, 2001).

We have performed classical Monte Carlo simulations for $T > 0$ in order to study the finite temperature behavior of $\{Mo_{72}Fe_{30}\}$ and $\{Mo_{72}Cr_{30}\}$. In Fig. 16 we show finite temperature simulations for the magnetization M (inset) and differential susceptibility dM/dB versus B for the classical Heisenberg model of $\{Mo_{72}Fe_{30}\}$. As the temperature T is raised from

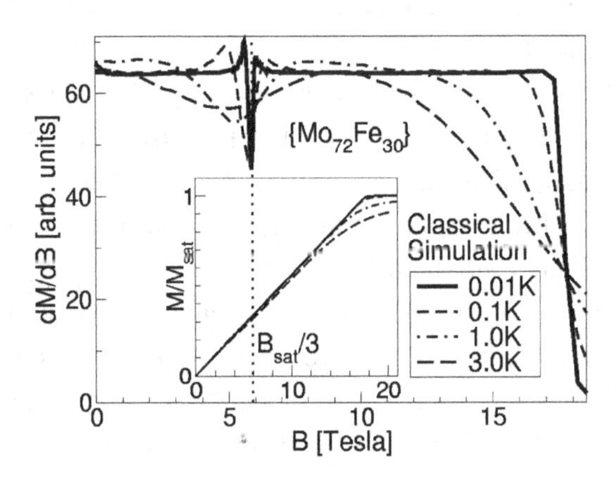

Fig. 16. Differential susceptibility dM/dB versus B for the classical Heisenberg model of $\{Mo_{72}Fe_{30}\}$ obtained by Monte Carlo simulations for temperatures given in the legend. The inset shows the corresponding curves for the magnetization M/M_{sat} with M_{sat} the saturation magnetization (Schröder *et al.*, 2005b).

0 K we find a striking anomaly in the differential susceptibility dM/dB versus magnetic field B. A deep narrow minimum in dM/dB emerges in the vicinity of one-third the saturation field B_{sat}, which upon increasing T extends over a larger field interval and its sharp features progressively deteriorate. In the classical case the drop in dM/dB can be understood as a result of the interplay of two effects: In the immediate vicinity of $B_{sat}/3$ a family of 'up-up-down' (*uud*) spin configurations are energetically competitive with the continuous family of spin configurations of lowest energy (Kawamura and Miyashita, 1985). However, the *uud* spin configurations are magnetically "stiff", i.e. $dM/dB \approx 0$ for low temperatures, and thus reduce the susceptibility of the system.

Moreover, we have found that this phenomenon is a common topological property of *all* structures that are assembled from corner-sharing triangles, i.e. the octahedron, dodecahedron, icosidodecahedron and even the two-dimensional Kagomé lattice (Schröder *et al.*, 2005a). Furthermore, one can show that this phenomenon is not at all an artifact of the classical treatment. It reflects a general intrinsic property of the very building block of these specific polytopes, namely, the simple AF equilateral Heisenberg spin triangle, and emerges for both classical and quantum spins as shown in Fig. 11.

The relevance of these theoretical results to real magnetic materials is demonstrated by measurements of the differential susceptibility for both molecules $\{Mo_{72}Fe_{30}\}$ and $\{Mo_{72}Cr_{30}\}$ (Schröder *et al.*, 2005a, 2008). In Fig. 17 experimental results of $\partial M/\partial B$ versus B for $\{Mo_{72}Fe_{30}\}$ for $T = 0.42$ K and 60 mK (upper panel, inset) and for $\{Mo_{72}Cr_{30}\}$ for $T = 1.3$ K and 0.5 K (lower panel, inset) are shown as grey squares. The dashed curves are the results of the single-J model for these temperatures using the numerical values given above for nearest-neighbor exchange constant J_C and the spectroscopic splitting factor g. One can clearly see that the experimental data show a minimum of the differential susceptibility at about one third of the saturation field as proposed by the classical simulations.

However, the inadequacy of the single-J model in this low temperature regime is striking in that the simulational data differ from the experimental data in four important ways. *First*, the experimental data exhibit a steep

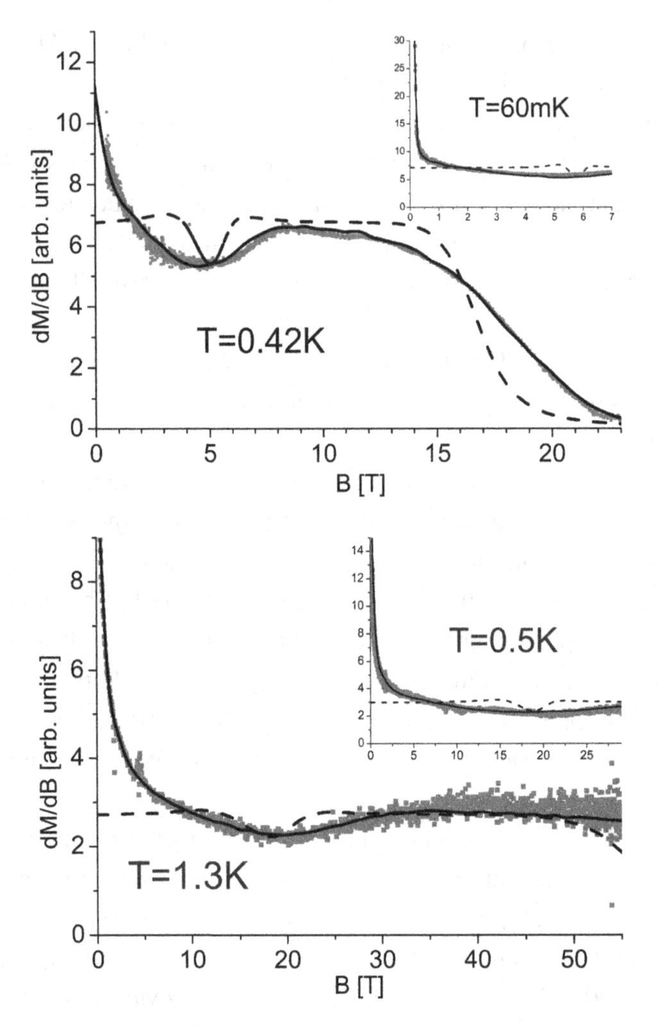

Fig. 17. Magnetic field dependence of the measured differential susceptibility $\partial M/\partial B$, shown as grey squares, for $\{Mo_{72}Fe_{30}\}$ for $T = 0.42$ K and 60 mK (upper panel, inset) and for $\{Mo_{72}Cr_{30}\}$ for $T = 1.3$ K and 0.5 K (lower panel, inset). The dashed curves are the results of the single-J model for these temperatures. Simulation results based on exchange disorder are show as solid curves (Schröder *et al.*, 2008).

rise for decreasing low fields, and the lower the temperature the steeper the rise. These features are entirely absent for the single-J model; in particular, $\partial M/\partial B$ is essentially independent of field in the low-field regime. *Second*, the local minimum in the experimental data is significantly broader than that predicted by the single-J model. *Third*, according to the single-J model a local minimum in $\partial M/\partial B$ versus B emerges at $T = 0$ K at a field $B = B_{sat}/3$ (approx. 6 T and 20 T for $\{Mo_{72}Fe_{30}\}$ and $\{Cr_{72}Fe_{30}\}$, respectively) and it is enhanced with increasing T as shown in Fig. 16. By contrast, the experimental data for fields in the vicinity of $B_{sat}/3$ differ insignificantly with temperature. *Fourth*, for $\{Mo_{72}Fe_{30}\}$ the experimental data for 0.42 K show a decrease with increasing field above 10 T, quite distinct from the pattern of the single-J model.

These discrepancies were known for a number of years and there were several attempts to explain them, specifically the broadening of the minimum versus B for $\{Mo_{72}Fe_{30}\}$. One attempt assumed an elevated spin temperature during the pulsed field measurements, however this could be ruled out since a subsequent steady-field measurement reproduced the results obtained by the pulsed-field technique (Schröder *et al.*, 2005a). Second, in a simulational study based on classical Monte Carlo calculations, effects of magnetic anisotropies, Dzyaloshinsky-Moriya, and dipole-dipole interactions have been considered (Hasegawa and Shiba, 2004). However, our own comprehensive simulational studies of these same mechanisms have shown that they give rise to only very minor corrections on the width of the minimum in $\partial M/\partial B$ versus B for any reasonable choices of model parameters.

Recently, it has been shown that the discrepancies described above at low temperatures ($T < 5$ K) can be resolved *all at once* and for *both* molecules in the *same* way by adopting a classical Heisenberg model where the 60 nearest-neighbor interactions are not identical; instead, the values of the exchange constants are described by a two-parameter probability distribution with a mean value as determined from experimental $\chi(T)$ data above 30 K using the single-J model (Schröder *et al.*, 2008).

In Fig. 17 we present results for $\partial M/\partial B$ versus B for $\{Mo_{72}Fe_{30}\}$ and $\{Mo_{72}Cr_{30}\}$ based on such an exchange disorder model (solid black curves). Apparently, there is an excellent agreement between the experimental data and the simulational results. Using the same set of parameters one can also calculate $\partial M/\partial B$ versus T for different magnetic field values B. Here, one finds again a very good agreement with low temperature experimental data (Schröder *et al.*, 2008).

The effect of exchange disorder as described above is strongly temperature dependent and becomes negligible for temperatures above ≈ 5 K, so that for higher temperatures the single-J model provides a satisfactory description of each molecule (Schröder *et al.*, 2008).

One can attribute the failure of the single-J model for low temperatures to the combined effect of a large number of diverse perturbing mechanisms. The effects of impurities, variations in the exchange-coupling geometry, weak magnetic exchange interactions of more-distant neighbors, Dzyaloshinsky-Moriya and dipole-dipole interactions in these magnetic molecules are some of the many effects that are excluded when one uses an idealized single-J model. On the other hand, it is at this stage an extremely difficult, essentially impossible task to realistically quantify the effects of the diverse mechanisms. A theoretical description based on a Heisenberg model where the nearest-neighbor exchange constant is chosen using a probability distribution provides a relatively simple, phenomenological platform for compromising between the need for microscopic realism versus practical limitations. Ultimately it is significant that a two-parameter probability description can actually provide the level of agreement that has been found.

The existence of a distribution of nearest-neighbor exchange constants can be expected to be responsible for a significant lifting of degeneracies of magnetic energy levels. To be specific, the quantum rotational band model (Schnack *et al.*, 2001), which is a solvable alternative to the nearest-neighbor single-J quantum Heisenberg model, predicts a discrete spectrum of energy levels, many of which have a very high degeneracy due to large multiplicity factors. Perturbing this model Hamiltonian by using a distribution of J-values would remove a major fraction of these degeneracies. The lifting of level degeneracies could provide a reasonable explanation for the classical characteristics of these molecules down to very low temperatures and the surprising fact that our simulational results based on the *classical* Heisenberg Hamiltonian are so successful in describing $\{Mo_{72}Cr_{30}\}$, despite the fact that the Cr^{III} ions have a small spin ($s = 3/2$). Stated differently, with the lifting of degeneracies and the fanning out of energy levels the effective temperature for the crossover from classical to quantum behavior can be anticipated to be considerably lower than that expected *a priori* for the single-J model.

Very recently a new spherical Keplerate and antiferromagnet abbreviated as $\{W_{72}Fe_{30}\}$ has been synthesized. First experimental and theoretical investigations once again show classical characteristics down to very low temperatures (Todea *et al.*, 2010).

3.4.2. *Metamagnetic phase transitions in magnetic polytopes*

As shown in the previous section frustration effects in the class of magnetic polytopes built from *corner*-sharing triangles lead to a variety of fascinating phenomena. Assuming a single antiferromagnetic nearest-neighbor exchange interaction all these polytopes exhibit a frustrated but *coplanar* classical groundstate. At $T = 0$ with increasing external magnetic field B the spins start tilting continuously until full saturation is reached. Consequently the magnetization $M(T = 0, B)$ is a linear function until the saturation field B_{sat} is reached.

If we direct our attention to magnetic polytopes built from *edge*-sharing polygons, we would refer for instance to the dodecahedron or the icosahedron. The dodecahedron is a polytope with 20 vertices built from edge-sharing pentagons, the icosahedron has 12 vertices and is built from edge-sharing equilateral triangles. In the case of a single antiferromagnetic exchange interaction between nearest neighbor spins these polytopes exhibit a frustrated complex *three dimensional* classical groundstate, however characterized by just a *single* value for all angles between adjacent spins in each molecule. It is interesting to note that the value of these angles has been determined first by numerical simulation using the stochastic Landau-Lifshitz equation Eq. (34). For such a simulation one applies a *time-dependent* temperature $T(t)$ which is slowly reduced over time so that $\lim_{t\to\infty} T(t) = 0$. This procedure allows the spin system to relax towards its groundstate. The application of a decreasing finite temperature creates sufficient thermal energy so that the system could not get stuck in local minima or metastable states. In order to check whether the final state at the end of such a simulation really represents the groundstate, one can repeat the simulations using different, e.g. random initial states and different functions for the temperature decay $T(t)$.

In Fig. 18 we show the result of such a simulation for the icosahedron. The simulation starts from random initial conditions and we have plotted

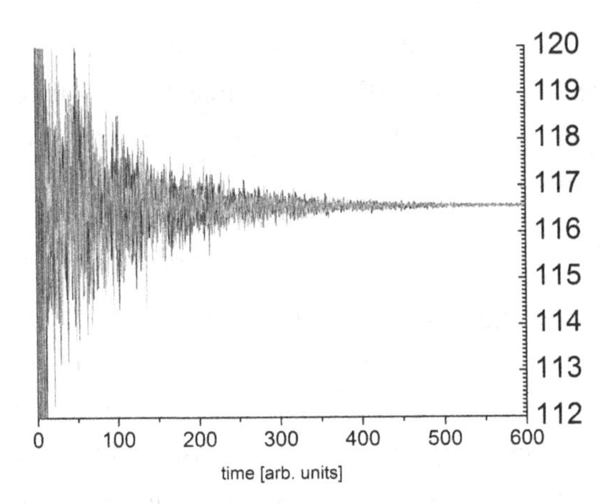

Fig. 18. Relaxation towards the groundstate of the antiferromagnetic icosahedron. Shown are all angles between adjacent spins as they evolve in time for an exponentially decaying temperature.

all angles between adjacent spins as they evolve in time for an exponentially decaying temperature. The convergence to one unique angle is clearly visible and we find $\alpha_I \approx 116.56°$ for the icosahedron and $\alpha_D \approx 138.19°$ for the dodecahedron, respectively. Later it has been shown that the classical groundstates of both polytopes can be completely determined by (semi-)analytical methods and the exact values for the angles are $\alpha_I = \arccos(-\sqrt{5}/5)$ and $\alpha_D = \arccos(-\sqrt{5}/3)$, respectively (Schmidt and Luban, 2003).

By theoretical investigations of the *field-dependent* behavior for both structures it was found that the icosahedron as well as the dodecahedron show another fascinating frustration effect (Schröder *et al.*, 2005b). In Fig. 19, we present results for the field-dependent classical spin configurations at $T = 0$ of the icosahedron as obtained by solving the stochastic Landau-Lifshitz equation starting from the groundstate at $B = 0$ described above. The external field B is slowly increased linearly until a certain value is reached and then reduced until it reaches again $B = 0$. Once again we represent the spin configurations by the angles between adjacent spins. On the up-cycle (following the black arrows and the black dashed lines) the angles start fanning out but smoothly group into only seven distinct angles. At $B \approx 3.3$ these seven angles suddenly collapse into just three distinct

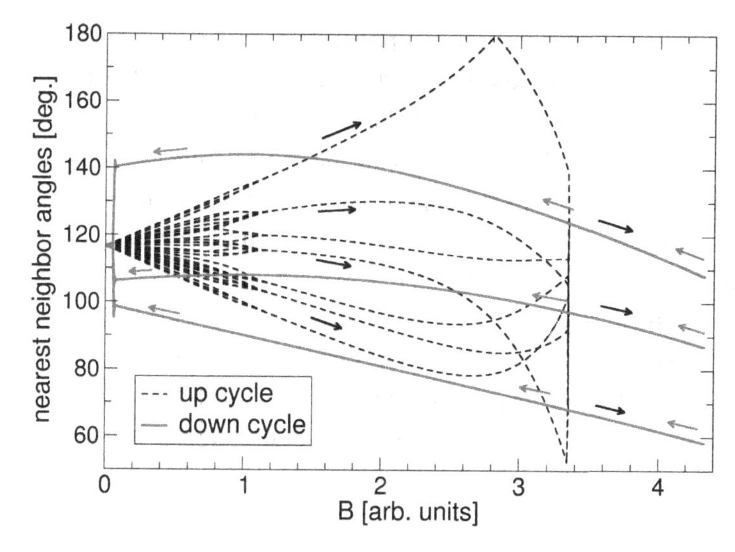

Fig. 19. Field-dependent classical spin configurations at $T = 0$ of the icosahedron. The external field B is slowly increased linearly until a certain value is reached and then reduced until it reaches again $B = 0$. On the up-cycle (following the black arrows and the black dashed lines) the angles start fanning out but smoothly group into only seven distinct angles. At $B \approx 3.3$ these seven angles suddenly collapse into just three distinct angles. On the down cycle (following the grey arrows and solid line curves) when reaching again the value of $B \approx 3.3$ but from above the system does not reverse to the seven-angle configuration but remains in the three-angle configuration until it collapses close to $B = 0$ and finally returns back to the groundstate.

angles. On the down cycle (following the grey arrows and grey curves) when reaching again the value of $B \approx 3.3$ but from above, the system does not revert to the seven-angle configuration but remains in the three-angle configuration until it collapses close to $B = 0$ and finally returns back to the groundstate. What we observe here is a very surprising phenomenon, namely the result of a first-order phase transition accompanied by metastability and hysteresis effects! Why is this surprising?

First, the system under study is a zero-dimensional polytope hosting just 12 spins. In contrast, it seems to be generally accepted that phase transitions only occur in the thermodynamic limit where the particle number N is going to infinity. The standard argument is that, under fairly general conditions, the (quantum or classical) partition function $\mathcal{Z}_N(\beta)$ will be an analytical function of the inverse temperature $\beta = \frac{1}{k_B T}$, and hence no thermodynamical function could show any discontinuity for finite N. However, this argument does not exclude phase transitions at zero temperature ($T = 0$, $\beta = \infty$). Indeed, there exist many thermodynamical functions of finite quantum systems which show jumps at $T = 0$. For example, the magnetization M at $T = 0$ of a finite spin system will always be a step function of the magnetic field B due to the discrete structure of the system's eigenstates. It may be that just because of the abundance of such examples of discontinuous thermodynamical functions one is hesitating to speak of genuine phase transitions for finite quantum systems at $T = 0$, since this term should be reserved to denote something extra-ordinary and not the standard case.

For classical systems the situation is different because the 'intrinsic phase transitions' due to the quantization of ground states are absent by definition. In order to avoid conflicts and confusions we refer to ($T = 0$)-phase transitions as so-called 'metamagnetic phase transition' (Coffey and Trugman, 1992; Schröder *et al.*, 2005b; Konstantinidis, 2007).

In Fig. 20 we plot the magnetization $M(B)$ of the icosahedron as a function of the external field B for $T = 0$ (left panel) and $T > 0$ (right panel). At the transition field $B \approx 3.3$ one finds a "magnetization jump"

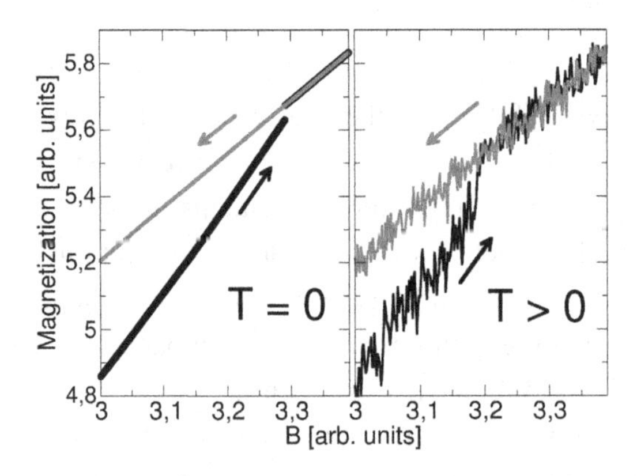

Fig. 20. Magnetization $M(B)$ of the icosahedron as a function of the external field B for $T = 0$ (left panel) and $T > 0$ (right panel). At the transition field $B \approx 3.3$ one finds a "magnetization jump" on the up cycle (black curves) whereas on the down-cycle (grey curves) the system remains in this metastable configuration.

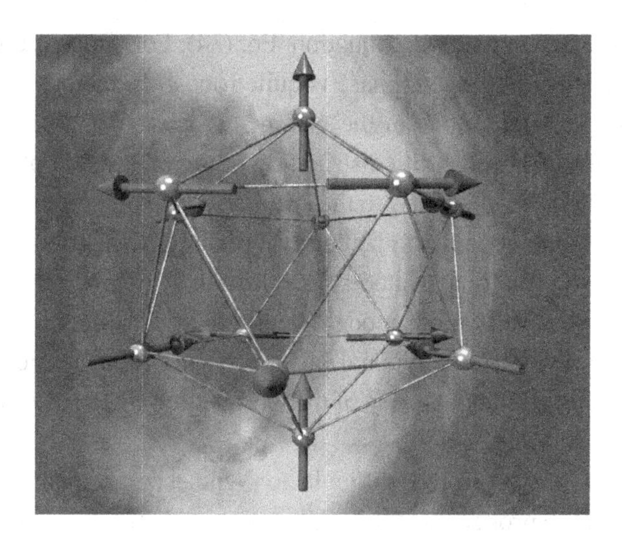

Fig. 21. Visualization of the metastable configuration described in the text.

on the up cycle (black curves) whereas on the down-cycle (grey curves) the system remains in this metastable configuration. The visualization of this metastable configuration right after the magnetization jump for $T = 0$ is shown in Fig. 21. This configuration can be characterized by two spin vectors which are "locked" in the direction of the external field and 10 spin vectors with a common polar angle θ and uniformly spaced relative azimuthal angles. The end points of the latter 10 spin vectors form a regular decagon.

It is actually quite interesting and rather educating to visualize a field sweep as shown in Fig. 19 or Fig. 20 by a movie and *watch* the metamagnetic phase transition over time.[13] Furthermore, as a response to an oscillating external field the two spins that are locked in the field direction simultaneously flip between up and down reminiscent of a nanomagnetic switch.

Using classical methods we can even study the stability of the metastable phase by performing simulations at finite temperatures. In the right panel of Fig. 20 we show a magnetic field sweep with the system coupled to a heat

[13]We invite the reader to examine movies of our simulations at http://spin.fh-bielefeld.de

bath according to the Landau-Lifshitz equation Eq. (34). For temperatures that are not too high, the metastable phase is quite robust. In fact, by performing a sequence of simulations in the parameter space of (T, B) one can determine the survival probability distribution of the metastable phase (Schröder *et al.*, 2005b). This leads us to the second surprising property of this system. Although one obtains similar survival probability distributions for systems showing thermally activated magnetization switching (Brown *et al.*, 2000), we emphasize that our model Hamiltonian does not contain any additional energy term (like an uniaxial anisotropy) providing an energy barrier. In fact, it is the special geometry of the icosahedron that causes the system to show metastability.

Unfortunately, it seems to be difficult to synthesize a magnetic molecule that reflects the properties of the antiferromagnetic icosahedron in supramolecular chemistry. As soon as the magnetic structure differs from that of a perfect icosahedron the magnetic properties are rather unspectacular (Tolis *et al.*, 2006).

3.4.3. *Critical slowing-down in Heisenberg magnetic molecules*

As pointed out in Sec. 3.2.1 the equilibrium spin correlation functions $C_{ij}^{\alpha\beta}(t; T)$ of Eq. (22) are of fundamental importance for studying the spin dynamics of magnetic molecules since these quantities enter the equations for the calculation of inelastic neutron scattering spectra and magnetic resonance experiments. In this context it is interesting to note that for *finite* systems of classical and quantum spins $(s > 1)$ that are interacting via antiferromagnetic Heisenberg exchange the equilibrium spin correlation function itself exhibits a common generic characteristic, i.e. it shows critical slowing-down for $T \to 0$ K. The corresponding Fourier time transform of the spin equilibrium correlation function exhibits a growing peak that shifts towards zero frequency on cooling (Schröder *et al.*, 2004).

By explicit calculations we have found that slowing-down effects occur for a variety of different systems including rings of equally-spaced spins and for a three-dimensional array of spins located at the sites of diverse polyhedra, including the tetrahedron, octahedron, dodecahedron, icosahedron, and the icosidodecahedron. The latter case is of particular interest since it serves as a model of the Keplerate type molecues $\{Mo_{72}Fe_{30}\}$ and $\{Mo_{72}Cr_{30}\}$ as described in a previous section.

Whereas for smaller systems the equilibrium spin correlation function can be calculated by quantum mechanical methods larger systems like the icosidodecahedron still can only be treated by classical spin dynamics methods. Calculations of $C_{ij}^{\alpha\beta}(t; T)$ for the classical Heisenberg model are conveniently performed using a numerical simulational method based on classical Monte Carlo methods as described in Sec. 3.2.1.

As one example we show in Fig. 22 data for the autocorrelation function $C_{ii}(t; T)$ as a function of time for a simple ring system $\{Fe_6\}$ (Waldmann, 2001) as well as for the icosidodecahedron as a model system for $\{Mo_{72}Fe_{30}\}$ using a single exchange constant. Although these two systems are geometrically totally different, their autocorrelation function $C_{ii}(t; T)$ shows similar behaviour, i.e. the decay time of the autocorrelation function becomes larger on cooling. By taking the Fourier transform $\Gamma_{ii}(\Omega, T)$ of the autocorrelation function one finds that on cooling the system's dynamic is dominated by low frequency modes, i.e. a peak in $\Gamma_{ii}(\Omega, T)$ arises and shifts to lower

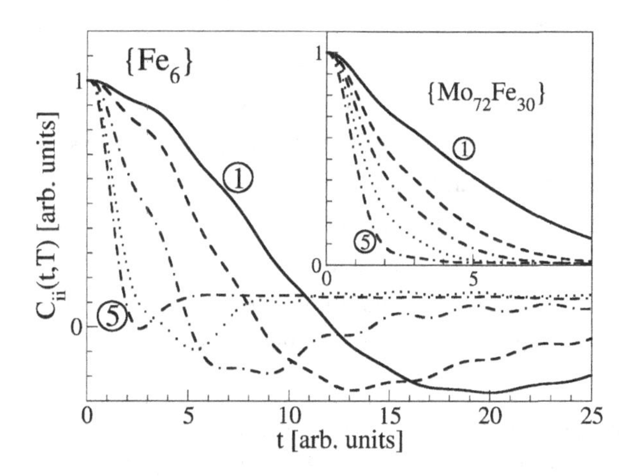

Fig. 22. Simulated autocorrelation functions $C_{ii}(t; T)$) for finite arrays of 6 and 30 classical spins interacting via antiferromagnetic Heisenberg exchange. $C_{ii}(t; T)$ exhibits effects of critical slowing-down as the temperature T is decreased towards 0 K (labeling: ⑤ high temperature, ① low temperature behavior) (Schröder *et al.*, 2004).

frequencies Ω while the magnitude of the peak increases. It is these dual properties that we identify as critical slowing-down.

For the quantum Heisenberg systems the function $\Gamma_{ii}(\Omega, T)$ consists of a finite 'comb' of Dirac delta functions which are singular for values of Ω equal to the excitation frequencies for pairs of energy eigenstates linked via angular momentum selection rules and whose temperature-dependent coefficients are extracted from the classical function by sampling at the excitation frequencies. As a result, slowing-down characteristics at low temperatures are therefore manifested in both quantum and classical Heisenberg systems (Schröder *et al.*, 2004).

4. Summary

To summarize, we have shown that there is a whole "arsenal" of powerful quantum and classical methods available to simulate and characterize magnetic molecules based on the rather simple Heisenberg model. Matrix-based calculations — while important — are limited to relatively small magnetic molecule systems. For more complex systems, the quantum Monte Carlo (QMC) method can be used to quickly obtain numerically exact results for the full (unapproximated) quantum Hamiltonian. As we have demonstrated, this method can be used for arbitrarily large systems; however, for strongly "frustrated" magnetic molecules the QMC method does not work at low temperatures. These systems can instead be studied using classical spin dynamics methods, which are neither limited by the size of the system nor by the complexity of interactions. Classical models also have the appeal that they allow us to "visualize" the behavior of the microscopic magnetic moments, though the results are approximations that lack the features of quantum mechanics.

Each method has its strengths and weaknesses, but in combination they give us the unique opportunity to reveal the underlying physics of these fascinating nanosystems. Moreover, as these methods continue to be improved and enhanced, we are optimistic that they will provide a seamless, multiscale approach to address future challenges such as the development and understanding of *functional* nanosystems based on magnetic molecules.

References

Anderson, D.P., Cobb, J., Korpela, E., Lebofsky, M. and Werthimer, D. (2002). SETI@home: An experiment in public-resource computing, *Commun. ACM* **45**(11), 56–61.

Antropov, V.P., Tretyakov, S.V. and Harmon, B.N. (1997). Spin dynamics in magnets: Quantum effects and numerical simulations (invited), *J. Appl. Phys.* **81**(8), 3961–3965.

Axenovich, M. and Luban, M. (2001). Exact ground state properties of the classical Heisenberg model for giant magnetic molecules, *Phys. Rev. B* **63**(10), 100407.

Borrás-Almenar, J.J., Clemente-Juan, J.M., Coronado, E. and Tsukerblat, B.S. (1999). High-nuclearity magnetic clusters: Generalized spin hamiltonian and its use for the calculation of the energy levels, bulk magnetic properties, and inelastic neutron scattering spectra, *Inorg. Chem.* **38**, 6081–6088.

Borrás-Almenar, J.J., Clemente-Juan, J.M., Coronado, E. and Tsukerblat, B.S. (2001). MAGPACK A package to calculate the energy levels, bulk magnetic properties, and inelastic neutron scattering spectra of high nuclearity spin clusters, *J. Comp. Chem.* **22**, 985–991.

Brown, G., Novotny, M.A. and Rikvold, P.A. (2000). Micromagnetic simulations of thermally activated magnetization reversal of nanoscale magnets, *J. Appl. Phys.* **87**(9), 4792–4794.

Caneschi, A., Cornia, A., Fabretti, A. and Gatteschi, D. (1999). Structure and magnetic properties of a dodecanuclear twisted-ring iron(III) cluster, *Angew. Chem. Int. Ed.* **38**, 1295–1297.

Coffey, D. and Trugman, S.A. (1992). Magnetic properties of undoped C_{60}, *Phys. Rev. Lett.* **69**(1), 176–179.

Cooper, G.J.T., Newton, G.N., Kögerler, P., Long, D.-L., Engelhardt, L., Luban, M. and Cronin, L. (2007). Structural and compositional control in $\{M_{12}\}$ cobalt and nickel coordination clusters detected magnetochemically and with cryospray mass spectrometry, *Angew. Chem.* **119**, 1362–1366.

de Graaf, C. and Sousa, C. (2006). Assessing the zero-field splitting in magnetic molecules by wave function-based methods, *Int. J. Quantum Chem.* **106**, 2470–2478.

Engelhardt, L. and Luban, M. (2006). Low-temperature magnetization and the excitation spectrum of antiferromagnetic Heisenberg spin rings, *Phys. Rev. B* **73**, 054430.

Engelhardt, L., Luban, M. and Schröder, C. (2006). Finite quantum Heisenberg spin models and their approach to the classical limit, *Phys. Rev. B* **74**(5), 054413.

Engelhardt, L., Martin, C., Prozorov, R., Luban, M., Timco, G.A. and Winpenny, R.E.P. (2009). High-field magnetic properties of the magnetic molecule $\{Cr_{10}Cu_2\}$, *Phys. Rev. B* **79**, 014404.

Engelhardt, L.P. (2006). *Quantum Monte Carlo Calculations Applied to Magnetic Molecules*, Ph.D. thesis, Iowa State University, Ames, IA, USA.

Engelhardt, L.P., Muryn, C.A., Pritchard, R.G., Timco, G.A., Tuna, F. and Winpenny, R.E.P. (2008). Octa-, deca-, trideca- and tetradeca-nuclear heterometallic cyclic chromium-copper cages, *Angew. Chem. Int. Ed.* **47**, 294–297.

Ferrenberg, A.M. and Swendsen, R.H. (1988). New monte carlo technique for studying phase transitions, *Phys. Rev. Lett.* **61**(23), 2635–2638.

Fisher, M.E. (1964). Magnetism in one-dimensional systems — the Heisenberg model for infinite spin, *Am. J. Phys.* **32**, 343–346.

García-Palacios, J.L. and Lázaro, F.J. (1998). Langevin-dynamics study of the dynamical properties of small magnetic particles, *Phys. Rev. B* **58**(22), 14937–14958.

Handscomb, D.C. (1964). Monte-Carlo method and the Heisenberg ferromagnet, *Proc. Cambridge Philos. Soc.* **60**, 115.

Hasegawa, M. and Shiba, H. (2004). Magnetic frustration and anisotropy effects in giant magnetic molecule $Mo_{72}Fe_{30}$, *J. Phys. Soc. Jpn* **73**(9), 2543–2549.

Inagaki, Y., Asano, T., Ajiro, Y., Narumi, Y., Kindo, K., Cornia, A. and Gatteschi, D. (2003). High field magnetization process in a dodecanuclear Fe(III) ring cluster, *J. Phys. Soc. Jpn.* **72**, 1178–1183.

Kawamura, H. and Miyashita, S. (1985). Phase transition of the Heisenberg antiferromagnet on the triangular lattice in a magnetic field, *J. Phys. Soc. Jpn* **54**(12), 4530–4538.

Kloeden, P.E. and Platen, E. (1992). *Numerical Solution of Stochastic Differential Equations* (Springer-Verlag, Berlin; New York), ISBN 3540540628 0387540628.

Konstantinidis, N.P. (2007). Unconventional magnetic properties of the icosahedral symmetry antiferromagnetic Heisenberg model, *Phys. Rev. B* **76**(10), 104434.

Landau, D. and Binder, K. (2009). *A Guide to Monte Carlo Simulations in Statistical Physics* (Cambridge University Press, Cambridge), ISBN 9780521768481.

Luban, M. and Luscombe, J.H. (1999). Equilibrium time correlation functions and the dynamics of fluctuations, *Am. J. Phys.* **67**(12), 1161–1169.

Martin, C., Engelhardt, L., Baker, M.L., Timco, G.A., Tuna, F., Winpenny, R.E.P., Tregenna-Piggott, P.L.W., Luban, M. and Prozorov, R. (2009). Radio frequency spectroscopy of the low energy spectrum of a magnetic molecule, *Phys. Rev. B* **80**, 100407(R).

Mentrup, D., Schmidt, H.J., Schnack, J. and Luban, M. (2000). Transition from quantum to classical Heisenberg trimers: Thermodynamics and time correlation functions, *Physica A* **278**(1-2), 214–221.

Mentrup, D., Schnack, J. and Luban, M. (1999). Spin dynamics of quantum and classical Heisenberg dimers, *Physica A* **272**(1-2), 153–161.

Metropolis, N., Rosenbluth, A.W., Rosenbluth, M.N., Teller, A.H. and Teller, E. (1953). Equation of state calculations by fast computing machines, *J. Chem. Phys.* **21**(6), 1087–1092.

Milstein, G.N. and Tretyakov, M.V. (2004). *Stochastic Numerics for Mathematical Physics*, Scientific Computation (Springer, Berlin, Heidelberg, New York), ISBN 3-540-21110-1.

Müller, A., Luban, M., Schröder, C., Modler, R., Kögerler, P., Axenovich, M., Schnack, J., Canfield, P., Bud'ko, S. and Harrison, N. (2001). Classical and quantum magnetism in giant Keplerate magnetic molecules, *ChemPhysChem* **2**(8-9), 517–521.

Müller, A., Sarkar, S., Shah, S.Q., Bögge, H., Schmidtmann, M., Kögerler, P., Hauptfleisch, B., Trautwein, A.X. and Schünemann, V.V. (1999). Archimedean synthesis and magic numbers: "Sizing" giant molybdenum-oxide-based molecular spheres of the Keplerate type, *Angew. Chem. Int. Ed. Engl.* **38**(21), 3238–3241.

Müller, A., Todea, A.M., van Slageren, J., Dressel, M., Bögge, H., Schmidtmann, M., Luban, M., Engelhardt, L. and Rusu, M. (2005). Triangular geometrical and magnetic motifs uniquely linked on a spherical capsule surface, *Angew. Chem.* **117**, 3925–3929.

Schmidt, H.J. and Luban, M. (2003). Classical ground states of symmetric Heisenberg spin systems, *J. Phys. A: Math. Gen.* **36**(23), 6351–6378.

Schnack, J., Luban, M. and Modler, R. (2001). Quantum rotational band model for the Heisenberg molecular magnet $\{Mo_{72}Fe_{30}\}$, *Europhys. Lett.* **56**(6), 863–869.

Schnalle, R. and Schnack, J. (2009). Numerically exact and approximate determination of energy eigenvalues for antiferromagnetic molecules using irreducible tensor operators and general point-group symmetries, *Phys. Rev. B* **79**, 104419.

Schröder, C. (1999). *Numerical Simulations of Thermodynamic Properties of Magnetic Systems Using Deterministic and Stochastic Heat Bath Couplings*, Dissertation, University of Osnabrück, Department of Physics, in German.

Schröder, C., Nojiri, H., Schnack, J., Hage, P., Luban, M. and Kögerler, P. (2005a). Competing spin phases in geometrically frustrated magnetic molecules, *Phys. Rev. Lett.* **94**(1), 017205.

Schröder, C., Prozorov, R., Kögerler, P., Vannette, M.D., Fang, X., Luban, M., Matsuo, A., Kindo, K., Müller, A. and Todea, A.M. (2008). Multiple nearest-neighbor exchange model for the frustrated magnetic molecules $\{Mo_{72}Fe_{30}\}$ and $\{Mo_{72}Cr_{30}\}$, *Phys. Rev. B* **77**(22), 224409.

Schröder, C., Schmidt, H.-J., Schnack, J. and Luban, M. (2005b). Metamagnetic phase transition of the antiferromagnetic Heisenberg icosahedron, *Phys. Rev. Lett.* **94**(20), 207203.

Schröder, C., Schnack, J., Mentrup, D. and Luban, M. (2004). Critical slowing-down in classical and quantum Heisenberg magnetic molecules, *J. Mag. Mag. Mater.* **272–276**(1), E721–E723.

Shanmugam, M., Engelhardt, L.P., Larsen, F.K., Luban, M., McInnes, E.J., Muryn, C.A., Overgaard, J., Rentschler, E., Timco, G.A. and Winpenny, R.E.P. (2006). Studies of a molecular hourglass: Synthesis and magnetic characterization of a cyclic dodecanuclear $\{Cr_{10}Cu_2\}$ complex, *Chem. Eur. J.* **12**, 8267–8275.

Syljuåsen, O.F. and Sandvik, A.W. (2002). Quantum Monte Carlo with directed loops, *Phys. Rev. E* **66**, 046701.

Todea, A.M., Merca, A., Bögge, H., Glaser, T., Engelhardt, L., Prozorov, R., Luban, M. and Müller, A. (2009). Polyoxotungstates now also with pentagonal units: Supramolecular chemistry and tuning of magnetic exchange in $\{(M)M_5\}_{12}V_{30}$ Keplerates (M = Mo, W), *Chem. Commun.* 3351–3353.

Todea, A.M., Merca, A., Bogge, H., Glaser, T., Pigga, J.M., Langston, M.L., Liu, T., Prozorov, R., Luban, M., Schröder, C., Casey, W.H. and Müller, A. (2010). Porous capsules $\{(M)M_5\}_{12}Fe^{III}_{30}$ (M=Mo^{VI}, W^{VI}): Sphere surface supramolecular chemistry with 20 ammonium ions, related solution properties, and tuning of magnetic exchange interactions, *Angew. Chem. Int. Ed. Engl.* **49**(3), 514–519.

Todea, A.M., Merca, A., Bögge, H., van Slageren, J., Dressel, M., Engelhardt, L., Luban, M., Glaser, T., Henry, M. and Müller, A. (2007). Extending the $\{(Mo)Mo_5\}_{12}M_{30}$ capsule Keplerate sequence: A $\{Cr_{30}\}$ cluster of $s = 3/2$ metal centers with a $\{Na(H_2O)_{12}\}$ encapsulate, *Angew. Chem. Int. Ed.* **46**, 6106–6110.

Tolis, E.I., Engelhardt, L.P., Mason, P.V., Rajaraman, G., Kindo, K., Luban, M., Matsuo, A., Nojiri, H., Raftery, J., Schröder, C., Timco, G.A., Tuna, F., Wernsdorfer, W. and Winpenny, R.E. (2006). Studies of an $\{Fe_9\}$ tridiminished icosahedron, *Chemistry* **12**(35), 8961–8968.

Troyer, M. and Wiese, U.-J. (2005). Computational complexity and fundamental limitations to fermionic quantum Monte Carlo simulations, *Phys. Rev. Lett.* **94**, 170201.

Tsai, S.-H., Lee, H.K. and Landau, D.P. (2005). Molecular and spin dynamics simulations using modern integration methods, *American Journal of Physics* **73**(7), 615–624.

Vannette, M.D., Safa-Sefat, A., Jia, S., Law, S.A., Lapertot, G., Bud'ko, S.L., Canfield, P.C., Schmalian, J. and Prozorov, R. (2008). Precise measurements of radio-frequency magnetic susceptibility in ferromagnetic and antiferromagnetic materials, *J. Magn. Magn. Mater.* **320**, 354.

Waldmann, O. (2001). Spin dynamics of finite antiferromagnetic Heisenberg spin rings, *Phys. Rev. B* **65**(2), 024424.

Wang, F. and Landau, D.P. (2001). Efficient, multiple-range random walk algorithm to calculate the density of states, *Phys. Rev. Lett.* **86**(10), 2050–2053.

Index